Green Energy and Technology

Ayhan Demirbas

Biofuels

Securing the Planet's Future Energy Needs

 Springer

Ayhan Demirbas, Professor of Energy Technology
Sila Science and Energy
Trabzon
Turkey

ISBN 978-1-84882-010-4   e-ISBN 978-1-84882-011-1

DOI 10.1007/978-1-84882-011-1

Green Energy and Technology ISSN 1865-3529

A catalogue record for this book is available from the British Library

Library of Congress Control Number: 2008940429

Cover design: WMXDesign, Heidelberg, Germany

Printed on acid-free paper

9 8 7 6 5 4 3 2 1

springer.com

Preface

Today's world is facing two critical problems: (1) high fuel prices, and (2) climatic changes. Experts suggest that current oil and gas reserves would suffice to last only a few more decades. It is well known that transport is almost totally dependent on fossil fuels, particularly petroleum-based fuels such as gasoline, diesel fuel, liquefied petroleum gas, and natural gas. Of special concern are the liquid fuels used in automobiles. Hence, there has been widespread recent interest in learning more about obtaining liquid fuels from non-fossil sources. The combination of rising oil prices, issues of security, climate instability, and pollution, and deepening poverty in rural and agricultural areas, is propelling governments to enact powerful incentives for the use of these fuels, which is in turn sparking investment. In fact, the world is on the verge of an unprecedented increase in the production and use of biofuels for transport. Production of grain-based ethanol and vegetable-oil-based biodiesel is today facing difficulties due to competition with food supply. This book unifies the production of various usable liquid fuels from biomass by using a variety of technologies.

Biofuels appear to be a potential alternative "greener" energy substitute for fossil fuels. They are renewable and available throughout the world. Biomass can contribute to sustainable development and globally environmental preservation since it is renewable and carbon neutral.

This book on biofuels attempts to address the needs of energy researchers, chemical engineers, chemical engineering students, energy resources specialists, engineers, agriculturists, crop cultivators, and others interested in a practical tool for pursuing their interests in relation to bioenergy. Each chapter in the book starts with basic/fundamental explanations suitable for general readers and ends with in-depth scientific details suitable for expert readers. General readers will include people interested in learning about solutions for current fuel and environmental crises. Expert readers will include chemists, chemical engineers, fuel engineers, agricultural engineers, farming specialists, biologists, fuel processors, policy makers, environmentalists, environmental engineers, automobile engineers, college students, research faculties, *etc*. The book may even be adopted as a text book for college courses that deal with renewable energy and/or sustainability.

The Introduction already comprises one seventh of the book; in these pages emphasis is laid in detail on global energy sources, fossil fuels, and renewables, *i.e.*, biomass, hydro, wind, solar, geothermal, and marine energy sources. The second chapter is entitled "Biomass Feedstocks" and includes main biomass sources, characterization, and valorization. The third chapter is an introduction to biofuels. Furthermore, processing conditions are discussed briefly, as well as alternative applications of biorenewable feedstocks in the following chapters. The fourth and fifth chapters on "Liquid and Gaseous Biofuels", including main liquid biofuels such as bioethanol, biodiesel, biogas, biohydrogen, liquid and gaseous fuels from the Fischer–Tropsch synthesis are addressed in detail. The sixth chapter on "Thermochemical Conversion Processes" covers the utilization of biorenewables for engine fuels and chemicals. The seventh and eighth chapters include "Biofuel Economy and Biofuel Policy".

Trabzon, Turkey, July 2008   *Ayhan Demirbas*

Contents

Chapter 1
Introduction

1.1 Introduction to Energy Sources

Energy plays a vital role in our everyday lives. Energy is one of the vital inputs to the socio-economic development of any country. There are different ways in which the abundance of energy around us can be stored, converted, and amplified for our use. Energy production has always been a concern for researchers as well as policy makers.

Energy sources can be classified into three groups: fossil, renewable, and nuclear (fissile). Fossil fuels were formed in an earlier geological period and are not renewable. The fossil energy sources include petroleum, coal, bitumens, natural gas, oil shales, and tar sands. Today fuels and chemicals are predominately derived from unsustainable mineral resources, petroleum, and coal, which leads to environmental pollution, greenhouse gas emissions, and problems with energy security. The renewable energy sources include biomass, hydro, wind, solar (both thermal and photovoltaic), geothermal, and marine energy sources. The main fissile energy sources are uranium and thorium (Demirbas, 2008). The energy reserves of the world are shown in Table 1.1 (Demirbas, 2006).

The world is presently being confronted with the twin crises of fossil fuel depletion and environmental degradation. To overcome these problems, recently renewable energy has been receiving increasing attention due to its environmental benefits and the fact that it is derived from renewable sources such as virgin or cooked vegetable oils (both edible and non-edible). The world's over-demand of

Table 1.1 Energy reserves of the world

Deuterium	Uranium	Coal	Shale oil	Crude oil	Natural gas	Tar sands
7.5×10^9	1.2×10^5	320.0	79.0	37.0	19.6	6.1

Each unit $= 1 \times 10^{15}$ MJ $= 1.67 \times 10^{11}$ bbl crude oil
Source: Demirbas, 2006a

A. Demirbas, *Biofuels*,
© Springer 2009

energy, the oil crisis, and the continuous increase in oil prices have led countries to investigate new and renewable fuel alternatives. Hence, energy sources, like sun, wind, geothermal, hydraulic, nuclear, hydrogen, and biomass have been taken into consideration (Karaosmanoglu and Aksoy, 1988).

Fissile materials are those that are defined to be materials that are fissionable by neutrons with zero kinetic energy. In nuclear engineering, a fissile material is one that is capable of sustaining a chain reaction of nuclear fission. Nuclear power reactors are mainly fueled with uranium, the heaviest element occurring in nature in more than trace quantities. The principal fissile materials are uranium-235, plutonium-239, and uranium-233.

Petroleum is the largest single source of energy consumed by the world's population; exceeding coal, natural gas, nuclear and renewables, as shown in Table 1.2 for the year 2005. In fact today, over 80% of the energy we use comes from three fossil fuels: petroleum, coal, and natural gas. While fossil fuels are still being created today by underground heat and pressure, they are being consumed much more rapidly than they are created. Hence, fossil fuels are considered to be non-renewable; that is, they are not replaced as fast as they are consumed. Unfortunately, petroleum oil is in danger of becoming short in supply. Hence, the future trend is towards using alternate energy sources. Fortunately, technological developments are making the transition possible.

About 98% of carbon emissions result from fossil fuel combustion. Reducing the use of fossil fuels would considerably reduce the amount of carbon dioxide and other pollutants produced. This can be achieved by either using less energy altogether or by replacing fossil fuel by renewable fuels. Hence, current efforts focus on advancing technologies that emit less carbon (e.g., high efficiency combustion) or no carbon such as nuclear, hydrogen, solar, wind, geothermal, or on using energy more efficiently and on developing sequestering carbon dioxide emitted during fossil fuel combustion.

Another problem with petroleum fuels are their uneven distribution in the world; for example, the Middle East has 63% of the global reserves and is the dominant supplier of petroleum. This energy system is unsustainable because of equity issues as well as environmental, economic, and geopolitical concerns that have far reaching implications. Interestingly, the renewable energy resources are more evenly

Table 1.2 Energy consumption in the world (2005)

Energy source	% of total
Petroleum	40
Natural gas	23
Coal	23
Nuclear energy power	8
Renewable energy	6

Source: Demirbas, 2008

distributed than fossil or nuclear resources. Also the energy flows from renewable resources are more than three orders of magnitude higher than current global energy need. Today's energy system is unsustainable because of equity issues as well as environmental, economic, and geopolitical concerns that will have implications far into the future. Hence, sustainable renewable energy sources such as biomass, hydro, wind, solar (both thermal and photovoltaic), geothermal, and marine energy sources will play an important role in the world's future energy supply. For example, it is estimated that by year 2040 approximately half of the global energy supply will come from renewables, and the electricity generation from renewables will be more than 80% of the total global electricity production. Table 1.3 shows the estimated global renewable energy scenario by 2040.

In recent years, recovery of the liquid transportation biofuels from biorenewable feedstocks has become a promising method for the future. The biggest difference between biorenewable and petroleum feedstocks is the oxygen content. Biorenewables have oxygen levels ranging from 10–44%, while petroleum has essentially none; making the chemical properties of biorenewables very different from petroleum. For example, biorenewable products are often more polar and some easily entrain water and can therefore be acidic.

According to the International Energy Agency (IEA), scenarios developed for the USA and the EU indicate that near-term targets of up to 6% displacement of petroleum fuels with renewable biofuels appear feasible using conventional biofuels, given available cropland. A 5% displacement of gasoline in the EU requires about 5% of the available cropland to produce ethanol while in the USA 8% is required. A 5% displacement of diesel requires 13% of cropland in the USA, and 15% in the EU (IEA, 2004).

Table 1.3 Estimated Global renewable energy scenario by 2040

	2001	2010	2020	2030	2040
Total consumption (million tons oil equivalent)	10,038	10,549	11,425	12,352	13,310
Biomass	1,080	1,313	1,791	2,483	3,271
Large hydro	22.7	266	309	341	358
Geothermal	43.2	86	186	333	493
Small hydro	9.5	19	49	106	189
Wind	4.7	44	266	542	688
Solar thermal	4.1	15	66	244	480
Photovoltaic	0.2	2	24	221	784
Solar thermal electricity	0.1	0.4	3	16	68
Marine (tidal/wave/ocean)	0.05	0.1	0.4	3	20
Total renewable energy sources	1,365.5	1,745.5	2,694.4	4,289	6,351
Renewable energy sources contribution (%)	13.6	16.6	23.6	34.7	47.7

Source: Demirbas, 2008

1.2   Short Supply of Fossil Fuels

Our modern way of life is intimately dependent upon fossil fuels, specifically hydrocarbons including petroleum, coal, and natural gas. For example, the plastic in keyboards and computers comes from crude oil or natural gas feedstocks. One of our most important sources of energy today is fossil fuels. Fossil fuels are found deposited in rock formations. Fossils are non-renewable and relatively rare resources. More importantly, the major energy demand is fulfilled by fossil fuels. Today, oil and natural gas are important drivers of the world economy. Oil and natural gas are also found in beds of sedimentary rock.

Fossil fuels or mineral fuels are fossil source fuels, that is, hydrocarbons found within the top layer of the Earth's crust. It is generally accepted that they formed from the fossilized remains of dead plants and animals by exposure to heat and pressure in the Earth's crust over hundreds of millions of years.

1.2.1   Petroleum in the World

Petroleum (derived from the Greek *petra* – rock and *elaion* – oil *or* Latin *oleum* – oil) or crude oil, sometimes colloquially called *black gold* or "Texas tea", is a thick, dark brown or greenish liquid. It is used to describe a broad range of hydrocarbons that are found as gases, liquids, or solids beneath the surface of the Earth. The two most common forms are natural gas and crude oil. Petroleum consists of a complex mixture of various hydrocarbons, largely of the alkane and aromatic compounds, but may vary much in appearance and composition. The physical properties of petroleum vary greatly. The color ranges from pale yellow through red and brown to black or greenish, while by reflected light it is, in the majority of cases, of a green hue. Petroleum is a fossil fuel because it was formed from the remains of tiny sea plants and animals that died millions of years ago, and sank to the bottom of the oceans. This organic mixture was subjected to enormous hydraulic pressure and geothermal heat. Over time, the mixture changed, breaking down into compounds made of hydrocarbons by reduction reactions. This resulted in the formation of oil-saturated rocks. The oil rises and is trapped under non-porous rocks that are sealed with salt or clay layers.

According to well accepted biogenic theory, crude oil, like coal and natural gas, is the product of compression and heating of ancient vegetation and animal remains over geological time scales. According to this theory, an organic matter is formed from the decayed remains of prehistoric marine animals and terrestrial plants. Over many centuries this organic matter, mixed with mud, is buried under thick sedimentary layers. The resulting high pressure and heat causes the remains to transform, first into a waxy material known as kerogen, and then into liquid and gaseous hydrocarbons by process of catagenesis. The fluids then migrate through adjacent rock layers until they become trapped underground in porous rocks termed reservoirs, forming an oil field, from which the liquid can be removed by drilling and pumping. The reservoirs are at different depths in different parts of the world, but the typical depth is 4–5 km. The thickness of the oil layer is about 150 m and is generally termed the "oil window". Three important elements of an oil reservoir are: a rich source rock, a migration conduit, and a trap (seal) that forms the reservoir.

According to the not well accepted abiogenic theory, the origin of petroleum is natural hydrocarbons. The theory proposes that large amounts of carbon exist naturally on the planet, some in the form of hydrocarbons. Due to it having a lower density than aqueous pore fluids, hydrocarbons migrate upward through deep fracture networks.

The first oil wells were drilled in China in the 4th century or earlier. The wells, as deep as 243 meters, were drilled using bits attached to bamboo poles. The oil was burned to produce heat needed in the production of salt from brine evaporation. By the 10th century, extensive bamboo pipelines connected oil wells with salt springs. Ancient Persian tablets indicate the medicinal and lighting uses of petroleum in the upper echelons of their society.

In the 8th century, the streets of the newly constructed Baghdad were paved with tar derived from easily accessible petroleum from natural fields in the region. In the 9th century, oil fields were exploited to produce naphtha in Baku, Azerbaijan. These fields were described by the geographer Masudi in the 10th century, and the output increased to hundreds of shiploads in 13th century as described by Marco Polo.

The modern history of petroleum began in 1846, with the discovery of the refining of kerosene from coal by Atlantic Canada's Abraham Pineo Gesner. Poland's Ignacy Łukasiewicz discovered a means of refining kerosene from the more readily available "rock oil" ("petroleum") in 1852; and in the following year the first rock oil mine was built in Bobrka, near Krosno in southern Poland. The discovery rapidly spread around the world, and Meerzoeff built the first Russian refinery in the mature oil fields of Baku in 1861, which produced about 90% of the world's oil. In fact, the battle of Stalingrad was fought over Baku (now the capital of the Azerbaijan Republic).

The first commercial oil well in North America was drilled by James Miller Williams in 1858 in Oil Springs, Ontario, Canada. In the following year, Edwin Drake discovered oil near Titusville, Pennsylvania, and pioneered a new method for producing oil from the ground, in which he drilled using piping to prevent borehole collapse, allowing for the drill to penetrate deeper into the ground. Previous methods for collecting oil had been limited. For example, ground collection of oil consisted of gathering it from where it occurred naturally, such as from oil seeps or shallow holes dug into the ground. The methods of digging large shafts into the ground also failed, as collapse from water seepage almost always occurred. The significant advancement that Drake made was to drive a 10 meter iron pipe through the ground into the bedrock below. This allowed Drake to drill inside the pipe, without the hole collapsing from the water seepage. The principle behind this idea is still employed today by many companies for petroleum drilling.

Drake's well was 23 meters deep, which is very shallow compared to today's well depth of 1000–4000 meters. Although technology has improved the odds since Edwin Drake's days, petroleum exploration today is still a gamble. For example, only about 33 in every 100 exploratory wells have oil, and the remaining 67 come up "dry".

For about 10 years Pennsylvania was the one great oil producer of the world, but since 1870 the industry has spread all over the globe. From the time of the completion of the first flowing well on the Baku field, Russia has ranked second on the list of producing countries, whilst Galicia and Romania became prominent in 1878 and 1880, respectively. Sumatra, Java, Burma, and Borneo, where active development began in 1883, 1886, 1890, and 1896, bid fair to rank before long among the chief sources of the oil supplies of the world.

Before the 1850s, Americans often used whale oil to light their homes and businesses. Drake refined the oil from his well into kerosene for lighting, which was used till the discovery of light bulbs. Gasoline and other products made during refining were simply discarded due to lack of use. In 1892, the "horseless carriage" solved this problem since it required gasoline. By 1920 there were nine million motor vehicles in USA and many gas stations to supply gasoline.

1.2.1.1 Properties of Petroleum, Crude Oil Refining, and World Petroleum Reserves

Crude oil is a complex mixture that is between 50% and 95% hydrocarbon by weight. The first step in refining crude oil involves separating the oil into different hydrocarbon fractions by distillation. An oil refinery cleans and separates the crude oil into various fuels and byproducts, including gasoline, diesel fuel, heating oil, and jet fuel. Main crude oil fractions are listed in Table 1.4. Since various components boil at different temperatures, refineries use a heating process called distillation to separate the components. For example, gasoline has a lower boiling point than kerosene, allowing the two to be separated by heating to different temperatures. Another important job of the refineries is to remove contaminants from

Table 1.4 Main crude oil fractions

Fraction	Boiling range (K)	Number of carbon atoms
Natural gas	<293	C_1 to C_4
Petroleum ether	293–333	C_5 to C_6
Ligroin (light naphtha)	333–373	C_6 to C_7
Gasoline	313–478	C_5 to C_{12}, and cycloalkanes
Jet fuel	378–538	C_8 to C_{14}, and aromatics
Kerosene	423–588	C_{10} to C_{16}, and aromatics
No. 2 diesel fuel	448–638	C_{10} to C_{20}, and aromatics
Fuel oils	>548	C_{12} to C_{70}, and aromatics
Lubricating oils	>673	>C_{20}
Asphalt or petroleum coke	Non-volatile residue	Polycyclic structures

the oil. For example, sulfur from gasoline or diesel to reduce air pollution from automobile exhausts. After processing at the refinery, gasoline and other liquid products are usually shipped out through pipelines, which are the safest and cheapest way to move large quantities of petroleum across land.

An important non-fuel use of petroleum is to produce chemical raw materials. The two main classes of petrochemical raw materials are olefins (including ethylene and propylene) and aromatics (including benzene and xylene isomers), both of which are produced in large quantities. A very important aspect of petrochemicals is their extremely large scale. The olefins are produced by chemical cracking by using steam or catalysts, and the aromatics are produced by catalytic reforming. These two basic building blocks serve as feedstock to produce a wide range of chemicals and materials including monomers, solvents, and adhesives. From the monomers, polymers or oligomers are produced for use as plastics, resins, fibers, elastomers, certain lubricants, and gels.

The oil industry classifies "crude" according to its production location (*e.g.*, "West Texas Intermediate, WTI" or "Brent"), relative density (API gravity), viscosity ("light", "intermediate", or "heavy"), and sulfur content ("sweet" for low sulfur, and "sour" for high sulfur). Additional classification is due to conventional and non-conventional oil as shown in Table 1.5.

Oil shale is a sedimentary rock that contains the solid hydrocarbon wax kerogen in tightly packed limy mud and clay. The kerogen may be decomposed at elevated temperatures (723 K), resulting in an oil suitable for refinery processing (Dorf, 1977). The oil shale layer is not hot enough to complete the oil generation. For the final step the kerogen must be heated up to 775 K and molecularly combine with additional hydrogen to complete the oil formation. This final process must be performed in the refinery and needs huge amounts of energy that otherwise have been provided by the geological environment during oil formation (Demirbas, 2000). The kerogen is still in the source rock and can not accumulate in oil fields. Typically, the ratio of kerogen to waste material is very low, making the mining of oil shales unattractive. Hence, due to a combination of environmental and economic concerns, it is very unlikely that oil shale mining will ever be performed at large scale, though in some places it has been utilized in small quantities. However, the shale oil reserves in the world are greater than those of crude oil or natural gas, as shown in Table 1.1.

Table 1.5 Classification of oils

Class	Viscosity of oil (measured in °API)
Light crude	>31.1
Medium oil	22.3–31.1
Heavy oil	<22.3
Deep sea oil above 500 meters water depth	–
Extra heavy oil below (including tar sands)	<10
Oil shale	–
Bitumen from tar sands	–

Tar sands are oil traps not deep enough in the Earth to allow for geological conversion into the conventional oil. This oil was not heated enough to complete the process of molecular breakage to reduce the viscosity. The oil has the characteristics of bitumen and is mixed with large amounts of sand due to the proximity to the Earth surface. The tar sand is mined, flooded with water in order to separate the heavier sand, and then processed in special refineries to reduce its high sulfur content (the original oil usually has 3–5% sulfur) and other components. This process needs huge amounts of energy and water. Only oil deposits in deep layers below 75 m are mined *in-situ* (COSO, 2007).

OPEC is the Organization of Oil Exporting Countries and its current members are Iran, Iraq, Kuwait, Saudi Arabia, Venezuela, Qatar, Indonesia, Libya, United Arab Emirates, Algeria, Nigeria, Ecuador and Gabon. OPEC members try to set production levels for petroleum to maximize their revenue. According to supply/demand economics, the less oil they produce, the higher the price of oil on the world market, and the more oil they produce, the lower the price. However, the OPEC countries do not always agree with each other. Some OPEC countries want to produce less oil to raise prices, whereas other OPEC countries want to flood the market with petroleum to reap immediate returns. In addition, the oil supply may be controlled for political reasons. For example, the 1973 OPEC oil embargo was a political statement against the US for supporting Israel in the Yom Kippur war. Such embargos or cuts in production cause drastic increases in the price of petroleum. Today, a significant portion of US oil import is from Canada and Mexico, which is more reliable and has a lower shipping cost. However, due to an internal law, Mexico can only export half the oil it produces to the US.

The US is a member of the Organization for Economic Co-operation and Development (OECD), which is an international organization of 30 countries that accept the principles of representative democracy and a free market economy. In the 1970s, as a counterweight to OPEC, OECD founded the International Energy Agency (IEA) which is regarded as the "energy watchdog" of the western world and is supposed to help to avoid future crises. IEA provides demand and supply forecasts in its annual World Energy Outlook (WEO) report, and the current situation of oil market in its monthly publication. WEO covers forecasts for the next two decades and is highly regarded by people related to the energy industry.

The price of a barrel (42 gallons or 159 liters) of crude oil is highly dependent on both its grade (*e.g.*, specific gravity, sulfur content, viscosity) and location. The price is highly influenced by the demand, current supply, and perceived future supply. Both demand and supply are highly dependent on global macroeconomic and political conditions. It is often claimed that OPEC sets a high oil price, and the true cost of oil production is only $2/barrel in the Middle East. These cost estimates ignore the cost of finding and developing the oil fields. In fact, the price is set by the cost of producing not the easy oil, but more difficult marginal oil. For example, by limiting production OPEC has caused development of more expensive areas of production such as the North Sea. On the other hand, investing in spare capacity is expensive and the low oil price environment of 1990s has led to cutbacks in investment. As a result, during the oil price rally seen since 2003, OPEC's spare capacity has not been sufficient to stabilize prices.

Petroleum is the most important energy source, as 35% of the world's primary energy needs is met by crude oil, 25% by coal, and 21% by natural gas, as shown in Table 1.6 (IEA, 2007). The transport sector (*i.e.*, automobiles, ships, and aircrafts) relies to well over 90% on crude oil. In fact, the economy and lifestyle of industrialized nations relies heavily upon a sufficient supply of crude oil at low cost.

Table 1.7 shows crude oil production data for various regions (IEA, 2007). The Middle East produces 32% of the world's oil (Table 1.8), but more importantly it has 64% of the total proven oil reserves in the world (Table 1.9). Also, its reserves are depleting at a slower rate than any other region in the world. The Middle East provides more than half of OPEC's total oil exports and has a major influence on worldwide crude oil prices, despite the fact that OPEC produces less than half the oil in the world.

The smaller petroleum reserves are on the verge of depletion, and the larger reserves are estimated to be depleted in less than 50 years at the present rate of consumption. Hence, the world is facing a bleak future of petroleum short supply. Figure 1.1 shows global oil production scenarios based on today's production. A peak

Table 1.6 1973 and 2005 fuel shares of total primary energy supply (TPES) (excludes electricity and heat trade)

	World		OECD	
	1973	2005	1973	2005
Oil	46.2	35.0	53.0	40.6
Coal	24.4	25.3	22.4	20.4
Natural gas	16.0	20.7	18.8	21.8
Combustible renewables and wastes	10.6	10.0	2.3	3.5
Nuclear	0.9	6.3	1.3	11.0
Hydro	1.8	2.2	2.1	2.0
Other (geothermal, solar, wind, heat, *etc.*)	0.1	0.5	0.1	0.7
Total (million tons oil equivalent)	6,128	11,435	3,762	5,546

Table 1.7 1973 and 2006 regional shares of crude oil production

Region	1973	2006
Middle East (%)	37.0	31.1
OECD (%)	23.6	23.2
Former USSR (%)	15.0	15.2
Africa (%)	10.0	12.1
Latin America (%)	8.6	9.0
Asia (excluding China) (%)	3.2	4.5
China (%)	1.9	4.7
Non-OECD Europe (%)	0.7	0.2
Total (million tons)	2,867	3,936

Table 1.8 Percentage of petroleum production by region

Middle East	Latin America	Eastern Europe	North America	Asia and Pacific	Africa	Western Europe
32	14	13	11	11	10	9

Table 1.9 Percentage of total proven reserves by region

Middle East	Latin America	Eastern Europe	North America	Asia and Pacific	Africa	Western Europe
64	12	6	3	4	9	2

Fig. 1.1 Global oil production scenarios based on today's production
Source: Demirbas, 2008

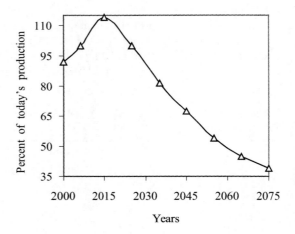

in global oil production is likely to occur by 2015 and thereafter the production will start to decline at a rate of several percent *per* year. By 2030, the global oil supply will be dramatically lower, which will create a supply gap that may be hard to fill by growing contributions from other fossil, nuclear, or alternative energy sources in that time frame.

1.2.2 *Natural Gas as the Fastest Growing Primary Energy Source in the World*

Natural gas was known in England as early as 1659. However, it did not replace coal gas as an important source of energy in the world until after World War II. The usefulness of natural gas (NG) has been known for hundreds of years. For example, the Chinese used NG to heat water. In the early days, NG was used to light lamps on the street and in houses.

Natural gas is a mixture of lightweight alkanes. Natural gas contains methane (CH_4), ethane (C_2H_6), propane (C_3H_8), butane and isobutane (C_4H_{10}), and pentanes (C_5H_{12}). The C_3, C_4, and C_5 hydrocarbons are removed before the gas is sold. The commercial natural gas delivered to the customer is therefore primarily a mixture of methane and ethane. The propane and butane removed from natural gas are usually liquefied under pressure and sold as liquefied petroleum gases (LPG). Natural gas is found to consist mainly of the lower paraffins, with varying quantities of carbon dioxide, carbon monoxide, hydrogen, nitrogen, and oxygen, in some cases also hydrogen sulfur and possibly ammonia. The chemical composition of NG is given in Table 1.10.

In recent years, NG has become the fastest growing primary energy source in the world, mainly because it is a cleaner fuel than oil or coal and not as controversial as nuclear power. NG combustion is clean and emits less CO_2 than all other petroleum derivate fuels, which makes it makes favorable in terms of the greenhouse effect. NG is used across all sectors, in varying amounts, including the industrial, residential, electric generation, commercial and transportation sectors.

NG is found in many parts of the world, but the largest reserves are in the former Soviet Union and Iran. Since the 1970s, world natural gas reserves have generally increased each year. World natural gas reserves by country are tabulated in Table 1.11.

Around the world, NG use is increasing for a variety of reasons including prices, environmental concerns, fuel diversification and/or energy security issues, market deregulation, and overall economic growth. Figure 1.2 shows production and consumption trends of natural gas in the last decades. In NG consumption, the United States ranks first, the former USSR region ranks second, and Europe ranks third. The largest NG producer is Russia, which is also the largest supplier of NG

Table 1.10 Chemical composition of NG

Component	Typical analysis (% by volume)	Range (% by volume)
Methane	94.9	87.0–96.0
Ethane	2.5	1.8–5.1
Propane	0.2	0.1–1.5
i-Butane	0.03	0.01–0.3
n-Butane	0.03	0.01–0.3
i-Pentane	0.01	trace–0.14
n-Pentane	0.01	trace–0.14
Hexanes plus	0.01	trace–0.06
Nitrogen	1.6	1.3–5.6
Carbon Dioxide	0.7	0.1–1.0
Oxygen	0.02	0.01–0.1
Hydrogen	trace	trace–0.02

Table 1.11 World natural gas reserves according to country

Country	Reserves (trillion cubic meters)	% of world total
Russian Federation	48.1	33.0
Iran	23.0	15.8
Qatar	8.5	5.8
United Arab Emirates	6.0	4.1
Saudi Arabia	5.8	4.0
United States	4.7	3.3
Venezuela	4.0	2.8
Algeria	3.7	2.5
Nigeria	3.5	2.4
Iraq	3.1	2.1
Turkmenistan	2.9	2.0
Malaysia	2.3	1.6
Indonesia	2.0	1.4
Uzbekistan	1.9	1.3
Kazakhstan	1.8	1.3
Rest of the world	23.8	16.5

Source: Demirbas, 2008

to Western Europe. Asia and Oceania import NG to satisfy their demands. Other regions are relatively minor producers and consumers of gas.

Compared to oil, only a moderate amount of NG is traded on world markets. Due to its low density, NG is more expensive to transport than oil. For example, a section of pipe in oil service can hold 15 times more energy than when used to transport high pressure NG. Hence, gas pipelines need to have a much larger diameter and/or fluid velocity for a given energy movement. In fact, pipeline transportation is not always feasible because of the growing geographic distance between gas reserves and markets. Many of the importing countries do not wish to solely rely on NG import due the potential political instabilities that may affect the long pipeline routes. The alternate transport routes are by ships or railcars. However, for economical transport, sufficient energy needs to be packaged in the containers, which is done by liquefaction. A full liquefied natural gas (LNG) chain consists of a liquefaction plant, low temperature, pressurized, transport ships, and a re-gasification terminal. World LNG trade is currently about 60 million metric tons *per* year, some 65% of which is imported by Japan.

The generation of electricity is an important use of NG. However, the electricity from NG is generally more expensive that from coal because of increased fuel costs. NG can be used to generate electricity in a variety of ways. These include (1) conventional steam generation, similar to coal fired power plants in which heating is used to generate steam, which in turns runs turbines with an efficiency of 30–35%; (2) centralized gas turbines, in which hot gases from NG combustion are used to turn the turbines; and (3) combine cycle units, in which both steam and hot combustion gases are used to turn the turbines with an efficiency of 50–60%.

Fig. 1.2 Production and consumption trends of natural gas in the world

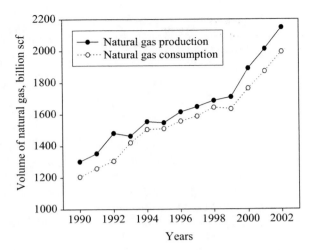

The use of NG in power production has increased due to the fact that NG is the cleanest burning alternative fossil fuel. Upon combustion, NG emits less CO_2 than oil or coal, virtually no sulfur dioxide, and only small amounts of nitrous oxides. CO_2 is a greenhouse gas, while the sulfur and nitrous oxides produced by oil and coal combustion cause acid rain. Both the carbon and hydrogen in methane combine with oxygen when NG is burned, giving off heat, CO_2 and H_2O. Coal and oil contain proportionally more carbon than NG, hence emit more CO_2.

Concerns about acid rain, urban air pollution, and global warming are likely to increase NG use in the future. NG burns far more cleanly than gasoline or diesel, producing fewer nitrous oxides, unburned hydrocarbons and particulates.

$$CH_4 \quad + \; 2O_2 \quad \rightarrow \quad CO_2 \; + \; 2H_2O \tag{1.1}$$
$$1.00\,g \qquad\qquad\qquad 2.75\,g$$

$$2C_4H_{10} \; + \; 13O_2 \quad \rightarrow \quad 8CO_2 + \; 10H_2O \tag{1.2}$$
$$1.00\,g \qquad\qquad\qquad 3.03\,g$$

$$C \qquad + \; O_2 \quad \rightarrow \quad CO_2 \tag{1.3}$$
$$1.00\,g \qquad\qquad\qquad 3.67\,g$$

From Eq. 1.1, among the fossil fuels, natural gas is the least responsible for CO_2 emissions. Liquefied petroleum gas (LPG) causes higher CO_2 than that of natural gas Eq. 1.2. The highest amount of CO_2 occurs according to Eq. 1.3. Thus responsibility of the fossil fuel increases with increasing its carbon number (Demirbas, 2005). The gases (they consist of three or more atoms like CO_2 and CH_4) with higher heat capacities than those of O_2 and N_2 cause a greenhouse effect.

Since, NG vehicles require large storage tanks, the main market may be for buses that are used within cities. Another use that may develop is the use of fuel cells for stationary and transportation application. The energy for fuel cells comes from hydrogen, which can be made from NG. Fuel cells eliminate the need for turbines or generators, and can operate at efficiencies as high as 60%. In addition, fuel cells can also operate at low temperatures, reducing the emissions of acid rain causing nitrous oxides, which are formed during high temperature combustion of any fuel.

1.2.2.1 Gas Hydrates

Natural gas (methane) can be obtained from gas hydrates. Gas hydrates are also called clathrates or methane hydrates. Gas hydrates are potentially one of the most important energy resources for the future. Methane gas hydrates are increasingly considered a potential energy resource. Methane gas hydrates are crystalline solids formed by combination of methane and water at low temperatures and high pressures. Gas hydrates have an ice-like crystalline lattice of water molecules with methane molecules trapped inside. Enormous reserves of hydrates can be found under continental shelves and on land under permafrost. The amount of organic carbon in gas hydrates is estimated to be twice that of all other fossil fuels combined. However, due to solid form of the gas hydrates, conventional gas and oil recovery techniques are not suitable. Table 1.12 shows worldwide amounts of organic carbon sources. The recovery of methane generally involves dissociating or melting *in-situ* gas hydrates by heating the reservoir above the temperature of hydrate formation, or decreasing the reservoir pressure below hydrate equilibrium (Lee and Holder, 2001).

The difficulty with recovering this source of energy is that the fuel is in solid form and is not amenable to conventional gas and oil recovery techniques (Lee and Holder, 2001). Proposed methods of gas recovery from hydrates generally deal with dissociating or melting *in-situ* gas hydrates by heating the reservoir beyond the temperature of hydrate formation, or decreasing the reservoir pressure below hydrate equilibrium. The models have been developed to evaluate natural gas production from hydrates by both depressurization and heating methods.

There are three methods to obtain methane from gas hydrates: (a) The depressurization method, (b) the thermal stimulation method, and (c) the chemical inhibition method. The thermal stimulation method is quite expensive. The chemical

Table 1.12 Worldwide amounts of organic carbon sources

Source of organic carbon	Amount (gigaton)
Gas hydrates (onshore and offshore)	10,000–11,000
Recoverable and non-recoverable fossil fuels (oil, coal, gas)	5,000
Soil	1,400
Dissolved organic matter	980
Land biota	880
Peat	500
Other	70

Source: Hacisalihoglu *et al.*, 2008

inhibitor injection method is also expensive. The depressurization method may prove useful for applying more than one production.

1.2.3 Coal as a Fuel and Chemical Feedstock

The first known and the oldest fossil fuel is coal. Coal has played a key role as a primary energy source as well as a primary source of organic chemicals. It is a complex, heterogeneous combustible material, made up of portions that are either useful (carbon and hydrogen), or useless (diluents such as moisture, ash and oxygen, or contaminants such as sulfur and heavy metals). Coal can be defined as a sedimentary rock that burns. It was formed by the decomposition of plant matter, and it is a complex substance that can be found in many forms. Coal is divided into four classes: anthracite, bituminous, subbituminous, and lignite. Elemental analysis gives empirical formulas such as $C_{137}H_{97}O_9NS$ for bituminous coal and $C_{240}H_{90}O_4NS$ for high-grade anthracite.

Coal is formed from plant remains that have been compacted, hardened, chemically altered, and metamorphosed underground by heat and pressure over millions of years. When plants die in a low-oxygen swamp environment, instead of decaying by bacteria and oxidation, their organic matter is preserved. Over time, heat and pressure remove the water and transform the matter into coal. The first step in coal formation yields peat, compressed plant matter that still contains leaves and twigs. The second step is the formation of brown coal or lignite. Lignite has already lost most of the original moisture, oxygen, and nitrogen. It is widely used as a heating fuel but is of little chemical interest. The third stage, bituminous coal, is also widely utilized as a fuel for heating. Bituminous coal is the most abundant form of coal and is the source of coke for smelting, coal tar, and many forms of chemically modified fuels. The chemical properties of typical coal samples are given in Table 1.13. Table 1.14 shows the world's recoverable coal reserves. Worldwide coal production and consumption in year 1998 were 4,574 and 4,548 million tons, respectively. The known world recoverable coal reserves in 1999 were 989 billion tons. Also, coal reserves are rather evenly spread around the globe: 25% are in the USA, 16% in Russia, and 11.5% in China. Although coal is much more abundant than oil or gas on a global scale, coalfields can easily become depleted on a regional scale.

Due to its abundance and wide distribution, coal accounts for 25% of the world's primary energy consumption and 37% of the energy consumed worldwide for electricity generation. For example, the known coal reserves in the world will be enough for consumption for over 215 years, while the known oil reserves are only about 39 times of the world's consumption, and the known natural gas reserves are about 63 times of the world's consumption level in 1998. With modern techniques coal can be mined, transported and stored efficiently and cost-effectively. International coal trade is growing steadily and the prices are kept low by the vigorous competition on supply. However, the future commercial develop-

Table 1.13 Chemical properties of typical coal samples

	Low rank coal	High volatile coal	High rank coal
Carbon, %	75.2	82.5	90.5
Hydrogen, %	6.0	5.5	4.5
Oxygen, %	17.0	9.6	2.6
Nitrogen, %	1.2	1.7	1.9
Sulfur, %	0.6	0.7	0.5
Moisture, %	10.8	7.8	6.5
Calorific value, MJ/kg	31.4	35.0	36.0

ment of coal depends critically on its environmental acceptability and in particular on the success of the power generation industry in reducing polluting emissions.

The major non-fuel use of coal is carbonization to make metallurgical coke. The production of activated carbon from coals has been of interest for years. The carbon in coal can be used as a source of specialty aromatic and aliphatic chemicals *via* processing, including gasification, liquefaction, direct conversion, and co-production of chemicals, fuels and electricity.

1.2.3.1   Conversion of Coal to Fuels and Chemicals

As early as 1800, coal gas, or town gas, was made by heating coal in the absence of air. Coal gas is rich in CH_4 and gives off up to 20.5 kJ *per* liter of gas burned. Coal gas became so popular that most major cities and many small towns had a local gas house in which it was generated, and gas burners were adjusted to burn fuel. Coal can be converted to water gas with steam (Demirbas, 2007).

$$C + H_2O \rightarrow CO + H_2 \tag{1.4}$$

Water gas burns to give CO_2 and H_2O, releasing roughly 11.2 kJ *per* liter of gas consumed.

Water gas formed by the reaction of coal with oxygen and steam is a mixture of CO, CO_2, and H_2. The ratio of H_2 to CO can be increased by adding water to this mixture, to take advantage of a reaction known as the water-gas shift reaction (Demirbas, 2007).

$$CO + H_2O \rightarrow CO_2 + H_2 \tag{1.5}$$

The concentration of CO_2 can be decreased by reacting of the CO_2 with coal at high temperatures to form CO.

$$C + CO_2 \rightarrow 2CO \tag{1.6}$$

Water gas from which the CO_2 has been removed is called synthesis gas because it can be used as a starting material for a variety of organic and inorganic compounds. It can be used as the source of H_2 for the synthesis of methanol, for example.

$$CO + 2H_2 \rightarrow CH_3OH \tag{1.7}$$

Table 1.14 The world's recoverable coal reserves

Country	Bituminous including anthracite	Subbituminous	Lignite
United States of America	115891	101021	33082
China	62200	33700	18600
India	82396	–	2000
South Africa	49520	–	–
Kazakhstan	31100	–	3000
Brazil	–	11929	–
Colombia	6267	381	–
Canada	3471	871	2236
Canada	3471	871	2236
Indonesia	790	1430	3150
Botswana	4300	–	–
Uzbekistan	1000	–	3000
Turkey	278	761	2650
Pakistan	–	2265	–
Thailand	–	–	1268
Chile	31	1150	–
Mexico	860	300	51
Peru	960	–	100
Kyrgyzstan	–	–	812
Japan	773	–	–
Korea (Democ. People's Rep.)	300	300	–
Zimbabwe	502	–	–
Venezuela	479	–	–
Philippines	–	232	100
Mozambique	212	–	–
Swaziland	208	–	–
Tanzania	200	–	–
Other	449	379	27

Source: IEA, 2006a

Methanol can then be used as a starting material for the synthesis of alkenes, aromatic compounds, acetic acid, formaldehyde, and ethyl alcohol (ethanol). Synthesis gas can also be used to produce methane, or synthetic natural gas (SNG) (Demirbas, 2007).

$$CO + 3H_2 \rightarrow CH_4 + H_2O \qquad (1.8)$$

$$2CO + 2H_2 \rightarrow CH_4 + CO_2 \qquad (1.9)$$

The first step toward making liquid fuels from coal involves the manufacture of synthesis gas (CO and H_2) from coal. In 1925, Franz Fischer and Hans Tropsch developed a catalyst that converted CO and H_2 at 1 atm and 250 to 300C into liquid hydrocarbons. By 1941, Fischer–Tropsch plants produced 740,000 tons of petroleum products *per* year in Germany (Dry, 1999). Fischer–Tropsch technology is based on a complex series of reactions that use H_2 to reduce CO to CH_2 groups linked to form long-chain hydrocarbons (Schulz, 1999).

$$nCO + (n + m/2)\,H_2 \rightarrow C_nH_m + nH_2O \qquad (1.10)$$

At the end of World War II, Fischer–Tropsch technology was under study in most industrial nations. Coal can be gasified to produce synthesis gas (syngas), which can be converted to paraffinic liquid fuels and chemicals by Fischer–Tropsch synthesis, which was developed in 1925 by Franz Fischer and Hans Tropsch to converted CO and H_2 at 1 atm and 575 K into liquid hydrocarbons using Fe/Co catalyst. The liquid product mainly contains benzene, toluene, xylene (BTX), phenols, alkylphenols and cresol. The low cost and high availability of crude oil, however, led to a decline in interest in liquid fuels made from coal.

Another approach to liquid fuels is based on the reaction between CO and H_2 to form methanol, CH_3OH.

$$CO + 2H_2 \rightarrow CH_3OH \qquad (1.11)$$

Methanol can be used directly as a transportation fuel, or it can be converted into gasoline with catalysts such as the ZSM-5 zeolite catalyst.

Methanol can also be produced from syngas with hydrogen and carbon monoxide in a 2 to 1 ratio. Coal-derived methanol typically has low sulfur and other impurities. Syngas from coal can be reformed by reacting with water to produce hydrogen. Ammonium sulfate from coal tar by pyrolysis can be converted to ammonia. Humus substances can be recovered from brown coal by alkali extraction.

1.3 Introduction to Renewable and Biorenewable Sources

Types of energy that are readily renewed are called renewable energy. Renewable energy sources (RES) can be readily renewed in a short time period. RES are also often called alternative sources of energy that use domestic resource and have the potential to provide energy service with zero or almost zero emission of both air pollutants and greenhouse gases. Renewable resources are more evenly distributed than fossil and nuclear resources. In 2005 the distribution of renewable energy consumption as a percentage of the total renewable energy in the world was: biomass, 46%; hydroelectric, 45%; geothermal, 6%; wind, 2%, and solar, 1% (EIA, 2006b). The most important benefit of renewable energy systems is the decrease of environmental pollution.

Worldwide research and development in the field of the RES and systems has been carried out during the last two decades. The types of energy that are readily renewed are called renewable energy. Renewable energies have been the primary energy source throughout the history of the human race. Examples of the RES include biomass, hydraulic, solar, wind, geothermal, wave, tidal, biogas, and ocean thermal energy. The RES are derived from those natural, mechanical, thermal and growth processes that repeat themselves within our lifetime and may be relied upon to produce predictable quantities of energy when required.

RES are also often called alternative sources of energy. The RES that use indigenous resources have the potential to provide energy services with zero or almost zero emissions of both air pollutants and greenhouse gases. Renewable energy is a clean or inexhaustible energy like hydrogen energy and nuclear energy. The most important benefit of renewable energy systems is the decrease of environmental pollution.

Renewable energy technologies produce marketable energy by converting natural phenomena into useful forms of energy. These technologies use the sun's energy and its direct and indirect effects on the Earth (solar radiation, wind, precipitation, and various plants, *i.e.*, biomass), gravitational forces (tides), and the heat of the Earth's core (geothermal) as the resources from which energy is produced. Currently, renewable energy sources supply 14% of the total world energy demand (Demirbas, 2005). RES are readily available in nature and the renewables are the primary energy resources. There is an urgency to develop use of the RES out of which biomass represents an important alternative. Various technologies exist through which biomass can be converted into the most preferred liquid form of the fuel including bioethanol, biodiesel, and biogasoline. The share of renewable energy sources is expected to increase significantly in coming decades.

Renewable technologies like water and wind power probably would not have provided the same fast increase in industrial productivity as fossil fuels (Edinger and Kaul, 2000). The share of renewable energy sources is expected to increase very significantly (to 30–80% in 2100). Biomass, wind and geothermal energy are commercially competitive and are making relatively fast progress (Fridleifsson, 2001). Renewable energy scenarios depend on environmental protection which is an essential characteristic of sustainable developments. Main renewable energy sources and their usage forms are given in Table 1.15.

Renewable energy is a promising alternative solution because it is clean and environmentally safe. These forms of energy also produce lower or negligible levels of greenhouse gases and other pollutants when compared with the fossil energy sources they replace. Approximately half of the global energy supply will be from renewables in 2040 according to European Renewable Energy Council

Table 1.15 Main renewable energy sources and their usage forms

Energy source	Energy conversion and usage options
Hydropower	Power generation
Modern biomass	Heat and power generation, pyrolysis, gasification, digestion
Geothermal	Urban heating, power generation, hydrothermal, hot dry rock
Solar	Solar home system, solar dryers, solar cookers
Direct solar	Photovoltaics, thermal power generation, water heaters
Wind	Power generation, wind generators, windmills, water pumps
Wave	Numerous designs
Tidal	Barrage, tidal stream

(EREC, 2006). The most significant developments in renewable energy production are observed in photovoltaics (from 0.2 to 784 Mtoe) and wind energy (from 4.7 to 688 Mtoe) between 2001 and 2040 (Table 1.3).

1.3.1 Non-combustible Renewable Energy Sources

Non-combustible renewable energy sources are hydro, geothermal, wind, solar, wave, tidal, and ocean thermal energy sources. Non-combustible renewable energies have been the primary energy source in the history of the human race. The potential of sustainable large hydro power is quite limited to some regions in the world. The potential for small hydro (<10 MW) power is still significant and will be even more significant in future. Photovoltaic (PV) systems and wind energy are technologies that have had annual growth rates of more than 30% during the last year, which will become more significant in future. Geothermal and solar thermal sources will be more important energy sources in future. PV will then be the largest renewable electricity source with a production of 25.1% of global power generation in 2040.

1.3.1.1 Hydropower

The water in rivers and streams can be captured and turned into hydropower, also called hydroelectric power. Large scale hydropower provides about one-quarter of the world's total electricity supply, virtually all of Norway's electricity and more than 40% of the electricity used in developing countries. The technically usable world potential of large-scale hydro power is estimated to be over 2200 GW, of which only about 25% is currently exploited.

There are two small-scale hydropower systems: microhydropower systems (MHP) with capacities below 100 kW and small hydropower systems (SHP) with capacity between 101 kW and 1MW. Large-scale hydropower supplies 20% of global electricity. In developing countries, considerable potential still exists, but large hydropower projects may face financial, environmental, and social constraints (UNDP, 2000). The two small-scale hydropower systems that are discussed in this section are the sites with capacities below 100 kW (referred to as microhydropower systems) and those with capacities between 101kW and 1MW (referred to as small hydropower systems). Microhydropower (MHP) systems, which use cross flow turbines and pelton wheels, can provide both direct mechanical energy (for crop processing) and electrical energy. However, due to design constraints, turbines up to a capacity of 30 kW are suitable for extracting mechanical energy. Of the total installed capacity of about 12MW of MHP systems, half is used solely for crop processing. The most popular of the MHP systems is the peltric set, which is an integrated pelton turbine and electricity generation unit with an average capacity of 1 kW. MHP systems are sometimes described as those having capacities below 100 kW, mini hydropower plants are those ranging from 100 to 1,000 kW, and small hydropower (SHP) plants are those that produce from 1 to 30 MW.

Dams are individually unique structures, and dam construction represents the biggest structure in the basic infrastructure in all nations. Today, nearly 500,000 square kilometers of land are inundated by reservoirs in the world, capable of storing 6,000 km^3 of water. As a result of this distribution of fresh water in the reservoirs, small but measurable changes have occurred in the world. The total intsalled capacity of hydropower is 640,000 MW (26% of the theoretical potential) generating an estimated 2,380 TWh/year in the world, producing nearly 20% of the world's total supply of electricity. 27,900 MW of the total hydropower is at small scale sites, generating 115 TWh/year (Penche, 1998; Gleick, 1999). The NAFTA countries are, now, the biggest producers, along with Latin America and EU/EFTA regions, but it is estimated that Asia will be generating more hydroelectricty than NAFTA countries at the end of the next decade.

There is no universal consensus on the definition of small hydropower. Some countries of the European Union, such as Portugal, Spain, Ireland, Greece, and Belgium, accept 10 MW as the upper limit for installed capacity. In Italy the limit is 3 MW, in France 8 MW, in UK 5 MW, in Canada 20–25 MW, in the United States 30 MW, however, a value of up to 10 MW total capacity is becoming generally accepted as small hydropower in the rest of the world. If the total installed capacity of any hydropower system is bigger than 10 MW, it is generally accepted as a large hydropower system (Demirbas, 2006). Small hydropower can be further subdivided into mini hydropower usually defined as <500 kW and microhydropower is <100 kW. The definition of microhydropower or small-scale hydropower varies in different countries. Small hydropower is one of the most valuable energy to be offered to the rural comminutes' electrification. Small hydroelectricity growth is to decrease the gap of decentralized production for private sector and municipal activity production. Small-scale hydropower systems supply the energy from the flowing or running water and convert it to electrical energy. The potential for small hydropower systems depends on the availability of water flow where the resource exists. If a well-designed small hydropower system is established anywhere, it can fit with its surroundings and will have minimal negative impacts on the environment. Small hydropower systems allow the achievement of self-sufficiency by use of scarce natural water resources. These systems supply low cost of energy production, which is applied in many developing countries in the world.

A water power plant is in general a highly effective energy conversion system. There is no pollution of the environment, but objections are raised relative to the flooding of valuable real estate and scenic areas. Whether a particular hydroelectric installation will be economically competitive with a fossil fuel power plant will depend upon a number of factors, in particular, fuel and construction costs. As far as non-fossil energy is concerned, hydropower and nuclear power resources are the principal assets, due to their high production potential and their economic efficiency.

There are two types of turbines: reaction turbines and impulse turbines. In reaction turbines, such as Francis and Kaplan turbines, water pressure applies force onto the face of the runner blades, which decreases as it proceeds through the turbine. Francis turbines are generally used in a head range of 5 to 250 meters and can be designed with either a vertical or horizontal shaft. Kaplan turbines are axial-flow reaction turbines, generally used for low-heads. In impulse turbines water pressure is converted into kinetic energy in the form of a high-speed jet that strikes buckets mounted on the periphery of the runner. The most common impulse type is the Pelton turbine. It is generally used in installations with a head of 50 to several hundred meters. By adjusting the flow through the nozzle, a Pelton turbine can operate at high efficiency over wide range of head and flow conditions.

Typically, larger turbines have higher efficiencies. For example, efficiency is usually above 90% for turbines producing several hundred kW or more, whereas the efficiency of a microhydropower turbine of 10 KW is likely to be in the order of 60% to 80%. Two main types of generators are used in the small hydropower industry: synchronous and asynchronous generators. Both generator types are very well known throughout the industry and have been steadily improved to meet the needs and demands of the hydropower sector.

Hydraulic ram (hydram) pumps are water-lifting or water pumping devices that are powered by filling water. Hydram pumps have been used for over two centuries in many parts of the world. The pump works by using the energy of a large amount of water lifting a small height to lift a small amount of that water to a much greater height. Wherever a fall of water can be obtained, the ram pump can be used as a comparatively cheap, simple and reliable means of raising water to considerable heights. The main and unique advantage of hydram pumps is that with a continuous flow of water, a hydram pump operates automatically and continuously with no other external energy source – be it electricity or hydrocarbon fuel. It uses a renewable energy source (stream of water), hence ensures low running cost (Demirbas, 2006).

Lifting water from the source to a higher location can usually be carried out through a number of potential water-lifting options, depending on the particular site conditions. One means of lifting water is the hydraulic ram pump or hydram. The hydram is a self-actuating pump operated by the same principle as a water hammer. If correctly installed and properly maintained, it is a dependable and useful device that can lift water to a great height without the use of any other source of energy or fuel than water itself. The hydram can be used for lifting water from a source lying at a lower elevation to a point of use located at a higher elevation for domestic use, drinking, cooking and washing, and irrigation of small areas, gardens, and orchards.

Energy that is stored in the gravitational field is called gravitational potential energy, or potential energy due to gravity. If the object is being lifted at constant velocity, then it is not accelerating, and the net force on it is zero. When lifting something at a constant velocity the force that you lift with equals the weight of the object. So, the work done lifting an object is equal to its mass times the acceleration due to gravity times the height of the lift. As the object falls it travels faster and faster, and thus, picks up more and more kinetic energy. This increase in kinetic energy during the fall is due to the drop in gravitational potential energy during the fall. The gravitational potential energy becomes the kinetic energy of the falling object. The water above receives energy as it falls down the short waterfall. This energy was stored as potential energy in the gravitational field of the Earth and came out of storage as the water dropped. This energy which came out of the gravitational field ended up being expressed as the kinetic energy of the water.

The efficiency of small hydropower depends mainly on the performance of the turbine. Today, generators commonly have efficiency rates of 98–99%. As a general rule, larger and newer plants have higher efficiencies of up to 90%. Efficiency can be as low as 60% for old and smaller plants. Hydropower is the most efficient way to generate electricity. Modern hydro turbines can convert as much as 90% of the available energy into electricity. The best fossil fuel plants are only about 50% efficient.

Hydropower provides unique benefits, rarely found in other sources of energy. These benefits can be attributed to the electricity itself, or to side-benefits, often associated with reservoir development. Principles of sustainable development of particular relevance to energy provision include: improving the well-being of entire populations, ensuring that development is people-centered, participatory and equitable; integrating environmental concerns into decision-making processes; and taking into account the full range of costs and benefits of development.

Investment costs for SHP plants vary according to site-specific and local characteristics. The most important system and cost elements are: (1) civil engineering, (2) equipment, and (3) turbine. The electrical generator represents less than 5% of the total cost of a power plant and the efficiency of generators for new plants is already close to 100%. Yet standardization of generator equipment for small hydropower may further reduce installation and maintenance costs.

Despite the recent debates, few would disclaim that the net environmental benefits of hydropower are far superior to fossil-based generation. Hydroelectricity is produced for an average of 0.85 cents *per* kWh. In comparison with hydropower, thermal plants take less time to design, obtain approval, build, and recover investment.

The remaining economically exploitable potential is about 5,400 TWh/yr. An investment of at least USD\$ 1,500 billion would be necessary to realize such a program. The mean level of hydropower plants capacity in the range of 50 MW to 100 MW, some 20,000 plants would need to be built. In order to implement a plant construction program of this magnitude, a great deal of work (technical, financial and political) would need to be accomplished by all the investments involved, particularly in Asia, South America and Africa (Demirbas, 2006).

1.3.1.2 Geothermal Energy

Geothermal energy can be utilized in various forms such as electricity generation, direct use, space heating, heat pumps, greenhouse heating, and industrial sectors. The electricity generation is improving faster in countries with rich geothermal energy. As an energy source, geothermal energy has come of age. The utilization has increased rapidly during the last three decades.

Geothermal energy for electricity generation has been produced commercially since 1913, and for four decades on the scale of hundreds of MW both for electricity generation and direct use. In Tuscany, Italy, a geothermal plant has been operating since the early 1900s. There are also geothermal power stations in the USA, New Zealand, and Iceland. In Southampton (UK) there is a district heating scheme based on geothermal energy. Hot water is pumped up from about 1,800 meters below ground. The utilization has increased rapidly during the last three decades. In 2000, geothermal resources have been identified in over 80 countries and there are quantified records of geothermal utilization in 58 countries in the world (Fridleifsson, 2001). The direct uses of geothermal energy in the world's top countries are given in Table 1.16.

Geothermal energy is heat from the Earth's interior. Nearly all geothermal energy refers to heat derived from the Earth's molten core. Some of what is often referred to as geothermal heat derives from solar heating of the surface of the Earth, although it amounts to a very small fraction of the energy derived from the Earth's core. For centuries, geothermal energy was apparent only through anomalies in the Earth's crust that permit the heat from Earth's molten core to venture close to the surface. Volcanoes, geysers, fumaroles, and hot springs are the most visible surface manifestations of these anomalies. The Earth's core temperature is estimated by most geologists to be around 5,000 K to 7,000 K. For reference, that is nearly as hot as the surface of the sun (although, substantially cooler than the sun's interior). Moreover, although the Earth's core is cooling, it is doing so very slowly in a geological sense, since the thermal conductivity of rock is very low and, fur-

Table 1.16 The direct uses of geothermal energy in the world's top countries

Country	Installed MWt	Production (GWh/a)
China	2282	10531
Japan	1167	7482
USA	3766	5640
Iceland	1469	5603
Turkey	820	4377
New Zealand	308	1967
Georgia	250	1752
Russia	308	1707
France	326	1360
Sweden	377	1147
Hungary	473	1135
Mexico	164	1089
Italy	326	1048
Romania	152	797
Switzerland	547	663

Source: Demirbas, 2008

ther, the heat being radiated from the Earth is being substantially offset by radioactive decay and solar radiation. Some scientists estimate that over the past three billion years, the Earth may have cooled several hundred degrees (Kutz, 2007).

Geothermal energy is clean, cheap and renewable, and can be utilized in various forms such as space heating and domestic hot water supply, CO_2 and dry-ice production process, heat pumps, greenhouse heating, swimming and balneology (therapeutic baths), industrial processes and electricity generation. The main types of direct use are bathing, swimming and balneology (42%), space heating (35%), greenhouses (9%), fish farming (6%), and industry (6%) (Fridleifsson, 2001). Geothermal energy can be utilized in various forms such as electricity generation, direct use, space heating, heat pumps, greenhouse heating, and industrial usage. Electricity is produced with geothermal steam in 21 countries spread over all continents. Low temperature geothermal energy is exploited in many countries to generate heat, with an estimated capacity of about 10,000 MW thermal.

Geothermal energy has been used for centuries, where it is accessible, for aquaculture, greenhouses, industrial process heat, and space heating. It was first used for production of electricity in 1904 in Lardarello, Tuscany, Italy, with the first commercial geothermal power plant (250 kWe) was developed in 1913. Since then geothermal energy has been used for electric power production all over the world, but most significantly in the United States, the Philippines, Mexico, Italy, Japan, Indonesia, and New Zealand.

Direct application of geothermal energy can involve a wide variety of end uses, such as space heating and cooling, industry, greenhouses, fish farming, and health spas. It uses mostly existing technology and straightforward engineering. The technology, reliability, economics, and environmental acceptability of direct use of geothermal energy have been demonstrated throughout the world.

The world's total installed geothermal electric capacity was 7304 MWe in 1996. In much of the world electricity from fossil fuel-burning electricity plants can be provided at half the cost of new geothermal electricity. A comparison of the renewable energy sources shows the current electrical energy cost to be 2–10 US¢/kWh for geothermal and hydro, 5–13 US¢/kWh for wind, 5–15 US¢/kWh for biomass, 25–125 US¢/kWh for solar photovoltaic, and 12–18 US¢/kWh for solar thermal electricity. Of the total electricity production from renewables of 2826 TWh in 1998, 92% came from hydropower, 5.5% from biomass, 1.6% from geothermal and 0.6% from wind. Solar electricity contributed 0.05% and tidal power 0.02%. Geothermal energy is a power source that produces electricity with minimal environmental impact. Geothermal energy, with its proven technology and abundant resources, can make a significant contribution towards reducing the emission of greenhouse gases (Demirbas, 2006).

1.3.1.3 Wind Energy

Wind power has long been used for grain-milling and water-pumping applications. Significant technical progress since the 1980s, however, driven by advances in aerodynamics, materials, design, controls, and computing power, has led to economically competitive electrical energy production from wind turbines. Technology development, favorable economic incentives (due to its early development status and environmental benefits), and increasing costs of power from traditional fossil sources have led to significant worldwide sales growth since the early 1980s (Kutz, 2007).

Renewable energy from the wind has been used for centuries to power windmills to mill wheat or pump water. More recently, large wind turbines have been designed that are used to generate electricity. This source of energy is non-polluting and freely available in many areas. Wind turbines are becoming more efficient. The cost of the electricity they generate is falling.

Wind energy is renewable, mostly distributed generation characterized by large variations in the production. The wind energy sector is one of the fastest-growing energy sectors in the world. The world wind power engineering entered during the stage of industrial development. From 1991 until the end of 2002, global installed capacity has increased from about 2GW to over 31GW, with an average annual growth rate of about 26%. During this period, both prices of wind turbines and the cost of wind-generated electricity have been reduced. Wind power would need to be produced 1% more to compensate for the losses of hydropower production, when wind power production. Wind power production, on an hourly level for 1–2 days ahead, is more difficult to predict than other production forms, or the load (Demirbas, 2006).

There are wind farms around the world. Because the UK is on the edge of the Atlantic Ocean it has one of the best wind resources in Europe. Offshore wind farms in coastal waters are being developed because winds are often stronger blowing across the sea. Turbines can produce between 500kW and 1MW of electricity. Production of wind-generated electricity has risen from practically zero in the early 1980s to more than 7.5 TWh *per* year in 1995. Cumulative generating capacity worldwide has topped 6500 MW in late 1997 (Garg and Datta, 1998). Figure 1.3 shows the growth in world wind turbine installed capacity.

Wind energy is a significant resource; it is safe, clean, and abundant. Wind energy is an indigenous supply permanently available in virtually every nation in the world. Wind energy is abundant, renewable, widely distributed, clean, and mitigates the greenhouse effect if it is used to replace fossil-fuel-derived electricity. Wind energy has limitations based on geography and meteorology, plus there may be political or environmental problems (*e.g.*, dead birds) with putting turbines in (Garg and Datta, 1998). On the other hand, wind can cause air pollution by degradation and distribution of pieces of pollutants such as waste paper, straw, *etc.* Figure 1.4 shows the growth scenarios for global installed wind power (IEA, 2006b).

An advantage of wind turbines over some forms of renewable energy is that they can produce electricity whenever the wind blows (at night and also during the day). In theory, wind systems can produce electricity 24 hours every day, unlike PV systems that cannot make power at night. However, even in the windiest places,

Fig. 1.3 Growth in world wind turbine installed capacity *Source:* Demirbas, 2006

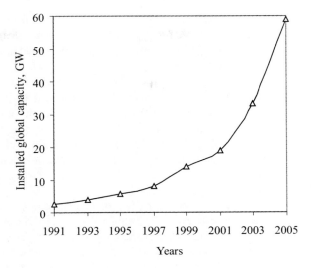

Fig. 1.4 Growth scenarios for global installed wind power *Source:* Demirbas, 2006

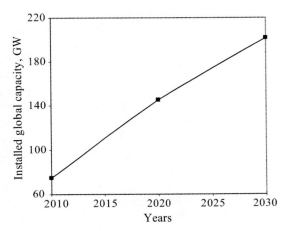

the wind does not blow all the time. So while wind farms do not need batteries for backup storage of electricity, small wind systems do need backup batteries.

Wind power in coastal and other windy regions is promising as well. By any measure the power in the wind is no longer an alternative source of energy. Wind energy has limitations based on geography and meteorology, plus there may be political or environmental problems (*e.g.*, dead birds) with putting turbines in.

1.3.1.4 Solar Energy

Solar energy is defined as that radiant energy transmitted by the sun and intercepted by the Earth. It is transmitted through space to the Earth by electromagnetic radiation with wavelengths ranging between 0.20 and 15μm. The availability of solar flux for terrestrial applications varies with season, time of day, location, and collecting surface orientation. In this chapter we shall treat these matters analytically (Kutz, 2007).

Energy comes from processes called solar heating (SH), solar home heating (SHH), solar dryers (SD), and solar cookers (SC), solar water heating (SWH), solar photovoltaic systems (SPV: converting sunlight directly into electricity), and solar thermal electric power (STEP: when the sun's energy is concentrated to heat water and produce steam, which is used to produce electricity). The major component of any solar system is the solar collector. Solar energy collectors are a special kind of heat exchanger that transform solar radiation energy to internal energy of the transport medium.

Solar dryers are used for drying fruits and spices. The three most popular types of SD are the box type, the cabinet type, and the tunnel type. The box type uses direct heat for dehydration. In cabinet type dryers, air heated by the collector dehydrates the food product, whereas in the tunnel type forced air circulation is used to distribute heat for dehydration. Cabinet and tunnel type dryers yield a high quality of dried products but they are very bulky and costly compared to box type dryers. Of about 800 dryers disseminated so far, 760 are of the box type (Pokharel, 2003).

Solar energy systems are solar home systems, solar photovoltaic (SPV) systems, solar water heating (SWH) systems, solar dryers, and solar cookers. These systems are installed and managed by a household or a small community. A solar home system is a PV system with a maximum capacity of 40 W. These systems are installed and managed by a household or a small community.

One of the most abundant energy resources on the surface of the Earth is sunlight. Today, solar energy has a tiny contribution in the world total primary energy supply of less than 1. Photovoltaic (PV) systems, other than solar home heating systems, are used for communication, water pumping for drinking and irrigation, and electricity generation. The total installed capacity of such systems is estimated at about 1000 kW. A solar home heating system is a solar PV system with a maximum capacity of 40 W. These systems are installed and managed by a household or a small community (Garg and Datta, 1998).

Like wind power markets, PV markets have seen rapid growth, and costs have fallen dramatically. The total installed capacity of such systems is estimated at about 1000 kW. The PV installed capacities are growing at a rate of 30% a year. Solar PV system has been found to be a promising energy in future. One of the most significant developments in renewable energy production is observed in PVs (EWEA, 2005; Reijnders, 2006; IEA, 2004). The PV will then be the largest renewable electricity source with a production of 25.1% of global power generation in 2040 (EWEA, 2005).

Photovoltaic (PV) systems, other than SHH systems, are used for communication, water pumping for drinking and irrigation, and electricity generation. Like wind power markets, PV markets have seen rapid growth and costs have fallen dramatically. The total installed capacity of such systems is estimated at about

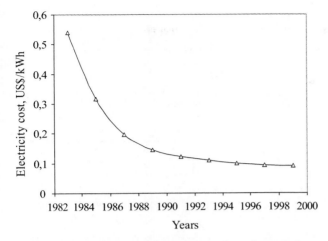

Fig. 1.5   Plot for electricity costs from solar thermal power plants

1000 kW. Solar photovoltaics and grid-connected wind installed capacities are growing at a rate of 30% a year (UNDP, 2000).

The solar thermal electricity power system is a device by the use of solar radiation for the generation of electricity through the solar thermal conversion (Xiao *et al.*, 2004). Figure 1.5 shows the plot for electricity costs from solar thermal power plants. Solar thermal electricity may be defined as the result of a process by which directly collected solar energy is converted to electricity through the use of some sort of heat to electricity conversion device (Mills, 2004). The last three decades have witnessed a trend in solar thermal electricity generation of increasing the concentration of sunlight (Kribus, 2002). There are three main systems of solar thermal electricity: solar towers, dishes and the parabolic troughs (Zhang *et al.*, 1998). Solar thermal power stations based on parabolic and heliostat trough concentrating collectors can soon become a competitive option on the world's electricity market (Trieb, 2000). Table 1.17 shows the economics and emissions of conventional technologies compared with solar power generation.

It has been estimated that a proposed solar system has a net fuel-to-electricity efficiency higher than 60% even when both the energy to produce high pressure oxygen and that to liquefy the captured CO_2 are taken into account (Kosugi and

Table 1.17   Economics and emissions of conventional technologies compared with solar power generation

Electricity generation technology	Carbon emissions g C/kWh	Generation costs US¢/kWh
Solar thermal and solar PV systems	0	9–40
Pulverized coal–natural gas turbine	100–230	5–7

Source: Demirbas, 2006

Table 1.18   Properties of common solar collectors

Type	Absorber	Motion	Temperature range, K
Flat plate collector	Flat	Stationary	300–350
Compound parabolic collector	Tubular	Stationary	330–510
Cylindrical trough collector	Tubular	Single-axis tracking	330–575
Parabolic trough collector	Tubular	Single-axis tracking	330–575
Parabolic dish collector	Point	Two-axes tracking	275–775
Heliostat trough collector	Point	Two-axes tracking	425–1260

Pyong, 2003). Based on futuristic trends, it is foreseen that by the year 2025, PV electricity may be more economical than fossil fuel electricity (Muneer *et al.*, 2005). A mathematical model for simulating an innovative design of a solar-heated anaerobic digester has been developed by Axaopoulos *et al.* (2001).

The major component of any solar system is the solar collector. Solar energy collectors are a special kind of heat exchanger that transform solar radiation energy to internal energy of the transport medium. A historical introduction into the uses of solar energy has been attempted followed by a description of the various types of collectors including flat-plate, compound parabolic, evacuated tube, parabolic trough, the Fresnel lens, parabolic dish, and heliostat field collectors (Kalogirou, 2004). Solar thermal electricity systems utilize solar radiation to generate electricity by the photo-thermal conversion method. There are basically two types of solar collectors: non-concentrating (stationary type) and concentrating. The properties of common solar collectors are given in Table 1.18. Temperatures exceeding about 2300 K should be feasible, given the appropriate concentration of the incident sunlight. The theoretical limit on the concentration of sunlight is about 40,000 (Kribus, 2002). A solar thermal conversion system with heliostat trough collector can achieve temperatures in excess of 2000 K.

Solar supported power plants (SSPPs) use technology that is similar to that used in coal-fired power plants. For example, SSPPs use similar steam-turbine generators and fuel delivery systems. Electricity costs are in the 9–40 US¢/kWh range (Table 1.17). The feasibility of combining gas and steam expansion in a power cycle has been extensively explored (Sorensen, 1983). Concentrated solar energy is used to produce steam the steam turns a turbine and drives a generator, producing electricity. Solar radiation can be used as a primary energy source or as a secondary energy source to power gas turbines.

1.3.1.5   Other Non-combustible Renewable Energy Sources

Wave energy, tidal energy, and ocean thermal energy conversion (OTEC) are the other non-combustible RES. Water energy sources are hydro, tidal and wave technologies. Marine energy sources are current, tidal, OTEC and wave technologies. The world wave resource is between 200 and 5000 GW mostly found in offshore locations (Garg and Datta, 1998). Wave energy converters fixed to the shoreline are likely to be the first to be fully developed and deployed, but waves are typically 2–3 times more powerful in deep offshore waters than at the shoreline. Wave energy can be harnessed in coastal areas, close to the shore. The first patent for a wave energy device was filed in Paris in 1799, and by 1973 there were 340 British patents for wave energy devices. By comparison to wind and PV, wave energy and tidal stream are very much in their infancy. Currently, around 1 MW of wave energy devices is installed worldwide, mainly from demonstration projects.

OTEC is an energy technology that converts solar radiation to electric power. OTEC systems use the ocean's natural thermal gradient to drive a power-producing cycle. As long as the temperature between the warm surface water and the cold deep water differs by about 20 K, an OTEC system can produce a significant amount of power. The oceans are thus a vast renewable resource, with the potential to help us produce billions of watts of electric power.

Water is a renewable energy source which can be used in electricity generation by using its lifting force (buoyant force). It is important an electricity generating apparatus using gravity and buoyancy can curtail costs of power generation and prevent environmental pollution and prevent destruction of an ecosystem. The hydraulic ram is an attractive solution for electricity generation where a large gravity flow exists. The wave conversion plant using buoyancy chambers is another solution for electricity generation using water lifting force. There are a good many reasons that water lifting force will be used future at many ranges for electricity generation. Hydraulic ram (hydram) pumps are water lifting or water pumping devices that are powered by filling water. Hydram pumps have been used for over two centuries in many parts of the world. It is a useful device that can pump water uphill from a flowing source of water above the source with no power requirement except the force of gravity.

1.3.2 Biorenewable Energy Sources

Biorenewable materials such as lignocellulosic materials, crops, grasses, animal wastes, and biogas are combustible renewable energy sources. Biomass resources include various natural and derived materials, such as woody and herbaceous species, wood wastes, bagasse, agricultural and industrial residues, waste paper, municipal solid waste, sawdust, biosolids, grass, waste from food processing, animal wastes, aquatic plants and algae, *etc*. The major organic components of biomass can be classified as cellulose, hemicelluloses and lignin. The average majority of biomass energy is produced from wood and wood wastes (64%), followed by municipal solid waste (24%), agricultural waste (5%) and landfill gases (5%) (Demirbas, 2000b). World production of biomass is estimated at 146 billion metric tons a year, mostly wild plant growth. Some farm crops and trees can produce up to 20 metric tons *per* acre of biomass a year (Cuff and Young, 1980).

There are three ways to use biomass. It can be burned to produce heat and electricity, changed to gas-like fuels such as methane, hydrogen and carbon monoxide or converted to a liquid fuel. Liquid fuels, also called biofuels, are predominantly two forms of alcohol: ethanol and methanol. Because biomass can be converted directly into a liquid fuel, it may someday supply much of our transportation fuel needs for cars, trucks, buses, airplanes, and trains. This is very important because nearly one-third of our nation's energy is now used for transportation.

The term biofuel is referred to as liquid or gaseous fuels for the transport sector that are predominantly produced from biomass. Biofuels generally offer many benefits including sustainability, reduction of greenhouse gas emissions, regional development, reduction of rural poverty, and fuel security. The biofuel economy, and its associated biorefineries, will be shaped by many of the same forces that shaped the development of the hydrocarbon economy and its refineries over the past century. The biggest difference between biofuels and petroleum fuels is the oxygen content. Biofuels have 10–45 wt.% oxygen while petroleum fuels have essentially none, making the chemical properties of biofuels very different than those from petroleum. There are two biomass-based liquid transportation fuels that might replace petroleum. Bioethanol can replace gasoline and biodiesel can replace diesel. Biomass can be converted into liquid and gaseous fuels through thermochemical and biological routes. A variety of fuels can be produced from biomass resources including liquid fuels, such as ethanol, methanol, biodiesel, Fischer–Tropsch diesel, and gaseous fuels, such as hydrogen and methane.

The most commonly used biofuel is ethanol, which is produced from sugarcane, corn, and other grains. A blend of gasoline and ethanol is already used in cities with high air pollution. However, ethanol made from biomass is currently more expensive than gasoline on a gallon-for-gallon basis. So, it is very important for scientists to find less expensive ways to produce ethanol from other biomass crops. Today, researchers have found new ways to produce ethanol from grasses, trees, bark, sawdust, paper, and farming wastes.

Diesel fuel can also be replaced by biodiesel made from vegetable oils. This fuel is now mainly being produced from canola oil and soybean oil. However, any plant oil such as palm, cottonseed, peanut, or sunflower may be used to produce biodiesel.

The replacement of petroleum with biorenewables for fuels and chemicals requires the identification of feedstock, intermediate, and product species that replace their fossil counterparts. Main biorenewable feedstocks are presented in Table 1.19. The properties of biorenewable feedstocks are compared to petroleum in Table 1.20.

Biofuels originate from plant oils, sugar beets, cereals, organic waste and the processing of biomass. Biological feedstocks that contain appreciable amounts of sugar – or materials that can be converted into sugar, such as starch or cellulose – can be fermented to produce ethanol to be used in gasoline engines. Ethanol feedstocks can be conveniently classified into three types: (1) sucrose-containing feedstocks, (2) starchy materials, and (3) lignocellulosic biomass. Feedstock for bioethanol is essentially comprised of sugar cane and sugar beet. The two are produced in geographically distinct regions. Sugar cane is grown in tropical and subtropical countries, while sugar beet is only grown in temperate climate countries.

Table 1.19 Main biorenewable feedstocks

Biorenewable feedstocks	Description
Sucrose-containing crops	Sugar cane, sugar beet, beet molasses, sweet sorghum
Starchy crops	Corn, potato, starchy food wastes
Lignocellulosic materials	Corn stover, wheat straw, perennial crops
Vegetable oils	Produced from soybeans, canola, palm
Animal fats	Tallow, lard, fish oil
Recycled products	Yellow grease, brown grease, cooking oil
Pyrolysis oil	Wood residues, waste biomass, organic solid wastes

Table 1.20 Typical properties of petroleum and biorenewable feedstocks and biodiesel

Property	Petroleum	Biorenewable		
	Crude oil	Soybean oil	Soybean oil methyl ester (biodiesel)	Pyrolysis oil
%C	83–86	77.7	78.1	60.8
%H	11–14	11.7	12.0	6.6
%O	–	10.5	9.8	32.5
%N	0.1–1	0.0011	0.001	0.3
%S	0.1–4	0.0006	0.0004	0.0008
Density (kg/L at 293 K)	0.86	0.92	0.87	1.23
Viscosity (mm^2/s at 313 K)	1.9–3.4	34.4	4.1	175
Higher heating value (MJ/kg)	45.3	39.6	41.2	26.7

Today the production cost of bioethanol from lignocellulosic materials is still too high, which is the major reason why bioethanol has not made its breakthrough yet. When producing bioethanol from corn or sugar cane the raw material constitutes about 40–70% of the production cost.

In European countries, beet molasses are the most utilized sucrose-containing feedstock. Sugar beet crops are grown in most of the EU-25 countries, and yield substantially more ethanol *per* hectare than wheat. The advantages with sugar beet are a lower cycle of crop production, higher yield, and high tolerance of a wide range of climatic variations, low water and fertilizer requirement. Sweet sorghum is one of the most drought resistant agricultural crops as it has the capability to remain dormant during the driest periods.

The conversion of sucrose into ethanol is easier compared to starchy materials and lignocellulosic biomass because previous hydrolysis of the feedstock is not required since this disaccharide can be broken down by the yeast cells; in addition, the conditioning of the cane juice or molasses favors the hydrolysis of sucrose.

Another type of feedstock that can be used for bioethanol production is starch-based materials. Starch is a biopolymer and defines as a homopolymer consisting only one monomer, D-glucose. To produce bioethanol from starch it is necessary to break down the chains of this carbohydrate for obtaining glucose syrup, which can be converted into ethanol by yeasts. The single greatest cost in the production

of bioethanol from corn, and the cost with the greatest variability, is the cost of
the corn.

Lignocellulosic biomass, such as agricultural residues (corn stover and wheat
straw), wood and energy crops, is attractive materials for ethanol fuel production
since it is the most abundant reproducible resource on Earth. Lignocellulosic perennial crops (*e.g.*, short rotation coppices and grasses) are promising feedstock
because of high yields, low costs, good suitability for low quality land (which is
more easily available for energy crops), and low environmental impacts.

1.3.2.1 Biomass

Biomass is the name given to all the Earth's living matter (Garg and Datta, 1998).
Biomass is organic material that has stored sunlight in the form of chemical energy. All biomass is produced by green plants converting sunlight into plant material through photosynthesis (Hall *et al.*, 1993).

Biomass provides a clean, renewable energy source that may dramatically improve our environment, economy, and energy security. Biomass energy generates
far less air emissions than fossil fuels, reduces the amount of waste sent to landfills, and decreases our reliance on foreign oil. Biomass energy also creates thousands of jobs and helps revitalize rural communities.

Biomass appears to be an attractive feedstock for three main reasons. First, it is
a renewable resource that could be sustainably developed in the future. Second, it
appears to have formidably positive environmental properties, reduced GHG
emissions, possibly reduced NO_x and SO_x depending on the fossil-fuels displaced.
However, also negative impacts, such as polycyclic aromatic hydrocarbons including polycyclic aromatic hydrocarbons, dioxins, furans, volatile organic compounds, and heavy metals especially when combusted in traditional stoves (Pastircakova, 2004). Third, it appears to have significant economic potential provided
that fossil fuel prices increase in the future.

Biomass now mainly represents only 3% of primary energy consumption in industrialized countries. However, much of the rural population in developing countries, which represents about 50% of the world's population, is reliant on biomass,
mainly in the form of wood, for fuel. Biomass accounts for 35% of primary energy
consumption in developing countries, raising the world total to 14% of primary
energy consumption (Demirbas, 2005).

Biomass is not an ideal form for fuel use. The heat content calculated on a dry
mass basis must be corrected for the natural water content that can reduce the net
heat available by as much as 20% in direct combustion applications. Faaij (2006)
suggested that gasification is commercially available with high overall efficiency
of about 40–50% when used with a combined cycle or 15–30% when used with
combined heat and power gas engine. Pyrolysis is less well developed than gasification. Faaij (2006) suggested that pyrolysis is not commercially available. It
produces 60–70% heat content of bio-oil/feed stock; smaller capacities compared

to gasification and anaerobic digestion. Pyrolysis is receiving attention as a pre-treatment step for long-distance transport of bio-oil.

Biomass energy generates far less air emissions than fossil fuels, reduces the amount of waste sent to landfills and decreases our reliance on foreign oil. Emission benefits depend on fossil-fuels replaced. GHG benefits (primarily CO_2) compared with all fossil fuels, and has multiple benefits compared with coal. However, acid gases and polycyclic aromatic hydrocarbons, *etc.*, may be increased, compared with natural gas in particular (especially if burnt in traditional stoves). Biomass energy also creates thousands of jobs and helps revitalize rural communities.

Biomass can be thermochemically converted into liquid fuel, gases, such as methane, carbon monoxide, or hydrogen by pyrolysis. Bioethanol can be obtained from cellulosic biomass by fermenting and distilling sugar solutions. Vegetable oils such as soybean and canola oils can be chemically converted into liquid fuel known as biodiesel. These fuels can be used as diesel fuel and gasoline in conventional engines with little modification to the system. Certain organic compounds, specifically municipal biosolids (sewage) and animal wastes (manures) can be biochemically converted into methane by anaerobic digestion. Energy crops, especially liquid biofuel (vegetable oils and biodiesels) crops have the potential to substitute a fraction of petroleum distillates and petroleum-based petrochemicals in the near future (Hamelinck and Faaij, 2002).

In the future, biomass has the potential to provide a cost-effective and sustainable supply of energy, while at the same time aiding countries in meeting their greenhouse gas reduction targets. In the short to medium term, biomass is expected to dominate energy supply. For the generation of electricity and heat, while using advanced combustion technology, organic wastes can be used as modern biomass. Also a number of crops and crop residues may fit modern bioenergy chains (Pimentel *et al.*, 1981; Haberl and Geissler, 2000; Hoogwijk *et al.*, 2003).

In industrialized countries, the main biomass processes utilized in the future are expected to be the direct combustion of residues and wastes for electricity generation, bio-ethanol and biodiesel as liquid fuels, and combined heat and power production from energy crops. In the short to medium term, biomass waste and residues are expected to dominate biomass supply, to be substituted by energy crops in the longer term. The future of biomass electricity generation lies in biomass integrated gasification/gas turbine technology, which offers high-energy conversion efficiencies. The electricity is produced by direct combustion of biomass, advanced gasification and pyrolysis technologies, which are almost ready for commercial scale use. Biomass is burned to produce steam, the steam turns a turbine and drives a generator, producing electricity. Because of potential ash build up (which fouls boilers, reduces efficiency and increases costs), only certain types of biomass materials are used for direct combustion. Gasifiers are used to convert biomass into a combustible gas (biogas). The biogas is then used to drive a high-efficiency, combined-cycle gas turbine. Heat is used to thermochemically convert biomass into a pyrolysis oil. The oil, which is easier to store and transport than solid biomass material, is then burned like petroleum to generate electricity. Pyrolysis also can convert biomass into phenolic oil, a chemical used to make wood adhesives, molded plastics and foam insulation. Wood adhesives are used to glue together polywood and other composite wood products. Biomass can also be converted into transportation fuels such as ethanol, methanol, biodiesel, and additives for reformulated gasoline. Biofuels are used in pure form or blended with gasoline (Demirbas, 2004).

1.3.2.2 Bioalcohols

Ethanol is the most widely used liquid biofuel. It is produced by fermentation of sugars, which can be obtained from natural sugars (*e.g.*, sugar cane, sugar beet), starches (*e.g.*, corn, wheat), or cellulosic biomass (*e.g.*, corn stover, straw, grass, wood). The most common feedstock is sugar cane or sugar beet, and the second common feedstock is corn starch. Currently, the use of cellulosic biomass is very limited due to expensive pretreatment the crystalline structure of cellulose required for breaking. Bioethanol is already an established commodity due to its ongoing non-fuel uses in beverages, and in the manufacture of pharmaceuticals and cosmetics. In fact, ethanol is the oldest synthetic organic chemical used by mankind. Table 1.21 shows ethanol production in different continents.

Bioethanol can be used as a 10% blend with gasoline without need for any engine modification. However, with some engine modification, bioethanol can be used at higher levels, for example, E85 (85% bioethanol).

Methanol is another possible replacement for conventional motor fuels. In fact, it has been considered as a possible large-volume motor fuel substitute at various times during gasoline shortages. In fact, it was often used in the early 20th century to power automobiles before the introduction of inexpensive gasoline. Later, synthetically produced methanol was widely used as a motor fuel in Germany during the Second World War. Again, during the oil crisis of 1970s, methanol blending with motor fuel received attention due to its availability and low cost. Methanol is poisonous and burns with an invisible flame. Similar to ethanol, methanol has a high octane rating and hence is suitable for Otto engines. Today, methanol is commonly used in biodiesel production for its reactivity with vegetable oils. Methanol can be used as one possible replacement for conventional motor fuels. Many tests have shown promising results using 85–100% by volume methanol as a transportation fuel in automobiles, trucks, and buses.

Before the 1920s, methanol was obtained from wood as a co-product of charcoal production, and hence was commonly known as wood alcohol. Methanol is

Table 1.21 Ethanol production in different continents (billion liters/year) in 2006

America	Asia	Europe	Africa	Oceania
22.3	5.7	4.6	0.5	0.2

Source: Demirbas, 2008

currently manufactured worldwide from syngas, which is derived from natural gas, refinery off-gas, coal, or petroleum, as

$$2H_2 + CO \rightarrow CH_3OH \tag{1.12}$$

The above reaction can be carried out in the presence of a variety of catalysts including Ni, Cu/Zn, Cu/SiO$_2$, Pd/SiO$_2$, and Pd/ZnO. In the case of coal, it is first pulverized and cleaned, then fed to a gasifier bed where it is reacted with oxygen and steam to produce the syngas. A 2:1 mole ratio of hydrogen to carbon monoxide is fed to a fixed catalyst bed reactor for methanol production. Also, the technology for making methanol from natural gas is already in place and in wide use. Current natural gas feedstocks are so inexpensive that even with tax incentives renewable methanol has not been able to compete economically.

The composition of biosyngas from biomass is shown in Table 1.22. The hydrogen to CO ratio in biosyngas is less than that from coal or natural gas, hence additional hydrogen is needed for full conversion to methanol. The gases produced from biomass can be steam reformed to produce hydrogen and followed by water–gas shift reaction to further enhance hydrogen content. Wet biomass can be easily gasified using supercritical water conditions. The main production pathways for methanol and biomethanol are compared in Table 1.23.

Methanol can be produced from biomass essentially any primary energy source. Thus, the choice of fuel in the transportation sector is to some extent determined by the availability of biomass. As regards the difference between hydrogen and methanol production costs, conversion of natural gas, biomass and coal into hydrogen is generally more energy efficient and less expensive than conversion into methanol.

Table 1.22   Composition of biosyngas from biomass gasification

Constituents	% by volume (dry and nitrogen free)
Carbon monoxide (CO)	28–36
Hydrogen (H$_2$)	22–32
Carbon dioxide (CO$_2$)	21–30
Methane (CH$_4$)	8–11
Ethene (C$_2$H$_4$)	2–4

Table 1.23   Main production facilities of methanol and biomethanol

Methanol	Biomethanol
Catalytic synthesis from CO and H$_2$	Catalytic synthesis from CO and H$_2$
Natural gas	Distillation of liquid from wood pyrolysis
Petroleum gas	Gaseous products from biomass gasification
Distillation of liquid from coal pyrolysis	Synthetic gas from biomass and coal

1.3.2.3 Bio-oil

The term bio-oil is used mainly to refer to liquid fuels. There are several reasons for bio-oils to be considered as relevant technologies by both developing and industrialized countries. They include energy security reasons, environmental concerns, foreign exchange savings, and socioeconomic issues related to the rural sector. Bio-oils are liquid or gaseous fuels made from biomass materials, such as agricultural crops, municipal wastes and agricultural and forestry byproducts *via* biochemical or thermochemical processes.

Bio-oil has a higher energy density than biomass, and can be obtained by quick heating of dried biomass in a fluidized bed followed by cooling. The byproducts char and gases can be combusted to heat the reactor. For utilization of biomass in remote locations, it is more economical to convert into bio-oil and then the transport the bio-oil. Bio-oil can be used in vehicle engines – either totally or partially in a blend.

Biomass is dried and then converted to an oily product as known bio-oil by very quick exposure to heated particles in a fluidized bed. The char and gases produced are combusted to supply heat to the reactor, while the product oils are cooled and condensed. The bio-oil is then shipped by truck from these locations to the hydrogen production facility. It is more economical to produce bio-oil at remote locations and then ship the oil, since the energy density of bio-oil is higher than biomass. For this analysis, it was assumed that the bio-oil would be produced at several smaller plants that are closer to the sources of biomass, such that lower cost feedstocks can be obtained.

1.3.2.4 Biodiesel

Vegetable oil (m)ethyl esters, commonly referred to as "biodiesel", are prominent candidates as alternative diesel fuels. The name biodiesel has been given to transesterified vegetable oil to describe its use as a diesel fuel. There has been renewed interest in the use of vegetable oils for making biodiesel due to its less polluting and renewable nature as against conventional diesel, which is a fossil fuel that may potentially become exhausted. Biodiesel is technically competitive with or offers technical advantages compared to conventional petroleum diesel fuel. Vegetable oils can be converted to their (m)ethyl esters *via* a transesterification process in the presence of a catalyst. Methyl, ethyl, 2-propyl and butyl esters have been prepared from vegetable oils through transesterification using potassium and/or sodium alkoxides as catalysts. The purpose of the transesterification process is to lower the viscosity of the oil. Ideally, transesterification is a potentially less expensive way of transforming the large, branched molecular structure of the bio-oils into smaller, straight-chain molecules of the type required in regular diesel combustion engines.

Biodiesel is obtained by reacting methanol with vegetable oils using a process called transesterification. The purpose of the transesterification process is to lower the viscosity and oxygen content of the vegetable oil. In this process, an alcohol (*e.g.*, methanol, ethanol, butanol) is reacted with fatty acids in the presence of an alkali catalyst (*e.g.*, KOH, NaOH) to produce biodiesel and glycerol. Being immiscible, biodiesel is easily separated from glycerol. Transesterification is an inexpensive way of transforming the large, branched molecular structure of the vegetable oils into smaller, straight-chain molecules of the type required in regular diesel combustion engines. Most biodiesel is currently made from soybean or palm oil. The high value of soybean oil as a food product makes production of a cost-effective biodiesel very challenging. However, there are large amounts of low-cost oils and fats (*e.g.*, restaurant waste, beef tallow, pork lard, yellow grease) that may possibly be converted to biodiesel. The problem with processing these low cost oils and fats is that they often contain large amounts of free fatty acids (FFAs) that cannot be converted to biodiesel using an alkaline catalyst.

Biodiesel is characterized by its physical and fuel properties including density, viscosity, iodine value, acid value, cloud point, pour point, gross heat of combustion, and volatility. Biodiesel produces slightly lower power and torque, hence the fuel consumption is higher than No. 2 diesel. However, biodiesel is better than diesel in terms of sulfur content, flash point, aromatic content, and biodegradability.

The cost of biodiesels varies depending on the base stock, geographic area, variability in crop production from season to season, the price of crude petroleum, and other factors. Biodiesel has over double the price of petroleum diesel. The high price of biodiesel is in large part due to the high price of the feedstock. However, biodiesel can be made from other feedstocks, including beef tallow, pork lard, and yellow grease.

Biodiesel is an environmentally friendly alternative liquid fuel that can be used in any diesel engine without modification. There has been renewed interest in the use of vegetable oils for making biodiesel due to its less polluting and renewable nature as against conventional petroleum diesel fuel. If biodiesel were valorized efficiently for energy purposes, this would be of benefit for the environment and the local population, job creation, provision of modern energy carriers to rural communities, avoidance of urban migration, and the reduction of CO_2 and sulfur levels in the atmosphere.

1.3.2.5   Fischer–Tropsch Products

In 1923, Franz Fischer and Hans Tropsch converted syngas into larger, useful organic compounds. With further refinement using a transition metal catalyst, they converted CO and H_2 mixtures into liquid hydrocarbons. The process is now widely utilized and is known as Fischer–Tropsch (FT) synthesis (FTS). The first FT plants began operation in Germany in 1938 and operated till the Second World War. Since 1955, Sasol, a world-leader in the commercial production of liquid fuels and chemicals from coal, has been operating FT processes. A major advantage of the FTS is the flexibility in feedstocks as the process starts with syngas, which can be obtained from a variety of sources (*e.g.*, natural gas, coal, biomass).

A major disadvantage of FTS is the polymerization-like nature of the process, yielding in a wide product spectrum, ranging from compounds with low molecular mass, methane, to products with very high molecular mass like heavy waxes.

Hydrocarbon synthesis from biomass-derived syngas (biosyngas) has been investigated as a potential way to use biomass. FTS can be utilized to produce chemicals, gasoline and diesel fuel. FT products are predominantly linear molecules hence the quality of the diesel fuel is very high. Since purified synthesis gas is used in FTS, all the products are free of sulfur and nitrogen.

Depending upon the operating conditions (*e.g.*, temperature, feed gas composition, pressure, catalyst type and promoters), FTS can be tuned to produce a wide range of olefins, paraffins and oxygenated products (*e.g.*, alcohols, aldehydes, acids, ketones). For example, the high-temperature fluidized bed FT reactors with iron catalysts are ideal for the production of large amounts of linear-olefins. FTS started with one mole of CO reacting with two moles of H_2 to produce hydrocarbon chain extension ($-CH_2-$), as follows

$$CO + 2H_2 \rightarrow -CH_2- + H_2O \qquad \Delta H = -165 \text{ kJ/mol} \qquad (1.13)$$

$-CH_2-$ is a building block for longer hydrocarbons. The overall reactions can be written as:

$$nCO + 2nH_2 \rightarrow (-CH_2-)_n + nH_2O \qquad (1.14)$$

$$nCO + (2n+1)H_2 \rightarrow C_nH_{2n+2} + nH_2O \qquad (1.15)$$

$$nCO + (n+m/2)H_2 \rightarrow C_nH_m + nH_2O \qquad (1.16)$$

From the above exothermic reactions, a mixture of paraffins and olefins is obtained. The typical operating temperature range is 200–350°C and the pressure range is 15–40 bar. Iron catalysts have a higher tolerance for sulfur, are cheaper, and produce more olefins and alcohols. However, the lifetime of the Fe catalysts is short (*e.g.*, typically 8 weeks in commercial installations).

1.3.2.6   Biogas

Biogas is an environmentally friendly, clean, cheap, and versatile gaseous fuel. It is mainly a mixture of methane and carbon dioxide obtained by anaerobic digestion of biomass, sewage sludge, animal wastes, and industrial effluents. Anaerobic digestion occurs in the absence of air and is typically carried out for a few weeks. Typical compositions of biogas and landfill gas are given in Table 1.24. CH_4 and CO_2 make up around 90% of the gas volume produced, both of which are greenhouse gases. However, CO_2 is recycled back by the plants, but methane can contribute to global warming. Hence, the capture and fuel use of biogas is beneficial in two ways: (a) fuel value, and (b) conversion of CH_4 into CO_2, a plant recyclable carbon.

Anaerobic digestion in landfills occurs in a series of stages, each of which is characterized by the increase or decrease of specific bacterial populations and the

Table 1.24 Composition of biogas and landfill gas

Component	Chemical formula	Landfill gas	Biogas
Methane (% by vol.)	CH_4	40–60	55–70
Carbon dioxide (% by vol.)	CO_2	20–40	30–45
Nitrogen (% by vol.)	N_2	2–20	0–2
Oxygen (% by vol.)	O_2	<1	<1
Heavier hydrocarbons (% by vol.)	C_nH_{2n+2}	<1	0
Hydrogen sulfide (ppm)	H_2S	20–200	100–500
Ammonia (NH_3) (ppm)	NH_3	0	80–100
Volatile organic compounds (% by vol.)	–	0.25–0.50	0

Source: Demirbas, 2008

formation and utilization of certain metabolic products. By volume landfill gas contains about 50% methane. Producing energy from landfill gas improves local air quality, eliminates a potential explosion hazards, and reduces greenhouse gas emissions to the atmosphere. Hydrogen, produced by passing an electrical current through water, can be used to store solar energy and regenerate it when needed for nighttime energy requirements. The solid waste management practices are collection, recovery and disposal, together with the results of cost analyses.

1.3.2.7 Biohydrogen

Hydrogen is not a primary fuel. It must be manufactured from water using energy from either fossil or non-fossil sources. The use of hydrogen fuel has the potential to improve the global climate and air quality. In addition, using a fuel cell, hydrogen can be utilized with high energy efficiency, which is likely to increase its share as an automotive fuel. The share of hydrogen in total automotive fuel in the future is depicted in Fig. 1.6. A fuel cell is a device or an electrochemical engine that converts the energy of a fuel directly to electricity and heat without combustion. It consists of two electrodes sandwiched around an electrolyte. When hydrogen passes over one electrode and oxygen over the other, electricity is generated. The reaction product is water vapor and the fuel cells are clean, quiet, and efficient A variety of technologies are available to produce hydrogen from biomass. Some biomass materials and conversion technologies used for hydrogen production are listed in Table 1.25.

The gases produced can be steam reformed to produce hydrogen and followed by a water–gas shift reaction to further enhance hydrogen production. When the moisture content of biomass is higher than 35%, it can be gasified in a supercritical water condition. Supercritical water gasification is a promising process to gasify biomass with high moisture contents due to its high gasification ratio (100% achievable), and the high hydrogen volumetric ratio (50% achievable) hydrogen produced by biomass gasification was reported to be comparable to that obtained by natural gas reforming. The process is more advantageous than fossil

Fig. 1.6 Share of hydrogen in total automotive fuel in the future
Source: Demirbas, 2008

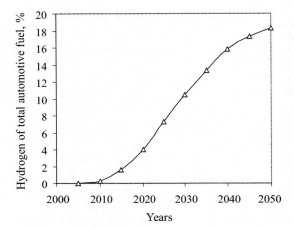

Table 1.25 List of some biomass materials and conversion technologies used for hydrogen production

Biomass species	Main conversion process
Bio-nut shell	Steam gasification
Olive husk	Pyrolysis
Tea waste	Pyrolysis
Crop straw	Pyrolysis
Black liquor	Steam gasification
Municipal solid waste	Supercritical water reforming
Crop grain residue	Supercritical water reforming
Pulp and paper waste	Microbial fermentation
Petroleum basis plastic waste	Supercritical water reforming
Manure slurry	Microbial fermentation

fuel reforming with respect to environmental benefits. It is expected that biomass thermochemical conversion will be one of the most economical large-scale renewable hydrogen technologies.

The strategy described herein is based on producing hydrogen from biomass pyrolysis using a co-product strategy to reduce the cost of hydrogen and it is concluded that only this strategy can compete with the cost of the commercial hydrocarbon-based technologies. This strategy will demonstrate how hydrogen and biofuel are economically feasible and can foster the development of rural areas when practiced on a larger scale. The processing of biomass to activated carbon is an alternative route to hydrogen with a valuable co-product that is practiced commercially.

References

Axaopoulos, P., Panagakis, P., Tsavdaris, A., Georgakakis, D. 2001. Simulation and experimental performance of a solar heated anaerobic digester. Solar Energy 70:155–164.

COSO (Crude Oil Supply Outlook). 2007. Report to the Energy Watch Group EWG-Series No 3/2007. http://www.energywatchgroup.org/fileadmin/global/pdf/EWG_Oilreport_10-2007.pdf.

Cuff, D.J., Young, W.J. 1980. US energy atlas. Free Press/McMillan, New York.

Demirbas, A. 2000. Biomass resources for energy and chemical industry. Energy Edu Sci Technol 5:21–45.

Demirbas, A. 2004. The importance of biomass. Energy Sources 26: 361–366.

Demirbas, A. 2005. Potential applications of renewable energy sources, biomass combustion problems in boiler power systems and combustion related environmental issues. Progress Energy Combus Sci 31:171–192.

Demirbas, A. 2006. Energy priorities and new energy strategies. Energy Edu Sci Technol 16:53–109.

Demirbas, A. 2007. Utilization of coals as sources of chemicals. Energy Sources 29:677–684.

Demirbas, A. 2008. Biodiesel: A realistic fuel alternative for diesel engines. Springer, London.

Dorf, R.C. 1977. Energy resources and policy. Addison-Wesley, Los Angeles, CA.

Dry, M.E. 1999. Fischer–Tropsch reactions and the environment. Appl Catal A: General 189:185–190.

Edinger, R., Kaul, S. 2000. Humankind's detour toward sustainability: Past, present, and future of renewable energies and electric power generation. Renew Sustain Energy Rev 4:295–313.

EREC. 2006. European Renewable Energy Council (EREC). Renewable energy scenario by 2040, EREC Statistics, Brussels.

EWEA. 2005. European Wind Energy Association (EWEA). Report: Large scale integration of wind energy in the European power supply: Analysis, issues and recommendations.

Faaij, A.P.C., 2006. Bio-energy in Europe: Changing technology choices. Energy Policy 34, 322–330.

Fridleifsson, I.B. 2001. Geothermal energy for the benefit of the people. Renew Sustain Energy Rev 5:299–312.

Garg, H.P., Datta, G. 1998. Global status on renewable energy, in Solar Energy Heating and Cooling Methods in Building, International Workshop: Iran University of Science and Technology. 19–20 May.

Gleick, P.H. 1999. The world's water. The biennial report on freshwater resources, Pacific Institute for Studies in Development, Environment, and Security, Oakland, CA.

Haberl, H., Geissler, S. 2000. Cascade utilization of biomass: strategies for a more efficient use of a scarce resource. Ecological Engineering 16:S111–S121.

Hacisalihoglu, B., Demirbas, A.H., Hacisalihoglu, S. 2008. Hydrogen from gas hydrate and hydrogen sulfide in the Black Sea. Energy Edu Sci Technol 21:109–115.

Hall, D.O., Rosillo-Calle, F., Williams, R.H., Woods, J. 1993. Biomass energy supply and prospects. In: Johansson, T.B, Kelly, H., Reddy, A.K.N., Williams, R.H., editors. Renewable energy: Sources for fuel and electricity. Island Press, Washington D.C., p. 593–651.

Hamelinck, C.N., Faaij, A.P.C. 2002. Future prospects for production of methanol and hydrogen from biomass. Journal of Power Sources 111:1–22.

Hoogwijk, M., Faaij, A., van den Broek, R., Berndes, G., Gielen, D., Turkenburg, W. 2003. Exploration of the ranges of the global potential of biomass for energy. Biomass and Bioenergy 25:119–133.

IEA (International Energy Agency). 2004. Biofuels for Transport: An International Perspective. 9, rue de la Fédération, 75739 Paris, cedex 15, France (available from: www.iea.org).

IEA. 2006a. The International Energy Agency (IEA). Key world energy statistics. Paris. http://www.iea.org/textbase/nppdf/free/2006/key2006.pdf.

EIA. 2006b. Annual Energy Outlook 2006. Energy Information Administration (EIA). E.I. Administration, U.S. Department of Energy.

IEA. 2007. The International Energy Agency (IEA). Key world energy statistics. Paris. http://www.iea.org/Textbase/nppdf/free/2007/key_stats_2007.pdf.

Karaosmanoglu, F., Aksoy, H.A. 1988. The phase separation problem of gasoline-ethanol mixture as motor fuel alternatives. J Thermal Sci Technol 11:49–52.

Kosugi,T., Pyong, S.P. 2003. Economic evaluation of solar thermal hybrid H_2O turbine. Energy 28:185–198.

Kribus, A. 2002. A high-efficiency triple cycle for solar power generation. Solar Energy 72:1–11.

Kutz, M. (ed.) 2007. Environmentally conscious alternative energy production. Wiley, Hoboken, NJ.

Lee, S.-Y., Holder, G.D. 2001. Methane hydrates potential as a future energy source. Fuel Proc. Technol. 71:181–186.

Mills, D. 2004. Advances in solar thermal electricity technology. Solar Energy 76:19–31.

Muneer, T., Asif, M., Munawwar, S. 2005. Sustainable production of solar electricity with particular reference to the Indian economy. Renew Sustain Energy Rew 9:444–473.

Pastircakova, K., 2004. Determination of trace metal concentrations in ashes from various biomass materials. Energy Education Science and Technology 13, 97–104.

Penche C. 1998. Layman's guidebook on how to develop a small hdro site. ESHA, European Small Hydropower Association, Directorate General for Energy (DG XVII).

Pimentel, D., Moran, M.A., Fast, S., Weber, G., Bukantis, R., Balliett, L., Boveng, P., Cleveland, C., Hindman, S., Young, M. 1981. Biomass energy from crop and forest residues. Science 212:1110–1115.

Pokharel, S. 2003. Promotional issues on alternative energy technologies in Nepal. Energy Policy 31:307–318.

Reijnders, L. 2006. Conditions for the sustainability of biomass based fuel use. Energy Policy 34:863–876.

Sorensen, H.A. Energy Conversion Systems. Wiley, New York, 1983.

Schulz, H. 1999. Short history and present trends of FT synthesis. Applied Catalysis A: General 186:1–16.

Trieb, F. 2000. Competitive solar thermal power stations until 2010 – The challenge of market introduction. Renew Energy 19:163–171.

UNDP. 2000. United Nations Development Programme (UNDP).World energy assessment 2000 – Energy and the challenge of sustainability. New York: UNDP; (ISBN 9211261260).

Xiao, C., Luo, H., Tang, R., Zhong, H. 2004. Solar thermal utilization in China. Renew Energy 29:1549–1556.

Zhang, Q.-C., Zhao, K., Zhang, B.-C., Wang, L.-F., Shen, Z.-L., Zhou, Z.-J., Lu, D.-Q., Xie, D.-L., Li, B.-F. 1998. New cermet solar coatings for solar thermal electricity applications. Solar Energy 64:109–114.

Chapter 2
Biomass Feedstocks

2.1 Introduction to Biomass Feedstocks

In the most biomass-intensive scenario, modernized biomass energy will by 2050 contribute about one half of total energy demand in developing countries (IPCC, 1997). The biomass intensive future energy supply scenario includes 385 million hectares of biomass energy plantations globally in 2050 with three quarters of this area established in developing countries (Kartha and Larson, 2000). Various scenarios have put forward estimates of biofuel from biomass sources in the future energy system. The availability of the resources is an important factor in the co-generative use of biofuel in the electricity, heat or liquid fuel market.

Biomass is a generic term for all vegetable material. It is generally a term for material derived from growing plants or from animal manure. Biomass has a unique characteristic compared with other forms of renewable energy: it can take various forms such as liquids, gases, and solids, and so can be used for electricity or mechanical power generation and heat. If biomass could be converted into useful energy, the consumption of fossil fuel and greenhouse gas emissions would be decreased. Furthermore, the use of biomass could lead to the creation of a new biomass industry, which would help revitalize agriculture and forestry, leading to social stability as well as economic stimulus (Saga *et al.*, 2008).

The first biomass sources used on Earth were wood and dry grass, and for a long time these were used for cooking and heating. Products with fuel characteristics that are obtained biotechnologically from plant sources are defined as biomass energy sources. Generally, biomass is an easily obtained energy source and therefore it is especially important for countries with forest and agriculture-based economics and those having limited sources of energy (Karaosmanoglu and Aksoy, 1988).

In the last decade, there has been rapid progress in the biofuel marketing trend. This has increased production capacity, international material flows, competition with conventional agriculture, competition with forest industries, and international

A. Demirbas, *Biofuels,*
© Springer 2009

trade flows. In turn, this has led to a strong international debate about the sustainability of biofuels production. Biomass is the most important bioenergy option at present and is expected to maintain that position during the first half of this century (Hamelinck and Faaij, 2006; IPCC, 2007).

There are three ways to use biomass. It can be burned to produce heat and electricity, changed to gas-like fuels such as methane, hydrogen and carbon monoxide or changed to a liquid fuel. Liquid fuels, also called biofuels, include mainly two forms of alcohol: ethanol and methanol. Because biomass can be changed directly into a liquid fuel, it may someday supply much of our transportation fuel needs for cars, trucks, buses, airplanes, and trains. This is very important because nearly one-third of our nation's energy is now used for transportation (Tewfik, 2004).

2.1.1   Definitions

The term biomass (Greek, bio, life + maza or mass) refers to non-fossilized and biodegradable organic material originating from plants, animals, and microorganisms derived from biological sources. Biomass includes products, byproducts, residues and waste from agriculture, forestry and related industries, as well as the non-fossilized and biodegradable organic fractions of industrial and municipal solid wastes. Biomass also includes gases and liquids recovered from the decomposition of non-fossilized and biodegradable organic material. Biomass residues mean biomass byproducts, residues and waste streams from agriculture, forestry, and related industries.

According to another definition, the term "biomass" refers to wood, short-rotation woody crops, agricultural wastes, short-rotation herbaceous species, wood wastes, bagasse, industrial residues, waste paper, municipal solid waste, sawdust, biosolids, grass, waste from food processing, aquatic plants and algae animal wastes, and a host of other materials (Demirbas, 2008a). Biomass is the plant and animal material, especially agricultural waste products, used as a source of fuel. Biomass as the solar energy stored in chemical form in plant and animal materials is among the most precious and versatile resources on Earth. It is a rather simple term for all organic materials that stem from plants, trees, crops and algae.

Biomass is organic material that has stored sunlight in the form of chemical energy. Biomass is commonly recognized as an important renewable energy, which is considered to be such a resource that during the growth of plants and trees; solar energy is stored as chemical energy *via* photosynthesis, which can be released *via* direct or indirect combustion. Figure 2.1 shows the carbon cycle, photosynthesis, and main steps of biomass technologies.

Chemical energy and organic carbon are obtained by organisms either directly or indirectly *via* the photosynthetic conversion of solar energy. These organisms have evolved metabolic machineries for the photochemical reduction of carbon dioxide to organic matter. The majority of the bioengineering strategies for biochemically derived fuels involve options for the disposition of organic matter

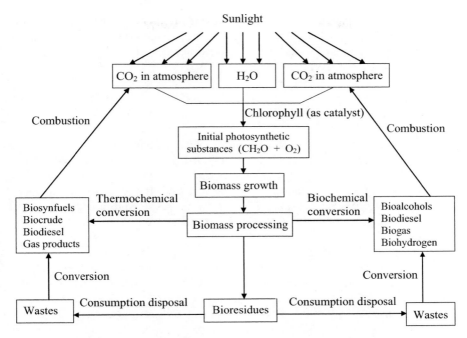

Fig. 2.1 Carbon cycle, photosynthesis, and main steps of biomass technologies

produced *via* photosynthate. The bulk of the presently exploited photosynthate is directed toward the production of wood, food, and feed. During processing and consumption, waste organic materials are generated, which can be used for energy production *via* combustion, pyrolysis, or biochemical conversions to ethanol, hydrogen, methane, and isopropanol.

All biomass is produced by green plants converting sunlight into plant material through photosynthesis. Photosynthesis is a carbon fixation reaction by reduction of carbon dioxide. The fixation or reduction of carbon dioxide is a light-independent process. Although some of the steps in photosynthesis are still not completely understood, the overall photosynthetic equation has been known since the 1800s.

The photosynthetic apparatus and the mechanisms by which it operates have been intensively investigated over the past 40 to 50 years. The photosynthesis is that of three series of interconnected oxidation-reduction reactions: The first involves the evolution of oxygen from water. The second is the transfer of H atoms to a primary hydrogen acceptor. The third is the reduction of CO_2 to carbohydrates by the primary hydrogen acceptor. The light energy required for photosynthesis is used to drive the H atoms against the potential gradient.

The photochemical stage of photosynthesis consists of two separate steps, I and II. The products of light reaction II are an intermediate oxidant and a strong oxidant that is capable of oxidizing water to oxygen. An intermediate oxidant and a strong reductant that can reduce carbon dioxide are produced in light reaction I. The two light reactions involve two pigment systems, photosystems I and II, interconnected by enzymatic reactions coupled with photophosphorylation yielding adenosine triphosphate (ATP).

2.1.1.1 Biomass Components

The components of biomass include cellulose, hemicelluloses, lignin, extractives, lipids, proteins, simple sugars, starches, water, hydrocarbons, ash, and other compounds. Two larger carbohydrate categories that have significant value are cellulose and hemicelluloses (holocellulose). The lignin fraction consists of non-sugar type macromolecules. Three structural components are cellulose, hemicelluloses, and lignin, which have the rough formulae $CH_{1.67}O_{0.83}$, $CH_{1.64}O_{0.78}$, and $C_{10}H_{11}O_{3.5}$, respectively.

The basic structure of all wood and woody biomass consists of cellulose, hemicelluloses, lignin, and extractives. Their relative composition is shown in Table 2.1. Softwoods and hardwoods differ greatly in wood structure and composition. Hardwoods contain a greater fraction of vessels and parenchyma cells. Hardwoods have a higher proportion of cellulose, hemicelluloses, and extractives than softwoods, but softwoods have a higher proportion of lignin. Hardwoods are denser than softwoods.

Cellulose is a linear polymer composed of repeating anhydroglucose units. Cellulose is a remarkably pure organic polymer, consisting solely of units of anhydroglucose held together in a giant straight chain molecule (Demirbas, 2000). These anhydroglucose units are bound together by β-(1,4)-glycosidic linkages. Due to this linkage, cellobiose is established as the repeat unit for cellulose chains (Fig. 2.2). Cellulose must be hydrolyzed to glucose before fermentation to ethanol. By forming intramolecular and intermolecular hydrogen bonds between OH groups within the same cellulose chain and the surrounding cellulose chains, the chains tend to be arranged parallel and form a crystalline supermolecular structure. Then, bundles of linear cellulose chains (in the longitudinal direction) form a microfibril which is oriented in the cell wall structure (Hashem *et al.*, 2007). Cellulose is insoluble in most solvents and has a low accessibility to acid and enzymatic hydrolysis.

The second major chemical species in wood are the hemicelluloses. They are amorphous polysaccharides, such as xylans, galactoglucomannans, arabinogalactans, glucans, and galactans. The hemicelluloses, unlike cellulose, not only contain glucose units, but they are also composed of a number of different pentose and

Table 2.1 Structural composition of wood (wt% of dry and ash free sample)

Wood species	Cellulose	Hemicelluloses	Lignin	Extractives
Hardwood	43–48	27–35	16–24	2–8
Softwood	40–44	24–29	26–33	1–5

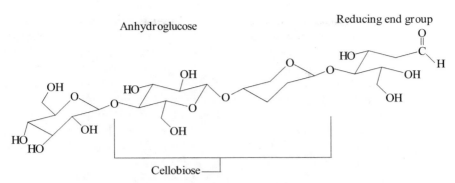

Fig. 2.2 The structure of cellulose. Anhydroglucose is the monomer of cellulose, cellobiose is the dimer
Source: Demirbas, 2008b

hexose monosaccharides. Hemicelluloses tend to be much shorter in length than cellulose, and the molecular structure is slightly branched.

Unlike cellulose, hemicelluloses consist of different monosaccharide units. In addition, the polymer chains of hemicelluloses have short branches and are amorphous. Because of the amorphous morphology, hemicelluloses are partially soluble or swellable in water. Hemicelluloses (arabinoglycuronoxylan and galactoglucomammans) are related to plant gums in composition, and occur in much shorter molecule chains than cellulose. Hemicelluloses are derived mainly from chains of pentose sugars, and act as the cement material holding together the cellulose micelles and fiber (Theander, 1985).

Among the most important sugar of the hemicelluloses component is xylose. In hardwood xylan, the backbone chain consists of xylose units which are linked by β-(1,4)-glycosidic bonds and branched by α-(1,2)-glycosidic bonds with 4-O-methylglucuronic acid groups (Hashem *et al.*, 2007). In addition, O-acetyl groups sometime replace the OH groups in position C_2 and C_3 (Fig. 2.3a). For softwood xylan, the acetyl groups are fewer in the backbone chain. However, softwood xylan has additional branches consisting of arabinofuranose units linked by α-(1,3)-glycosidic bonds to the backbone (Fig. 2.3b). Hemicelluloses are largely soluble in alkali and, as such, are more easily hydrolyzed (Timell, 1967; Wenzl *et al.*, 1970; Goldstein, 1981).

Lignin (sometimes "lignen") is a chemical compound that is most commonly derived from wood and is an integral part of the cell walls of plants, especially in tracheids, xylem fibers, and sclereids. It is one of most abundant organic compounds on Earth after cellulose and chitin. The empiric chemical formula of lignin is $C_{20}H_{19}O_{14}N_2$. Lignin is a complex, high molecular weight polymer built of hydroxyphenylpropane units. Lignin is a completely different polymeric material, being highly cross linked and having phenolic like structures as the monomeric base. It is the lignin that holds the wood cells together and provides the extraordinary composite strength characteristics of a piece of wood.

Lignin is a large, cross-linked macromolecule with molecular mass in excess of 10,000 amu. It is relatively hydrophobic and aromatic in nature. The molecular weight in nature is difficult to measure, since it is fragmented during preparation. The molecule consists of various types of substructures that appear to repeat in random manner. Lignin plays a crucial part in conducting water in plant stems. The polysaccharide components of plant cell walls are highly hydrophilic and thus permeable to water, whereas lignin is more hydrophobic. The crosslinking of polysaccharides by lignin is an obstacle for water absorption to the cell wall. Thus, lignin makes it possible for plants to form vascular systems that conduct water efficiently. Lignin is present in all vascular plants, but not in bryophytes, which supports the idea that the main function of lignin is related to water transport.

Lignins are polymers of aromatic compounds. Their functions are to provide structural strength, provide sealing of a water conducting system that links roots with leaves, and protect plants against degradation (Glasser and Sarkanen, 1989). Lignin is a macromolecule, which consists of alkylphenols and has a complex three-dimensional structure. Lignin is covalently linked with xylans in the case of hardwoods and with galactoglucomannans in the case of softwoods (Sarkanen and

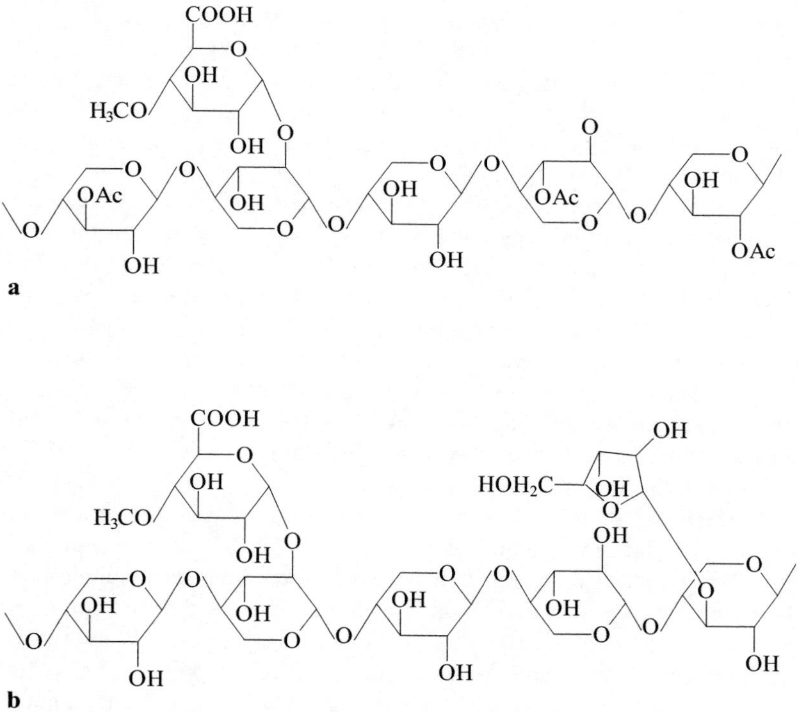

Fig. 2.3 Schematic illustration of xylans: **a** Partial xylan structure from hardwood and **b** Partial xylan structure from softwood
Source: Demirbas, 2008b

Ludwig, 1971). The basic chemical phenylpropane units of lignin (primarily sy-ringyl, guaiacyl and p-hydroxy phenol) as shown in Fig. 2.4 are bonded together by a set of linkages to form a very complex matrix. This matrix comprises a variety of functional groups, such as hydroxyl, methoxyl and carbonyl, which impart a high polarity to the lignin macromolecule (Demirbas and Kucuk, 1993; Hashem *et al.*, 2007). Cellulose and lignin structures have been investigated extensively in earlier studies (Young, 1986; Hergert and Pye, 1992; Mantanis *et al.*, 1995; Bridgwater, 2003; Garcia-Valls and Hatton, 2003; Mohan *et al.*, 2006).

Pyrolysis of lignin, *e.g.*, during combustion, yields a range of products, of which the most characteristic ones are methoxy phenols. Of those, the most important are guaiacol and syringol and their derivatives; their presence can be used to trace a smoke source to a wood fire. Lignin biosynthesis begins with the synthesis of monolignols. The starting material is the amino acid phenylalanine. The first reactions in the biosynthesis are shared with the phenylpropanoid pathway, and monolignols are considered to be a part of this group of compounds. There are three main types of monolignols: coniferyl alcohol, sinapyl alcohol, and paracou-maryl alcohol. Different plants use different monolignols. For example, gymno-sperms as Norway spruce have a lignin that consist almost entirely coniferyl alco-hol. Dicotyledonic lignin is a mixture of conifyl alcohol and sinapyl alcohol (normally more of the latter), and monocotylednic lignin is a mixure of all three monolignols. Some monocotyledons have mostly coniferyl alcohol (as many grasses), while other have mainly sinapyl alcohols, as some palms. Monolignols are synthesized in the cytosol as glucosides. The glucose is added to the mono-lignol to make them water soluble and to reduce their toxicity. The glucosides are transported through the cell membrane to the apoplast. The glucose is then re-moved and the monolignols are polymerized into lignin. Four main monolignols in the lignin structure are given in Fig. 2.5.

Wood and woody biomass also contain lesser amounts of tannins, simple sug-ars, starches, pectins, and organic soluble extractives. Extractives include terpenes, tall oil and the fatty acids, esters, and triglycerides, which contribute to paper mill pitch problems (Demirbas, 1991).

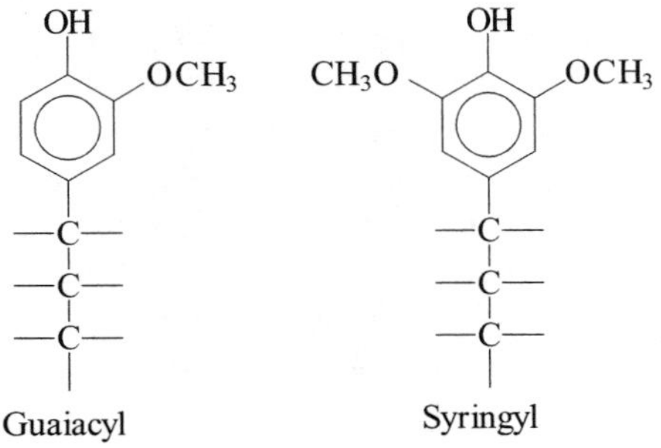

Fig. 2.4 Schematic illustration of building units of lignin *Source:* Demirbas, 2008b

Guaiacyl   Syringyl

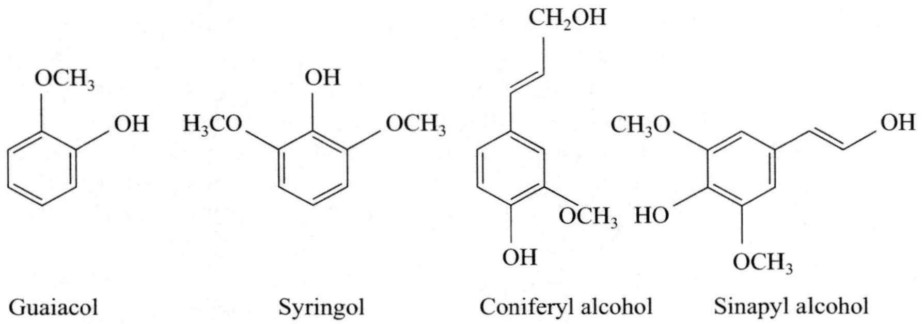

Guaiacol   Syringol   Coniferyl alcohol   Sinapyl alcohol

Fig. 2.5 The four main monolignols in the lignin structure

2.1.1.2 Modern Biomass, Bioenergy, and Green Energy

The term "modern biomass" is generally used to describe traditional biomass use through efficient and clean combustion technologies and sustained supply of biomass resources, environmentally sound and competitive fuels, and heat and electricity using modern conversion technologies. Modern biomass produced in a sustainable way excludes traditional uses of biomass as fuel-wood and includes electricity generation and heat production, as well as transportation fuels, from agricultural and forest residues and solid waste. On the other hand, "traditional biomass" is produced in an unsustainable way and it is used as a non-commercial source – usually with very low efficiencies for cooking in many countries (Goldemberg and Coelho, 2004). Modern biomass excludes traditional uses of biomass as fuelwood and includes electricity generation and heat production from agricultural and forest residues and solid waste.

Traditional biomass markets have been inefficient but technological developments have reduced energy, emission, and material flows through the system thus improving the efficiency of biomass energy systems. The energy market demands cost effectiveness, high efficiency, and reduced risk to future emission limits. Modernization of biomass conversion implies the choice of technologies: (a) offer the potential for high yields, (b) economic fuel availability, (c) low adverse environmental impacts, and (d) suitability to modern energy systems. A number of systems that meet the mentioned criteria for modernized biomass conversion can be identified (Larson, 1993).

Bioenergy is energy derived from organic sources such as all lignocellulosic materials, trees, agricultural crops, municipal solid wastes, food processing, and agricultural wastes and manure. Bioenergy is an inclusive term for all forms of biomass and biofuels. Biomass is renewable organic matter like crops, plant wastes, forest wood byproducts, aquatic plants, manure, and organic municipal wastes.

Bioenergy is one of the forms of renewable energy. It has been used bioenergy, the energy from biomass, for thousands of years, ever since people started burning wood to cook food and today, wood is still our largest biomass resource for bio-energy. The use of bioenergy has the potential to greatly reduce our greenhouse gas emissions. Replacing fossil fuels with energy from biomass has several distinct environmental implications. If biomass is harvested at a rate which is sustainable, using it for energy purposes does not result in any net increase in atmospheric carbon dioxide, a greenhouse gas.

Green energy is an alternate term for renewable energy that is generated from sources that are considered environmentally friendly (*e.g.*, hydro, solar, biomass (landfill) or wind). Green power is sometimes used in reference to electricity generated from "green" sources. Green energy production is the principal contributor in economic development of a developing country. Its economy development is based on agricultural production, and most people live in the rural areas. Implementation of integrated community development programs is therefore very necessary.

Green power refers to electricity supplied from more readily renewable energy sources than traditional electrical power sources. Green power products have become widespread in many electricity markets worldwide and can be derived from renewable energy sources. The environmental advantages of the production and use of green electricity by using green electrons seem to be clear. Market research indicates that there is a large potential market for green energy in Europe in general. Green power marketing has emerged in more than a dozen countries around the world.

Many green electricity products are based on renewable energy sources like wind, biomass, hydro, biogas, solar, and geothermal power (Murphya and Niitsuma, 1999). Green power products have become widespread in many electricity markets worldwide which can be derived from renewable energy sources. There has been interest in electricity from renewable sources named green electricity or green pools as a special market (Elliott, 1999). The term green energy is also used for green energy produced from cogeneration, energy from municipal waste, natural gas, and even conventional energy sources. Green power refers to electricity supplied from more readily renewable energy sources than traditional electrical power sources. The environmental advantages of the production and use of green electricity seem to be clear. The use of green energy sources like hydro, biomass, geothermal, and wind energy in electricity production reduces CO_2 emissions (Fridleifsson, 2003). Emissions such as SO_2, CO_2 and NO_x are reduced considerably, and the production and use of green electricity contributes to diminishing the green house effect (Arkesteijn and Oerlemans. 2005).

In general, a sustainable energy system includes energy efficiency, energy reliability, energy flexibility, energy development and continuity, combined heat and power (CHP) or cogeneration, fuel poverty, and environmental impacts. The environmental impacts of energy use are not new. For centuries, wood burning has contributed to the deforestation of many areas. On the other hand, the typical characteristics of a sustainable energy system can be derived from political definitions. A sustainable energy system can be defined also by comparing the performance of different energy systems in terms of sustainability indicators (Alanne and Sari, 2006). Because, by definition, sustainable energy systems must support both human and ecosystem health over the long term, goals on tolerable emissions should look well into the future. They should also take into account the public's tendency to demand more (UNDP, 2000).

2.1.2 Biomass Feedstocks

Biomass feedstocks are marked by their tremendous diversity, which makes them rather difficult to characterize as a whole. Feedstocks that can be utilized with conversion processes are primarily the organic materials now being landfilled. These include forest products wastes, agricultural residues, organic fractions of municipal solid wastes, paper, cardboard, plastic, food waste, green waste, and other waste. Non-biodegradable organic feedstocks, such as most plastics, are not convertible by biochemical processes. Biobased materials require pretreatment by chemical, physical, or biological means to open up the structure of biomass.

The major categories of biomass feedstock are as follows:

1. Forest products:

   - Wood
   - Logging residues
   - Trees, shrubs and wood residues
   - Sawdust, bark, *etc.*

2. Biorenewable wastes:

   - Agricultural wastes
   - Crop residues
   - Mill wood wastes
   - Urban wood wastes
   - Urban organic wastes

3. Energy crops:

   - Short rotation woody crops
   - Herbaceous woody crops
   - Grasses
   - Starch crops
   - Sugar crops
   - Forage crops
   - Oilseed crops

4. Aquatic plants:

   - Algae
   - Water weed
   - Water hyacinth
   - Reed and rushes

5. Food crops:

   - Grains
   - Oil crops

6. Sugar crops:

   - Sugar cane
   - Sugar beets
   - Molasses
   - Sorghum

7. Landfill
8. Industrial organic wastes
9. Algae, kelps, lichens and mosses.

   Biomass is the world's fourth largest energy source worldwide, following coal, oil and natural gas. Biomass appears to be an attractive feedstock for three main reasons. First, it is a renewable resource that may be sustainably developed in the future. Second, it appears to have formidably positive environmental properties, reduced GHG emissions, possibly reduced NO_x and SO_x depending on the fossil-fuels displaced. However, also negative impacts, such as polycyclic aromatic hydrocarbons including polycyclic aromatic hydrocarbons, dioxins, furans, volatile organic compounds, and heavy metals, especially when combusted in traditional stoves. Third, it appears to have significant economic potential provided that fossil fuel prices will increase in the future.

   Biomass is a sustainable feedstock for chemicals and energy products. Biomass feedstocks are more evenly distributed in the world. As an energy source that is highly productive, renewable, carbon neutral, and easy to store and transport, biomass has drawn worldwide attention recently.

   Biomass offers important advantages as a combustion feedstock due to the high volatility of the fuel and the high reactivity of both the fuel and the resulting char. However, it should be noticed that in comparison with solid fossil fuels, biomass contains much less carbon and more oxygen and has a low heating value.

   The waste products of a home include paper, containers, tin cans, aluminum cans, and food scraps, as well as sewage. The waste products of industry and commerce include paper, wood, and metal scraps, as well as agricultural waste products. Biodegradable wastes, such as paper fines and industrial biosludge, into mixed alcohol fuels (*e.g.*, isopropanol, isobutanol, isopentanol). The wastes are first treated with lime to enhance reactivity; then they are converted to volatile fatty acids (VFAs) such as acetic acid, propionic acid, and butyric acid – using a mixed culture of microorganisms derived from cattle rumen or anaerobic waste treatment facilities. Pulp and paper wastes may also be treated to produce methane. The contents of domestic solid waste are given in Table 2.2.

   Typical solid wastes include wood material, pulp and paper industry residues, agricultural residues, organic municipal material, sewage, manure, and food processing byproducts. Biomass is considered one of the main renewable energy resources

Table 2.2 Contents of domestic solid waste (percentage of total)

Component	Lower limit	Upper limit
Paper waste	33.2	50.7
Food waste	18.3	21.2
Plastic matter	7.8	11.2
Metal	7.3	10.5
Glass	8.6	10.2
Textile	2.0	2.8
Wood	1.8	2.9
Leather and rubber	0.6	1.0
Miscellaneous	1.2	1.8

Source: Demirbas, 2004

of the future due to its large potential, economic viability and various social and environmental benefits. It was estimated that by 2050 biomass may provide nearly 38% of the world's direct fuel use and 17% of the world's electricity. If biomass is produced more efficiently and used with modern conversion technologies, it can supply a considerable range and diversity of fuels at small and large scales. Municipal solid waste (MSW) is defined as waste durable goods, non-durable goods, containers and packaging, food scraps, yard trimmings, and miscellaneous inorganic wastes from residential, commercial, and industrial sources (Demirbas, 2004).

Forests are principal global economic as well as ecological resource. Forests have played a big role in the development of human societies. The prime direct or marketable product of most forests is wood for use as timber, fuelwood, pulp and paper, providing some 3.4 billion cubic meters of timber-equivalent a year globally. Asia and Africa use 75% of global wood fuels. The lumber, plywood, pulp, and paper industries burn their own wood residues in large furnaces and boilers to supply 60% of the energy needed to run factories (Demirbas, 2003). Figure 2.6 shows the use of world wood products; lumber, plywood, paper paperboard between 1970 and 2005. The availability of fuelwood from the forest is continually declining at an ever-increasing rate due to indiscriminate deforestation and slow regeneration, as well as afforestation (Jain and Singh, 1999). The fuelwoods generally used by local people were identified and analyzed quantitatively to select a few species with the best fuelwood characteristics so that plantation on non-agricultural lands could be undertaken (Jain, 1992).

Wood biomass involves trees with commercial structure and forest residues not being used in the traditional forest products industries. Available forest residues may appear to be an attractive fuel source. Collection and handling and transport costs are critical factors in the use of forest residues. Although the heat produced from wood wastes is less than that from oil or gas, its cost compared to fossil fuels makes it an attractive source of readily available heat or heat and power. The most effective utilization of wood wastes, particularly in the sawmilling and plywood industry, plays an important role in energy efficient production.

Fig. 2.6 Use of world
wood products: lumber,
plywood, paper, and
paperboard (1970–2005)

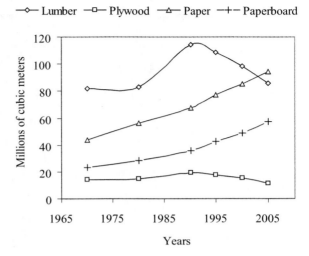

Table 2.3 shows the sources of available forest and wood manufacturing residues. Wood preparation involves the conversion of roundwood logs into a form suitable for pulping and includes processes for debarking, chipping, screening, handling, and storage.

Forest residues typically refer to those parts of trees such as treetops, branches, small-diameter wood, stumps and dead wood, as well as undergrowth and low-value species. The conversion of wood to biofuels and biochemicals has long been a goal of the forest products industry. Forest residues alone count for some 50% of the total forest biomass and are currently left in the forest to rot (Demirbas, 2001).

The importance of biomass in different world regions is given in Table 2.4. For large portions of the rural populations of developing countries, and for the poorest sections of urban populations, biomass is often the only available and affordable source of energy for basic needs such as cooking and heating. As shown in Table 2.4, the importance of biomass varies significantly across regions. In Europe, North America, and the Middle East, the share of biomass averages 2–3% of total final energy consumption, whereas in Africa, Asia and Latin Amer-

Table 2.3 Sources of available forest and wood manufacturing residues

Source of residue	Type of residue
Forest operations	Branches, bark chips, leaves/needles, stumps, roots, and sawdust
Lumber production	Bark, sawdust, clippings, split wood
Saw milling	Bark, trimmings, split wood, sawdust, planer shavings, sander dust
Plywood production	Bark, sawdust, veneer clippings and wastes, panel trim, sander dust
Paper production	Slab chips, pulping reject, sawdust, clippings
Paperboard production	Bark, sawdust, screening fines, panel trim, sander dust

Table 2.4   The importance of biomass in different world regions

Region	Share of biomass in final energy consumption
Africa (average)	62.0
Burundi	93.8
Ethiopia	85.6
Kenya	69.6
Somalia	86.5
Sudan	83.7
Uganda	94.6
South Asia (average)	56.3
East Asia (average)	25.1
China	23.5
Latin America (average)	18.2
Europe (average)	3.5
North America (average)	2.7
Middle East (average)	0.3

ica, which together account for three-quarters of the world's population, biomass provides a substantial share of the energy needs: one-third on average, but as much as 80–90% in some of the poorest countries of Africa and Asia (*e.g.*, Angola, Ethiopia, Mozambique, Tanzania, Democratic Republic of Congo, Nepal, and Myanmar). Indeed, for large portions of the rural populations of developing countries, and for the poorest sections of urban populations, biomass is often the only available and affordable source of energy for basic needs such as cooking and heating.

Agricultural residues, grasses, algae, kelps, lichens, and mosses are also important biomass feedstocks in the world. Algae can grow practically in every place where there is enough sunshine. Some algae can grow in saline water. The most significant difference of algal oil is in the yield and hence its biodiesel yield. According to some estimates, the yield (*per* acre) of oil from algae is over 200 times the yield from the best-performing plant/vegetable oils (Sheehan *et al.*, 1998). Microalgae are the fastest growing photosynthesizing organisms. They can complete an entire growing cycle every few days. Approximately 46 tons of oil/hectare/year can be produced from diatom algae. Different algae species produce different amounts of oil. Some algae produce up to 50% oil by weight.

2.2   Biomass Characterization

Characterization of biomass fuels has been reviewed by Bushnell *et al.* (1989). In general, combustion models of biomass can be classified as macroscopic or microscopic. The characterization of the biomass fuels is generally divided into three broad categories: (1) composition and structure, (2) reactivity measures, and (3) ash chemistry.

The macroscopic properties of biomass are given for macroscopic analysis, such as ultimate analysis, heating value, moisture content, particle size, bulk density, and ash fusion temperature. Properties for microscopic analysis include thermal, chemical kinetic and mineral data (Ragland *et al.*, 1991). Fuel characteristics such as ultimate analysis, heating value, moisture content, particle size, bulk density, and ash fusion temperature of biomass have been reviewed (Bushnell *et al.*, 1989). Fuel properties for the combustion analysis of biomass can be conveniently grouped into physical, chemical, thermal, and mineral properties.

Physical property values vary greatly and properties such as density, porosity, and internal surface area are related to biomass species, whereas bulk density, particle size, and shape distribution are related to fuel preparation methods. Important chemical properties for combustion are the ultimate analysis, proximate analysis, analysis of pyrolysis products, higher heating value, heat of pyrolysis, heating value of the volatiles, and heating value of the char.

Thermal property values such as specific heat, thermal conductivity, and emissivity vary with moisture content, temperature, and degree of thermal degradation by one order of magnitude. Thermal degradation products of biomass consist of moisture, volatiles, char, and ash. Volatiles are further subdivided into gases such as light hydrocarbons, carbon monoxide, carbon dioxide, hydrogen and moisture, and tars. The yields depend on the temperature and heating rate of pyrolysis. Some properties vary with species, location within the biomass, and growth conditions. Other properties depend on the combustion environment. Where the properties are highly variable, the likely range of the property is given (Ragland *et al.*, 1991).

2.2.1   Characterization of Biomass Feedstock and Products

Heterogeneity is an inherent characteristic of biomass materials. Since variability of any biomass cannot be controlled, processes that use biomass feedstocks must be able to monitor the chemical composition of the feedstock and compensate for variability during processing. Chemical changes during the processing of biomass must also be monitored to ensure that the process maintains a steady state in spite of the feedstock variability. Finally, process residues and products must be evaluated to assess overall process economics.

Physical property values vary greatly and properties such as density, porosity, and internal surface area are related to biomass species whereas bulk density, particle size, and shape distribution are related to fuel preparation methods.

Important chemical properties for combustion are the ultimate analysis, proximate analysis, analysis of pyrolysis products, higher heating value, heat of pyrolysis, heating value of the volatiles, and heating value of the char.

Thermal property values such as specific heat, thermal conductivity, and emissivity vary with moisture content, temperature, and degree of thermal degradation by one order of magnitude. Thermal degradation products of biomass consist of moisture, volatiles, char, and ash. Volatiles are further subdivided into gases such as light hydrocarbons, carbon monoxide, carbon dioxide, hydrogen and moisture, and tars. The yields depend on the temperature and heating rate of pyrolysis. Some properties vary with species, location within the biomass, and growth conditions. Other properties depend on the combustion environment. Where the properties are highly variable, the likely range of the property is given (Demirbas, 2004).

Current methods for chemical characterization of biomass feedstocks, process intermediates, and residues are not applicable in a commercial setting because they are very expensive and cannot provide analysis information in a time frame useful for process control.

2.2.2 Biomass Process Design and Development

Biomass process design and development requires acquiring the information needed to understand and characterize the fundamental chemical and physical processes that govern biomass conversion at high temperatures and pressures. The models can be used to determine operating conditions that optimize thermal efficiency and to examine design strategies for integrating combined cycles for the production of synthesis gas and electric power with minimum impact on the environment. Biomass gasification and pyrolysis both require precise characterization of the breakdown products being generated, so that processes can be fine-tuned to produce optimal end products.

Near-infrared spectrometry correlated by multi-variate analysis characterizes in minutes what would otherwise require three or four days and cost far more. Opportunities for use in the lumber and paper industries, let alone biorefineries, are almost limitless. Modern biotechnology can not only transform materials extracted from plants, but can transform the plants to produce more valuable materials. Plants can be developed to produce high-value chemicals in greater quantity than they do naturally, or even to produce compounds they do not naturally produce.

The syngas obtained from the processes can be used to produce hydrogen which, in turn, can be used as a fuel or to make plastics, fertilizers, and a wide variety of other products. The syngas can also be converted to sulfur-free liquid transportation fuels using a catalytic process (known as the Fischer–Tropsch Synthesis), or provide base chemicals for producing biobased products.

If biomass is heated to high temperatures in the total absence of oxygen, it pyrolyzes to a liquid that is oxygenated, but otherwise has similar characteristics to petroleum. This pyrolysis oil or bio-oil can be burned to generate electricity or it can be used to provide base chemicals for biobased products. As an example, phenolic compounds can be extracted from bio-oil to make adhesives and plastic resins.

Plant and animal fats and oils are long hydrocarbon chains, as are their fossil-fuel counterparts. Fatty acid methyl esters from transesterification of the fats and oils directly convert to substitutes for petroleum diesel. Known as biodiesel, it differs primarily in containing oxygen, so it burns more cleanly, either by itself or as an additive. Biodiesel use is small but growing rapidly in the United States. It is made mostly from soybean oil and used cooking oil. Soybean meal, the coproduct of oil extraction is now used primarily as animal feed, but could also be a base for making biobased products. Glycerin is already used to make a variety of products.

Another way to convert waste biomass into useful fuels and products is to have natural consortiums of anaerobic microorganisms decompose the material in closed systems. Anaerobic microorganisms digest organic material in the absence of oxygen and produce biogas as a waste product.

Direct combustion is the old way of using biomass. Biomass thermochemical conversion technologies such as pyrolysis and gasification are certainly not the most important options at present; combustion is responsible for over 97% of the world's bioenergy production.

In previous studies, the mechanisms of degradation by direct pyrolysis and cracking as well as catalytic, and liquefaction of biomass were not extensively identified (Demirbas, 2002). The reactions in biomass degradation process are much more complex. Many researchers have tried to investigate and propose reaction mechanisms but no definitive study has been conducted. These proposed reactions are general and the type of biomass will dictate the type of processes or reactions required to breakdown and rearrange molecules. The more complex the raw biomass is chemically, the more complex the reaction mechanisms required, and thus the increased difficulty in determining them (Balat, 2008a–c).

It is difficult to determine exactly what types of reactions occur during thermochemical conversion processes. The processes of carbonaceous materials take place through a sequence of structural and chemical changes that involve at least the following steps.

New techniques are needed to provide analytical support for large-scale processes that convert biomass to fuels and chemicals. One solution is to use established methods to calibrate rapid, inexpensive spectroscopic techniques, which can then be used for feedstock and process analysis. A complete chemical characterization is available in a time frame relevant for process control, meaning that the information can be used to make the process adjustments necessary for steady-state production. Process monitoring is one possible application of these rapid analytical techniques. The new techniques for biomass analysis can support and improve research by providing levels of information that would have been too costly to pursue using traditional wet chemical methods.

2.3 Biomass Fuel Analyses

Main biomass fuel analyses are (a) particle size and specific gravity, (b) ash content, (c) moisture content, (d) extractive content, (f) element (C, H, O and N) content, and (g) structural constituent (cellulose, hemicelluloses and lignin) content.

2.3.1   Particle Size and Specific Gravity

Particle size of biomass should be as much as 0.6 cm, sometimes more, in a profitable combustion process. Biomass is much less dense and has significantly higher aspect ratios than coal. It is also much more difficult to reduce to small sizes.

2.3.2   Ash Content

Ash or inorganic materials in plants depend on the type of the plant and the soil contamination in which the plant grows. On average wood contains about 0.5% ash. Ash contents of hard and soft woods are about 0.5% and 0.4%, respectively. Insoluble compounds act as a heat sink in the same way as moisture, lowering combustion efficiency, but soluble ionic compounds can have a catalytic effect on the pyrolysis and combustion of the fuel. The presence of inorganic compounds favors the formation of char. Ash content is an important parameter directly affecting the heating value. High ash content of a plant part makes it less desirable as fuel (Demirbas, 1998).

The composition of mineral matter can vary between and within each biomass sample. Mineral matter in fruit shells consists mostly of salts of calcium, potassium, silica, and magnesium, but salts of many other elements are also present in lesser amounts (Demirbas, 2002).

2.3.3   Moisture Content

Moisture in biomass generally decreases its heating value. Moisture in biomass is stored in spaces within the dead cells and within the cell walls. When the fuel is dried the stored moisture equilibrates with the ambient relative humidity. Equilibrium is usually about 20% in air dried fuel. Moisture percentage of the wood species varies from 41.27 to 70.20%. The heating value of a wood fuel decreases with an increase in the moisture content of the wood. The moisture content varies from one tree part to another. It is often the lowest in the stem and increases toward the roots and the crown. The presence of water in biomass influences its behavior during pyrolysis and affects the physical properties and quality of the pyrolysis liquid. The results obtained show that for higher initial moisture contents the maximum liquid yield on a dry feed basis occurs at lower pyrolysis temperatures between 691 K and 702 K (Demirbas, 2004).

2.3.4   Extractive Content

Again the heat content, which is a very important factor affecting utilization of any material as a fuel, is affected by the proportion of combustible organic components (called extractives) present in it. The HHVs of the extractive-free plant parts were found to be lower than those of the unextracted parts, which indicates a likely positive contribution of extractives towards the increase of HHV. The extractive content is an important parameter directly affecting the heating value. High extractive content of a plant part makes it desirable as fuel. Again, the heat content, which is a very important factor affecting utilization of any material as a fuel, is affected by the proportion of extractives present in it. Extractives raise the higher heating values of the wood fuels.

2.3.5 Element Content

Both the chemical and the physical composition of the fuel are important determining factors in the characteristics of combustion. Biomass can be analyzed by breaking it down into structural components (called proximate analysis) or into chemical elements (called ultimate analysis). The heat content is related to the oxidation state of the natural fuels in which carbon atoms generally dominate and overshadow small variations of hydrogen content. On the basis of literature values for different species of wood, Tillman (1978) also found a linear relationship between HHV and carbon content.

2.3.6 Structural Constituent Content

Biomass fuels are composed of biopolymers that consist of various types of cells and the cell walls are built of cellulose, hemicelluloses, and lignin. HHVs of biomass fuels increase with increasing their lignin contents. In general, the FC content of wood fuels increases with increase in their FC contents.

2.3.7 The Energy Value of Biomass

Again the heat content, which is a very important factor affecting utilization of any material as a fuel, is affected by the proportion of combustible organic components (called extractives) present in it. The HHVs of the extractive-free plant parts were found to be lower than those of the unextracted parts which indicate a likely positive contribution of extractives towards the increase of HHV. Extractive content is important parameter directly affecting the heating value. High extractive content of a plant part makes it desirable as fuel. Again the heat content, which is a very important factor affecting utilization of any material as a fuel, is affected by the proportion of extractives present in it.

Both the chemical and the physical composition of the fuel are important determining factors in the characteristics of combustion. Biomass can be analyzed by breaking it down into structural components (called proximate analysis) or into chemical elements (called ultimate analysis). The heat content is related to the oxidation state of the natural fuels in which carbon atoms generally dominate and overshadow small variations of hydrogen content.

The higher heating values (HHVs) or gross heat of combustion includes the latent heat of the water vapor products of combustion because the water vapor was allowed to condense to liquid water. The HHV (in MJ/kg) of the biomass fuels as a function of fixed carbon (FC, wt%) was calculated from Eq. 2.1 (Demirbas, 1997):

$$HHV = 0.196\,(FC) + 14.119 \tag{2.1}$$

In earlier work (Demirbas *et al.*, 1997), formulae were also developed for estimating the HHVs of fuels from different lignocellulosic materials, vegetable oils, and diesel fuels using their chemical analysis data. For biomass fuels such as coal, HHV was calculated using the modified Dulong's formula, as a function of the carbon, hydrogen, oxygen, and nitrogen contents from Eq. 2.2:

$$HHV = 0.335\,(CC) + 1.423\,(HC) - 0.154\,(OC) - 0.145\,(NC) \tag{2.2}$$

where (CC) is the carbon content (wt%), (HC) the hydrogen content (wt%), (OC) the oxygen content (wt%), and (NC) the nitrogen content (wt%).

The heat content is related to the oxidation state of the natural fuels in which carbon atoms generally dominate and overshadow small variations of hydrogen content. On the basis of literature values for different species of wood, Tillman (1978) also found a linear relationship between HHV and carbon content.

The HHVs of extractive-free samples reflect the HHV of lignin relative to cellulose and hemicelluloses. Cellulose and hemicelluloses (holocellulose) have a HHV 18.60 MJ/kg, whereas lignin has a HHV from 23.26 to 26.58 MJ/kg. As discussed by Baker (1982), HHVs reported for a given species reflect only the samples tested and not the entire population of the species. The HHV of a lignocellulosic fuel is a function of its lignin content. In general, the HHVs of lignocellulosic fuels increase with increase of their lignin contents and the HHV is highly correlated with lignin content. For the model including the lignin content, the regression equation was

$$HHV = 0.0889\,(LC) + 16.8218 \tag{2.3}$$

where LC was the lignin content (wt% daf and extractive-free basis).

Table 2.5 shows moisture, ash, and higher heating value (HHV) analysis of biomass fuels.

Biomass combustion is a series of chemical reactions by which carbon is oxidized to carbon dioxide, and hydrogen is oxidized to water. Oxygen deficiency leads to incomplete combustion and the formation of many products of incomplete combustion. Excess air cools the system. The air requirements depend on the chemical and physical characteristics of the fuel. The combustion of the biomass relates to the fuel burn rate, the combustion products, the required excess air for complete combustion, and the fire temperatures.

Table 2.5   Moisture, ash, and higher heating value (HHV) analysis of biomass fuels

Fuel common/scientific name	Moisture (wt% of fuel)	Ash (wt% of dry fuel)	HHV (MJ/kg, daf)
Almond shells/*Pranus dulcis*	7.5	2.9	19.8
Almond hulls/*Pranus dulcis*	8.0	5.8	20.0
Beech wood/*Fagus orientalis*	6.5	0.6	19.6
Hazelnut shells/*Corylus avellena*	7.2	1.4	19.5
Oak wood/*Quersus predunculata*	6.0	1.7	19.8
Oak bark/*Quersus predunculata*	5.6	9.1	22.0
Olive pits/*Olea europaea*	7.0	1.8	22.0
Olive husk/*Olea europaea*	6.8	2.1	21.8
Pistachio shells/*Pistocia vera*	8.1	1.3	19.9
Rice straw/*Oryza sativa*	11.2	19.2	18.7
Spruce wood/*Picea orientalis*	6.7	0.5	20.5
Switchgrass/*Panicum virgatum*	13.1	5.9	19.9
Wheat straw/*Triticum aestivum*	6.4	8.1	19.3

Source: Demirbas, 2004

2.4   Biomass Optimization and Valorization

Biomass is very important for implementing the Kyoto agreement to reduce carbon dioxide emissions by replacing fossil fuels. Developing biorenewable sources of energy has become necessary due to limited supply of fossil fuels. Global environmental concerns and decreasing resources of crude oil have prompted demand for alternative fuels. The global climate change is also the major environmental issue of current times. Global warming, the Kyoto Protocol, the emission of greenhouse gases, and the depletion of fossil fuels are the topics of environmental pleadings worldwide. Rapidly increasing energy requirements parallel technological development in the world, and research and development activities are forced to study new and biorenewable energy investigations. The purpose of the work is to optimize the system's operation. The main reason to build described system is to supply stand alone system using renewable energy sources. So the power plant has to produce energy independent of any weather fluctuations. An energy system is made up of an energy supply sector and energy end-use technologies. The object of the energy system is to deliver to consumers the benefits that energy offers. The energy system commonly consists of energy resources and production, security, conversion, use, distribution, and consumption.

Main wood valorization technologies include pulp and paper making, bio-oil by pyrolysis, synthesis gas by gasification, sugar by hydrolysis, ethanol by sugar fermentation, and adhesives by alkali liquefaction and polymerization.

The main research areas of biomass optimization and valorization are:

- Biogas and organic fertilizer production through anaerobic digestion
- Energy crops production
- Fractionation of biomass

- Biomass modernization
- Biomass cogeneration
- Biomass cofiring
- Biomass economy
- Biomass policy
- Sustainability of biorenewables
- Conversion of biomass into useful fuels and chemicals.

The basic structure of all woody biomass consists of three organic polymers: cellulose, hemicelluloses, and lignin in trunk, foliage, and bark. Added to these materials are extractives and minerals or ash. The proportion of these wood constituents varies between species, and there are distinct differences between hardwoods and softwoods. Hardwoods have a higher proportion of cellulose, hemicelluloses, and extractives than softwoods, but softwoods have a higher proportion of lignin. In general, hardwoods contain about 43% cellulose, 22% lignin, and 35% hemicelluloses, while softwoods contain about 43% cellulose, 29% lignin, and 28% hemicelluloses (Rydholm, 1965).

Biomass has been recognized as a major world renewable energy source to supplement declining fossil fuel resources. Biomass power plants have advantages over fossil-fuel plants, because their pollution emissions are less. Energy from biomass fuels is used in the electric utility, lumber and wood products, and pulp and paper industries. Wood fuel is a renewable energy source and its importance will increase in future. Biomass can be used directly or indirectly by converting it into a liquid or gaseous fuel. A large number of research projects in the field of thermochemical conversion of biomass, mainly on liquefaction pyrolysis, and gasification, have been undertaken.

When biomass is used directly in an energy application without chemical processing then it is combusted. Conversion may be effected by thermochemical, biological, or chemical processes. These may be categorized as follows: direct combustion, pyrolysis, gasification, liquefaction, supercritical fluid extraction, anaerobic digestion, fermentation, acid hydrolysis, enzyme hydrolysis, and esterification. Figure 2.7 shows main biomass conversion processes. Biomass can be converted to biofuels such as bioethanol and biodiesel, and thermochemical conversion products such as syn-oil, biosyngas, and biochemicals. Bioethanol is a fuel derived from renewable sources of feedstock; typically plants such as wheat, sugar beet, corn, straw, and wood. Bioethanol is a petrol additive/substitute. Biodiesel is better than diesel fuel in terms of sulfur content, flash point, aromatic content, and biodegradability.

The energy dimension of biomass use is importantly related to the possible increased use of this source as a critical option to address the global warming issue. Biomass is generally considered as an energy source completely CO_2-neutral. The underlying assumption is that the CO_2 released in the atmosphere is matched by the amount used in its production. This is true only if biomass energy is sustainably consumed, *i.e.*, the stock of biomass does not diminish in time. This may not be the case in many developing countries.

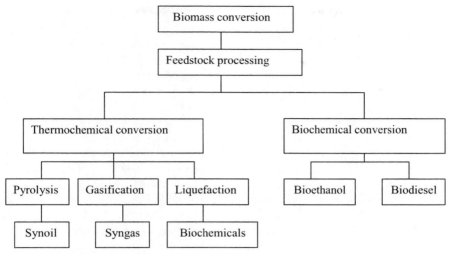

Fig. 2.7 Main biomass conversion processes

Biomass is burned by direct combustion to produce steam, the steam turns a turbine, and the turbine drives a generator, producing electricity. Gasifiers are used to convert biomass into a combustible gas (biogas). The biogas is then used to drive a high efficiency, combined cycle gas turbine. Biomass consumption for electricity generation has been growing sharply in Europe since 1996, with 1.7% of power generation in 1996.

2.4.1 Fuels from Biomass

Direct combustion is the old way of using biomass. Biomass thermochemical conversion technologies such as pyrolysis and gasification are certainly not the most important options at present; combustion is responsible for over 97% of the world's bioenergy production (Demirbas, 2004). Direct combustion and co-firing with coal for electricity production from biomass has been found to be a promising method in the nearest future. The supply is dominated by traditional biomass used for cooking and heating, especially in rural areas of developing countries. The traditional biomass cooking and heating produces high levels of pollutants.

Biomass energy currently represents approximately 14% of world final energy consumption, a higher share than that of coal (12%) and comparable to those of gas (15%) and electricity (14%). Biomass is the main source of energy for many developing countries and most of it is non-commercial. Hence there is enormous difficulty in collecting reliable biomass energy data. Yet good data are essential for analyzing tendencies and consumption patterns, for modeling future trends and for designing coherent strategies.

There are three ways to use biomass. It can be burned to produce heat and electricity, changed to gas-like fuels such as methane, hydrogen and carbon monoxide or changed to a liquid fuel. Liquid fuels, also called biofuels, include mainly two forms of alcohol: ethanol and methanol. The most commonly used biofuel is ethanol, which is produced from sugarcane, corn, and other grains. A blend of gasoline and ethanol is already used in cities with high air pollution.

There are several ways to make use of the energy contained in the biomass from old direct burning to pyrolysis, gasification, and liquefaction. Pyrolysis is the thermochemical process that converts organic materials into usable fuels. Pyrolysis produces energy fuels with high fuel-to-feed ratios, making it the most efficient process for biomass conversion and the method most capable of competing with and eventually replacing non-renewable fossil fuel resources.

The main transportation fuels that can be obtained from biomass using different processes are sugar ethanol, cellulosic ethanol, grain ethanol, biodiesel, pyrolysis liquids, green diesel, green gasoline, butanol, methanol, syngas liquids, biohydrogen, algae diesel, algae jet fuel, and hydrocarbons.

In the liquefaction process, biomass is converted to liquefied products through a complex sequence of physical structure and chemical changes. The feedstock of liquefaction is usually wet matter. In the liquefaction, biomass is decomposed into small molecules. These small molecules are unstable and reactive, and can repolymerize into oily compounds with a wide range of molecular weight distribution (Demirbas, 2000).

Liquefaction can be accomplished directly or indirectly. Direct liquefaction involves rapid pyrolysis to produce liquid tars and oils and/or condensable organic vapors. Indirect liquefaction involves the use of catalysts to convert non-condensable, gaseous products of pyrolysis or gasification into liquid products. Alkali salts, such as sodium carbonate and potassium carbonate, can break down the hydrolysis of cellulose and hemicellulose, into smaller fragments. The degradation of biomass into smaller products mainly proceeds by depolymerization and deoxygenation. In the liquefaction process, the amount of solid residue increased in proportion to the lignin content. Lignin is a macromolecule, which consists of alkylphenols and has a complex three-dimensional structure. It is generally accepted that free phenoxyl radicals are formed by thermal decomposition of lignin above 230°C and that the radicals have a random tendency to form a solid residue through condensation or repolymerization (Demirbas, 2000).

The changes during liquefaction process involve all kinds of processes such as solvolysis, depolymerization, decarboxylation, hydrogenolysis, and hydrogenation. Solvolysis results in micellar-like substructures of the biomass. The depolymerization of biomass leads to smaller molecules. It also leads to new molecular rearrangements through dehydration and decarboxylation. When hydrogen is present, hydrogenolysis and hydrogenation of functional groups, such as hydroxyl groups, carboxyl groups, and keto groups also occur.

Pyrolysis is the basic thermochemical process for converting biomass to a more useful fuel. Biomass is heated in the absence of oxygen, or partially combusted in a limited oxygen supply, to produce a hydrocarbon rich gas mixture, an oil-like liquid and a carbon rich solid residue. Rapid heating and rapid quenching produced the intermediate pyrolysis liquid products, which condense before further reactions break down higher-molecular-weight species into gaseous products. High reaction rates minimize char formation. Under some conditions, no char is formed. At higher fast pyrolysis temperatures, the major product is gas.

Pyrolysis is the simplest and almost certainly the oldest method of processing one fuel in order to produce a better one. Pyrolysis can also be carried out in the presence of a small quantity of oxygen ("gasification"), water ("steam gasification") or hydrogen ("hydrogenation"). One of the most useful products is methane, which is a suitable fuel for electricity generation using high-efficiency gas turbines.

Cellulose and hemicelluloses form mainly volatile products on heating due to the thermal cleavage of the sugar units. The lignin mainly forms char since it is not readily cleaved to lower molecular weight fragments. The progressive increase in the pyrolysis temperature of the wood led to the release of the volatiles thus forming a solid residue that is different chemically from the original starting material (Demirbas, 2000). Cellulose and hemicelluloses initially break into compounds of lower molecular weight. This forms an "activated cellulose", which decomposes by two competitive reactions: one forming volatiles (anhydrosugars) and the other char and gases. The thermal degradation of the activated cellulose and hemicelluloses to form volatiles and char can be divided into categories depending on the reaction temperature. In a fire all these reactions take place concurrently and consecutively. Gaseous emissions are predominantly a product of pyrolytic cracking of the fuel. If flames are present, fire temperatures are high, and more oxygen is available from thermally induced convection.

The biomass pyrolysis is attractive because solid biomass and wastes can be readily converted into liquid products. These liquids, such as crude bio-oil or slurry of charcoal of water or oil, have advantages in transport, storage, combustion, retrofitting, and flexibility in production and marketing.

Gasification is a form of pyrolysis, carried out in the presence of a small quantity of oxygen at high temperatures in order to optimize the gas production. The resulting gas, known as the producer gas, is a mixture of carbon monoxide, hydrogen and methane, together with carbon dioxide and nitrogen. The gas is more versatile than the original solid biomass (usually wood or charcoal): it can be burnt to produce process heat and steam, or used in gas turbines to produce electricity.

Biomass gasification technologies are expected to be an important part of the effort to meet these goals of expanding the use of biomass. Gasification technologies provide the opportunity to convert renewable biomass feedstocks into clean fuel gases or synthesis gases. Biomass gasification is the latest generation of biomass energy conversion processes, and is being used to improve efficiency, and to reduce the investment costs of biomass electricity generation through the use of gas turbine technology. High efficiencies (up to about 50%) are achievable using combined-cycle gas turbine systems, where waste gases from the gas turbine are recovered to produce steam for use in a steam turbine.

Various gasification technologies include gasifiers where the biomass is introduced at the top of the reactor and the gasifying medium is either directed co-currently (downdraft) or counter-currently up through the packed bed (updraft). Other gasifier designs incorporate circulating or bubbling fluidized beds. Tar yields can range from 0.1% (downdraft) to 20% (updraft) or greater in the product gases.

The process of synthetic fuels (synfuels) from biomass will lower the energy cost, improve the waste management, and reduce harmful emissions. This triple assault on plant operating challenges is a proprietary technology that gasifies biomass by reacting it with steam at high temperatures to form a clean burning synthetic gas (syngas: CO + H2). The molecules in the biomass (primarily carbon, hydrogen, and oxygen) and the molecules in the steam (hydrogen and oxygen) reorganize to form this syngas.

2.4.2   Chemicals from Biomass

Such features of larch wood, as an increased content of extractive compounds and its high density create some technological problems for pulping process. It seems that a production of high-value added chemicals is the most profitable way from an economical point of view of larch wood valorization. High value organic compounds such as arabinogalactan, quercitin dihydrate vanillin, microcrystalline cellulose, and levulinic acid have been obtained from larch wood. Both arabinogalactan and quercitin dihydrate have been extracted from larch wood with boiling water (Kuznetsov *et al.*, 2005).

Chemicals can be obtained from thermal depolymerization and decomposition of biomass structural components, well-known cellulose, hemicelluloses, lignin form liquids, and gas products, as well as a solid residue of charcoal. The bio-oils pyrolysis of biomass are composed of a range of cyclopentanone, methoxyphenol, acetic acid, methanol, acetone, furfural, phenol, formic acid, levoglucosan, guaiocol and their alkylated phenol derivatives. The structural components of the biomass samples mainly affect pyrolytic degradation products. A reaction mechanism is proposed, which describes a possible reaction route for the formation of the characteristic compounds found in the oils. The temperature and heating rate are most important parameters affecting the composition of chemicals. The supercritical water conditioning and liquefaction partial reactions also occur during the pyrolysis. Acetic acid is formed in the thermal decomposition of all three main components of biomass. In the pyrolysis reactions of biomass: water is formed by dehydration; acetic acid comes from the elimination of acetyl groups originally linked to the xylose unit; furfural is formed by dehydration of the xylose unit; formic acid proceeds from carboxylic groups of uronic acid; and methanol arises from methoxyl groups of uronic acid.

The pyrolysis process can mainly produce charcoal, condensable organic liquids, non-condensable gases, acetic acid, acetone, and methanol. Among the liquid products, methanol is one of the most valuable products. The liquid fraction of the pyrolysis products consists of two phases: an aqueous phase containing a wide variety of organo-oxygen compounds of low molecular weight and a non-aqueous phase containing insoluble organics of high molecular weight. This phase is called tar and is the product of greatest interest. The ratios of acetic acid, methanol, and acetone of the aqueous phase are higher than those of the non-aqueous phase.

Chemicals potentially derived from lignin conversions are syngas, methanol, dimethyl ether, ethanol, mixed alcohols, byproduct C_1 to C_4 gases, hydrocarbons, oxygenates, Fischer–Tropsch liquids, cyclohexane, styrenes, biphenyls, phenol, substituted phenols, catechols, cresols, resorcinols, eugenol, syringols, guaiacols, vanillin, vanilic acid, aromatic acids, aliphatic acids, syringaldhyde and aldehydes, quinones, cyclohexanol/al, cyclohexanal, beta keto adipate, benzene, toluene, xylene (BTX), and their derivates, higher alkylates, substituted lignins, drugs, mixed aromatic polyols, carbon fiber, fillers, *etc.*

As the temperature increases there is an increase in the yield of gas. The influence of residence time on the yield of products is small, with a slight decrease in oil yield and increase in char yield. The main gases produced are carbon dioxide, carbon monoxide, methane, and hydrogen, and there is significant production of oil and char. Pyrolysis of biomass produces high conversion rates to a gas composed mainly of hydrogen and carbon dioxide with in addition carbon monoxide and C_1–C_4 hydrocarbons. Similar to conventional gasification processes, an oil and char reaction products are produced. The oils are composed of a range of oxygenated compounds, including, cyclopentanone, methoxybenzene, acetic acid, furfural, acetophenone, phenol, benzoic acid, and their alkylated derivatives.

Biological processes are essentially microbic digestion and fermentation. High moisture herbaceous plants (vegetables, sugar cane, sugar beet, corn, sorghum, and cotton), marine crops, and manure are the most suitable for biological digestion. Intermediate-heat gas is methane mixed with carbon monoxide and carbon dioxide. Methane (high-heat gas) can be efficiently converted into methanol.

Cellulose is a remarkable pure organic polymer, consisting solely of units of anhydro glucose held together in a giant straight chain molecule. Cellulose must be hydrolyzed to glucose before fermentation to ethanol. Conversion efficiencies of cellulose to glucose may be dependent on the extent of chemical and mechanical pretreatments to structurally and chemically alter the pulp and paper mill wastes. The method of pulping, the type of wood, and the use of recycled pulp and paper products also could influence the accessibility of cellulose to cellulase enzymes. Hemicelluloses (arabinoglycuronoxylan and galactoglucomammans) are related to plant gums in composition, and occur in much shorter molecule chains than cellulose. The hemicelluloses, which are present in deciduous woods chiefly as pentosans and in coniferous woods almost entirely as hexosanes, undergo thermal decomposition very readily. Hemicelluloses are derived mainly from chains of pentose sugars, and act as the cement material holding together the cellulose micelles and fiber. Cellulose is insoluble in most solvents and has a low accessibility to acid and enzymatic hydrolysis. Hemicelluloses are largely soluble in alkali and, as such, are more easily hydrolyzed.

Hydrolysis (saccharification) breaks down the hydrogen bonds in the hemicellulose and cellulose fractions into their sugar components: pentoses and hexoses. These sugars can then be fermented into bioethanol.

Cellulose [hexosan, $(C_6H_{10}O_5)_n$] hydrolysis produces glucose (a hexose, $C_6H_{12}O_6$). The hydrolysis of cellulose is catalyzed by mineral acids and enzymes. Hemicelluloses hydrolysis produces both hexose and pentose sugars: mannose, galactose, xylose, and arabinose that are not all fermented with existing strains. The hemicelluloses fraction typically produces a mixture of sugars including xylose, arabinose, galactose, and mannose. These are both pentosans: xylose and arabinose, and hexosans: galactose and mannose. The hydrolysis of hemicelluloses is catalyzed by mineral acids and enzymes.

2.4.3   Char from Biomass

Agricultural and forestry byproducts may also offer an inexpensive and renewable additional source of activated carbons. These waste materials have little or no economic value and often present a disposal problem. Therefore, there is a need to valorize these low-cost byproducts. Their conversion into activated carbons would add economic value, help reduce the cost of waste disposal, and most importantly provide a potentially inexpensive alternative to the existing commercial activated carbons. Activated carbon has been prepared from dried municipal sewage sludge and batch mode adsorption experiments (Reddy et al., 2006).

Active carbons are carbonaceous materials with a highly developed internal surface area and porosity. Activated carbon is widely used as an effective adsorbent in many applications such as air separation and purification, vehicle exhaust emission control, solvent recovery, and catalyst support because of its high specific pore surface area, adequate pore size distribution, and relatively high mechanical strength. The large surface area results in high capacity for adsorbing chemicals from gases and liquids (Zanzi, 2001).

The starting materials used in commercial production of activated carbons are those with high carbon contents such as wood, lignite, peat, and coal of different ranks, or low-cost and abundantly available agricultural byproducts. Active carbons can be manufactured from virtually any carbonaceous precursor, but the most commonly used materials are wood, coal, and coconut shell. The development of activated carbons from agricultural carbonaceous wastes will be advantageous for environmental problems. In water contamination, wastewater contains many traces of organic compounds, which are a serious environmental problem. In the development of activated carbons, agricultural carbonaceous wastes will be used, as this will eliminate the problem of waste disposal while at the same time societies will derive great economic benefits from such commercialized products.

Activated carbons are used in the following applications:

1. They can be used as adsorbents for the removal or reduction of gaseous pollutants from the exhaust gases of industrial sources.
2. They can be used as adsorbents for the removal of volatile organic compounds, ozone from air, mercury and dioxin emissions from incinerator flue gas, and hydrogen sulfide emissions from sewage treatment facilities.

3. They can be used to remove chlorine and organic chemicals (such as phenols, polychlorinated biphenyls, trihalomethanes, pesticides and halogenated hydrocarbons), heavy metals, and organic contaminants from water.
4. They can be used to extract some harmful elements of cigarette smoke by incorporation in filter tips of the cigarettes.

Lignin gives higher yields of charcoal and tar from wood, although lignin has a threefold higher methoxyl content than wood (Demirbas, 2000). Phenolics are derived from lignin by cracking the phenyl-propane units of the macromolecule lattice. The formation of char from lignin under mild reaction conditions is a result of the breaking of relatively weak bonds, like the alkyl-aryl ether bonds, and the consequent formation of more resistant condensed structures, as has already been noted (Domburg *et al.*, 1974). One additional parameter that may also have an effect on the char formation is the moisture content of the kraft lignin used. It has been found that the presence of moisture increases the yield of char from the pyrolysis of wood waste at temperatures between 660 K and 730 K (Demirbas, 2006).

The destructive reaction of cellulose is started at temperatures lower than 425 K and is characterized by a decreasing polymerization degree. Thermal degradation of cellulose proceeds through two types of reaction: a gradual degradation, decomposition and charring on heating at lower temperatures, and a rapid volatilization accompanied by the formation of levoglucosan on pyrolysis at higher temperatures. The degradation of cellulose to a more stable anhydrocellulose, which gives higher biochar yield, is the dominant reaction at temperature <575 K (Shafizadeh, 1985). At temperatures >575 K, cellulose depolymerizes, producing volatiles. If the heating rate is very high, the residence time of the biomass at temperatures <575 K is insignificant. Thus a high heating rate provides a shorter time for the dehydration reactions and the formation of less reactive anhydrocellulose, which gives a higher yield of char (Zanzi, 2001). The result is that the rapid heating of the biomass favors the polymerization of cellulose and the formation of volatiles and suppresses the dehydration to anhydrocellulose and char formation. Hence the effect of heating rate is stronger in the pyrolysis of biomass than in that coal.

The initial degradation reactions include depolymerization, hydrolysis, oxidation, dehydration, and decarboxylation (Shafizadeh, 1982). The isothermal pyrolysis of cellulose in air and milder conditions, in the temperature range 623–643 K, has been investigated (Fengel and Wegener, 1983). Under these conditions, the pyrolysis reactions produced 62–72% aqueous distillate and left 10–18% charred residue. After the pyrolysis, the residue was found to consist of some water soluble materials, in addition to char and undecomposed cellulose.

The hemicelluloses undergo thermal decomposition very readily. The hemicelluloses reacted more readily than cellulose during heating. The thermal degradation of hemicelluloses begins above 373 K during heating for 48 h; hemicelluloses and lignin are depolymerized by steaming at high temperature for a short time. The methoxyl content of wet meals decreased at 493 K. The stronger effect of the heating rate on the formation of bi-char from biomass than from coal may be attributed to the cellulose content of the biomass (Demirbas, 2006). It is well known that heating rate has a significant effect on the pyrolysis of cellulose. The heating rate has a much greater effect on the pyrolysis of biomass than on that of coal. The quick devolatilization of the biomass in rapid pyrolysis favors the formation of char with high porosity and high reactivity (Zanzi *et al.*, 1996). The decreased formation of char at the higher heating rate is accompanied by an increased formation of tar. The net effect is a decrease in the volatile fuel production and an increased yield of biochar cellulose converted to levoglucosan at temperatures above 535 K (Freudenberg and Neish, 1968).

2.4.4   Adhesives from Biomass

Wood adhesives play an essential role in industry. Main wood valorization technologies include pulp and paper making, bio-oil by pyrolysis, synthesis gas by gasification, sugar by hydrolysis, ethanol by sugar fermentation and adhesives by alkali liquefaction and polymerization. Lignin is a complex, high molecular weight polymer built of hydroxyphenylpropane units. There are two types of phenolic resins: resol and novalac. A phenolic resin for partially replacing phenol was used with modified organosolv lignin in phenol-formaldehyde (PF) resin production. Organosolv lignin-phenol-formaldehyde (LPF) resins were produced in a two-step preparation with different additions of lignin.

Adhesion is the state in which two surfaces are held together by interfacial forces, which may be valence forces, interlocking action, or both. The adhesives used in commercial wood composite products are usually synthetic polymer resins, based on the condensation reaction of formaldehyde with phenol, urea, resorcinol or melamine (Cetin and Ozmen, 2003). Some structural, semistructural and nonstructural wood adhesives are listed in Table 2.6. Approximately 1 million metric tons of urea-formaldehyde resin are produced annually. More than 70% of this urea-formaldehyde resin is used by the forest products industry for a variety of purposes.

Fermentation residues were obtained by growing the anaerobic cellulolytic bacteria *Ruminococcus albus* 7 or *Clostridium thermocellum* ATCC 27405 on a fibrous fraction derived from lucerne (*Medicago sativa* L.) were converted to effective co-adhesive for phenol–formaldehyde (PF) bonding of aspen veneer sheets to one another (Weimer *et al.*, 2005).

The wood adhesive market is very large and problems due to volatile organic compounds and toxic chemicals in many adhesives and their production are significant. In addition, most of the adhesives are derived from depleting petrochemical resources. An environmentally friendly wood adhesive based on renewable resources and produced by microbial fermentation has been explored (Haag *et al.*, 2004).

A method is described for making adhesive from biomass. The liquefaction oil is prepared from lignin-bearing plant material, and a phenolic fraction is extracted there from. The phenolic fraction is reacted with formaldehyde to yield a phenol–formaldehyde resin. At present, the production of wood composites mainly relies on the petrochemical-based and formaldehyde–based adhesives such as PF resins and urea–formaldehyde (UF) resins (Liu and Li, 2007). Phenol–formaldehyde adhesives are used to manufacture plywood, flakeboard, and fiberglass insulation. Phenolic resins occur only in dark, opaque colors and can therefore be used only to manufacture dark-colored products.

Phenolic resins are some of the principal thermofixed synthetic polymers, the third most important polymeric matrix for composites (Lubin, 1969), and are also known for their high temperature resistance (Knop and Pilato, 1985; Kopf and Little, 1991). Pure phenolic resin can be obtained through the condensation reaction between phenol (C_6H_5OH) and formaldehyde (CH_2O), producing methylene bridges between the phenol molecules (Leite *et al.*, 2004). Thermosetting polymers make excellent structural adhesives because they undergo irreversible chemical change, and on reheating, they do not soften and flow again. They form cross-linked polymers that have high strength, have resistance to moisture and other chemicals, and are rigid enough to support high, long-term static loads without deforming. Phenolic, resorcinolic, melamine, isocyanate, urea, and epoxy are examples of types of wood adhesives that are based on thermosetting polymers.

Marine mussel adhesive protein is an excellent example of a formaldehyde-free adhesive from renewable resources. To cope with turbulent tides and waves, mussels stick to rocks or other substances in seawater by secreting an adhesive protein, commonly called marine adhesive protein. (Li *et al.*, 2004).

Resistance to chemical attack is generally improved by resin impregnation, which protects the underlying wood and reduces movement of liquid into the wood. Resistance to acids can be obtained by impregnating with phenolic resin and to alkalis by impregnating with furfural resin. The adhesion properties of

Table 2.6 Structural, semistructural, and non-structural wood adhesives

Structural	Phenol-formaldehyde
	Resorcinol-formaldehyde
	Phenol-resorcinol-formaldehyde
	Emulsion polymer/isocyanate
	Melamine-formaldehyde
	Melamine-urea-formaldehyde
	Epoxy
	Isocyanate
	Urea-formaldehyde
	Casein
Semistructural	Cross-linked polyvinyl acetate
	Polyurethane
Non-structural	Polyvinyl acetate
	Animal
	Soybean
	Elastomeric construction
	Elastomeric contact
	Starch

different natural fillers without the addition of coupling agents by considering the different filler morphology were investigated. The adhesion behavior has been determined in a qualitative way from microscopic observation. Moreover, it has been quantified that its influence on mechanical properties decreases, since in this type of materials, usually, tensile strength falls (Crespo *et al.*, 2007).

The wood-species dependent performance of polymeric isocyanate resin (PMDI) has been investigated by fracture analysis and solid-state NMR (Das *et al.*, 2007). The surface modification of cellulosic fibers was carried out using organofunctional silane coupling agents in an ethanol/water medium (Abdelmouleh *et al.*, 2004). A new extra-cellular polysaccharide-based adhesive with performance was carried out that may be useful in some wood product applications (Haag *et al.*, 2006). Polysaccharides are generally non-toxic, biodegradable, and produced from renewable resources. In a previous study (Yesil *et al.*, 2007) the bond strengths of three different composite resins bonding to different base substrates were tested.

Hydrogen bonding forces are important in the interfacial attraction of polar adhesive polymers for the hemicellulosics and cellulosics, which are rich with polar hydroxyl groups. These forces of attraction, sometimes referred to as specific adhesion, are particularly important in wetting of water carriers and adsorption of adhesive polymers onto the molecular structures of wood. Water is used as the carrier for most wood adhesives, primarily because water readily absorbs into wood. The relationships between the activation energy and moisture content were investigated. Both the reaction enthalpy and reaction rate increased with the increase in moisture content and remained almost unchanged or increased slightly after the moisture content reached 12% (He and Yan, 2005). The results of a dynamic mechanical analysis technique (DMA), previously developed for estimating the extent of residual cure of phenol-formaldehyde resol resins, depend on the moisture content of the resin (Lorenz and Christiansen, 1995).

There are two types of phenolic resins: resol and novalac. The first one is synthesis under basic conditions with pH conditions with excess of formaldehyde, and the latter is carried out at acidic pH with excess of phenol (Pérez, *et al.*, 2007). They are widely used in industry because of their chemical resistance, electrical insulation, and dimensional stability (Gardziella *et al.*, 2000). There are some published studies involving lignin-resol resins (Danielson *et al.*, 1998; Alonso *et al.*, 2001.) fewer employing lignin-novalac resins for different applications (El-Saied *et al.* 1984; Ysbrandy *et al.* 1992).

It has been demonstrated in the literature by its application in products ranging from wood adhesives to plastics (Hoyt and Goheen, 1971). In these applications the lignin degradation products/fragments are cross-linked to increase their molecular mass (plastic) or to form a rigid three-dimensional cross-linked structure (adhesive). Formaldehyde reacts with phenol at positions ortho- or para- to the aromatic hydroxyl group (Knop and Scheib, 1979). The alkali lignin produced by the alkaline pulping of softwood consequently contains a low content of positions reactive towards electrophiles such as formaldehyde. In natural lignin the positions on the aryl rings para- and ortho- to the hydroxyl groups are usually occupied by alkoxy or alkyl substituents (Sarkanen and Ludwig, 1971).

The reaction of formaldehyde with lignin model compounds in acidic medium is shown to give fast cross-linking of alkali-substituted phenolic and etherified phenolic lignin model compounds at positions meta- to the aromatic hydroxyl groups. This reaction differs from the reaction of formaldehyde with phenolic lignin model compounds in alkaline conditions, where the reaction with formaldehyde always occurs at position ortho-/para- to the aromatic hydroxyl group (Demirbas and Ucan, 1991).

The lignin degradation products and their sodium salts can be converted into very weak organic acids by treating mineral acids. Most phenols have Kas in the neighborhood of 10^{-10}, and are thus considerably weaker than the carboxylic acids (Kas about 10^{-5}). Most phenols are weaker than carbonic acid ($CO_2 + H_2O$), and hence, unlike carboxylic acids, do not dissolve in aqueous bicarbonate solutions. Indeed, phenols are conveniently liberated from their salts by action of carbonic acid (Ka's about 10^{-7}) (Morrison and Boyd, 1983).

Organosolv lignin-phenol-formaldehyde (LPF) resins were produced in a two-step preparation with different additions of lignin. The method selected for the manufacture of lignin resins dealt with modification of the lignin by the methylolation route. Organosolv lignin-based resins showed comparatively good strength and stiffness. The tensile strength properties of test samples made from organosolv lignin resins were equal to or better than those of test samples made from PF resin only (Cetin and Ozmen, 2003). The formaldehyde reactivity of the organosolv lignin is determined in accordance with the method described by Wooten *et al.* (1988). Methylolated lignin-phenol-formaldehyde resins are prepared using the procedure of Sellers (1993). Figure 2.8 shows the curve for lignin-formaldehyde methylolation *versus* reaction time.

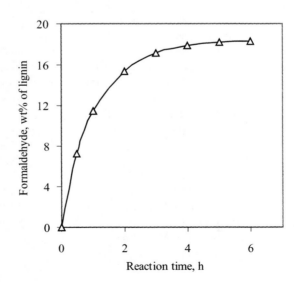

Fig. 2.8 Curve for lignin-formaldehyde methylolation *versus* reaction time

Table 2.7 Typical lignin-phenol methylol co-polymerization using organosolv lignin

Constituent	% by weight
Formaldehyde	42.8
Water	22.3
Lignin	19.1
Sodium hydroxide	15.8

Table 2.8 Typical lignin-formaldehyde-urea methylol co-condensation using organosolv lignin

Constituent	% by weight
Formaldehyde	43.3
Lignin phenolics	40.5
Sodium hydroxide	9.9
Urea	6.3

Typical lignin-phenol methylol co-polymerization using organosolv lignin is given in Table 2.7. Typical lignin-formaldehyde-urea methylol co-condensation using organosolv lignin is shown in Table 2.8.

2.4.5 Valorization of Wood

Forests are a principal global economic as well as ecological resource. Forests have played a big role in the development of human societies. The prime direct or marketable product of most forests is wood for use as timber, fuelwood, pulp, and paper, providing some 3.4 billion cubic meters of timber-equivalent a year globally. Asia and Africa use 75% of global wood fuels.

The lumber, plywood, pulp, and paper industries burn their own wood residues in large furnaces and boilers to supply 60% of the energy needed to run factories. The availability of fuelwood from the forest is continually declining at an ever-increasing rate due to indiscriminate deforestation and slow regeneration, as well as afforestation. The fuelwoods generally used by local people have been identified and analyzed quantitatively to select a few species with the best fuelwood characteristics so that plantation on non-agricultural lands could be undertaken (Jain, 1992).

The reduction of particle size and moisture content, together with the most appropriate storage and handling systems are necessary for an efficiently operated wood waste combustion system. Size reduction may be carried out in several stages in a hog or attrition mill, with screening before and in between. The moisture in residues may be reduced either by mechanical pressing, air-drying or the use of hot air dryers, or a combination of all three. Generally slabs, edgings, peeler cores, veneer waste, and trimmings are transported by mechanical conveyors.

2.4.5.1 Reuse of Wood Wastes

Forest energy involves the use of forest biomass, which is currently not being used in the traditional forest products industries. Essentially, this means the forest residues left after forest harvesting of residual trees and scrub or undermanaged woodland. Forest residues alone count for some 50% of the total forest biomass and are currently left in the forest to rot.

Wood biomass involves trees with commercial structure and forest residues not being used in the traditional forest products industries. Available forest residues appear to be an attractive fuel source. Collection and handling and transport costs are critical factors in the use of forest residues. Although the heat produced from wood wastes is less than that from oil or gas, its cost compared to fossil fuels makes it an attractive source of readily available heat, or heat and power. The most effective utilization of wood wastes, particularly in the sawmilling and plywood industry, plays an important role in energy efficient production. Wood preparation involves the conversion of roundwood logs into a form suitable for pulping and includes processes for debarking, chipping, screening, handling, and storage.

Forest residues typically refer to those parts of trees such as treetops, branches, small-diameter wood, stumps and dead wood, as well as undergrowth and low-value species. The conversion of wood to biofuels and biochemicals has long been a goal of the forest products industry. Forest residues alone count for some 50% of the total forest biomass and are currently left the forest to rot.

The principal sources of waste wood are two waste streams: municipal solid waste (MSW) and construction and demolition waste (C&DW). Municipal solid waste (MSW) is waste from residential, commercial, institutional, and industrial sources. Each generates distinctly different types of wood waste, with differing degrees and levels of recyclability. The primary components of wood waste are used lumber, shipping pallets, trees, branches, and other wood debris from construction and demolition clearing and grubbing activities. Construction and demolition (C&D) waste is defined as solid, largely inert waste resulting from the construction, repair, demolition, or razing of buildings, roads, and other structures. The term also includes debris from the clearing of land for construction. Other sources of waste wood include chemically treated wood from railroad ties, telephone and utility poles, miner poles, crossties, constructors, pier and dock timbers, untreated wood from logging and silvicultural operations, and industrial waste wood outside the MSW and C&DW streams. Chemical treatments and costs of collection make much of this material difficult to recover.

The wood waste preparation process generally involves hogging, dewatering, screening, size reduction, bulk storage, blending and drying prior to combustion so as to ensure a reliable and consistent supply of quality fuel to the burners. The handling, treatment, and storage of wood waste fuel are considerably more costly and troublesome than what is required for traditional fossil fuels.

The most limiting factors for using of wood waste as fuel for power generation are transportation costs and its energy content. The economics of wood waste energy generation becomes more attractive as traditional fuel prices increase.

2.4.5.2   Fractionation of Wood Residues and Wood Wastes

Wood related industries and households consume the most biomass energy. The lumber, pulp, and paper industries burn their own wood wastes in large furnaces and boilers to supply 60% of the energy needed to run the factories. Biomass includes 60% wood and 40% non-wood materials (Demirbas, 2000).

Wood biomass involves trees with commercial structure and forest residues not being used in the traditional forest products industries. Forest residues typically refer to those parts of trees such as treetops, branches, small-diameter wood, stumps and dead wood, as well as undergrowth and low-value species. The conversion of wood to biofuels and biochemicals has long been a goal of the forest

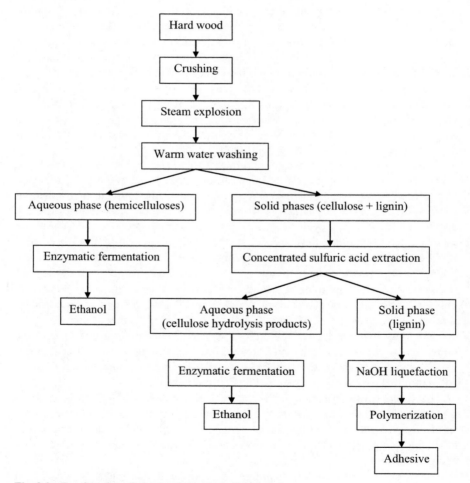

Fig. 2.9 Fractionation of wood and chemicals from wood

products industry. Forest residues alone count for some 50% of the total forest biomass and are currently left the forest to rot.

The opportunity to combine fuelwood production with effluent disposal has many potential environmental benefits. The economic feasibility of woody biomass plantations is difficult to justify at present but if full costing of externalities such as CO_2 emissions are to be applied in the future, then acceptable payback periods for boiler conversions and handling and storage facilities may be achieved (Demirbas, 2000).

Fractionation refers to the conversion of wood into its constituent components (cellulose, hemicelluloses, and lignin). Processes include steam explosion, aqueous separation and hot water systems. Commercial products of biomass fractionation include levulinic acid, xylitol and alcohols. Figure 2.9 shows the fractionation of wood and chemicals from wood.

Main fractionation chemicals from wood ingredients are:

1. Dissociation of cell components → lignin fragment + oligosaccharides + cellulose
2. Hydrolysis of cellulose (saccharification) → glucose
3. Conversion of glucose (fermentation) → ethanol + lactic acid
4. Chemical degradation of cellulose → levulinic acid + xylitol
5. Chemical degradation of lignin → phenolic products.

Wood fuel is a renewable energy source and its importance will increase in future. Three main determinants of the costs of operating and constructing a wood-fired power plant of a given size exist. They are: the availability of the required fuel, the delivered fuel prices, and the financing and construction of the desired power plant. Most previous research has focused on the feasibility of single-source fuel plants. One popular approach has been to examine the feasibility of short-rotation intensive-cultivation plantations.

References

Abdelmouleh, M., Boufi, S., Belgacem, M.N., Duarte, A.P., Salah, A., Gandini, A. 2004. Modification of cellulosic fibres with functionalised silanes: development of surface properties. Int J Adhesion Adhesives 24:43–54.

Alanne, K., Sari, A. 2006. Distributed energy generation and sustainable development. Renew Sust Energy Rev 10:539–558.

Alonso, M.V., Rodriguez, J.J., Oliet, M., Rodriguez, F., Garcia, J., Gilarranz, M.A. 2001. Characterization and structural modification of ammonic lignosulfonate by methylolation. J Appl Polym Sci 82:2661–2668.

Arkesteijn, K., Oerlemans, L. 2005. The early adoption of green power by Dutch households. An empirical exploration of factors influencing the early adoption of green electricity for domestic purposes. Energy Policy 33:183–196.

Balat, M. 2008a. Mechanisms of thermochemical biomass conversion processes. Part 1: Reactions of pyrolysis. Energy Sources Part A 30:620–635.

Balat, M. 2008b. Mechanisms of thermochemical biomass conversion processes. Part 2: Reactions of gasification. Energy Sources Part A 30:636–648.

Balat, M. 2008c. Mechanisms of thermochemical biomass conversion processes. Part 3: Reactions of liquefaction. Energy Sources Part A 30:649–659.

Baker, A.J. 1982. Wood fuel properties and fuel products from Woods, in Proc. Fuelwood, Management and Utilisation Seminar, Nov. 9–11, Michigan State Univ., East Lansing, MI.

Bridgwater, A.V. 2003. Renewable fuels and chemicals by thermal processing of biomass. Chem Ind J 91:87–102.

Bushnell, D.J., Haluzok, C., Dadkhah-Nikoo, A. 1989. Biomass fuel characterization, testing and evaluating the combustion characteristics of selected biomass fuels. Bonneville Power Administration, Corvallis, OR.

Cetin, N.S., Ozmen, N. 2003. Studies on lignin-based adhesives for particleboard panels. Turk J Agric For 27:183–189.

Crespo, J.E., Balart, R., Sanchez, L., Lopez, J. 2007. Mechanical behaviour of vinyl plastisols with cellulosic fillers. Analysis of the interface between particles and matrices. Int J Adhesion Adhesives 27:422–428.

Danielson, B., Simanson, R.. 1998. Kraft lignin in phenol formaldehyde resin. Part 1. Partial replacement of phenol by kraft lignin in phenol formaldehyde adhesives for plywood. J Adhesion Sci 12:923–939.

Das, S., Malmberg, M.J., Frazier, C.E. 2007. Cure chemistry of wood/polymeric isocyanate (PMDI) bonds: Effect of wood species. Int J Adhesion Adhesives. 27: 250–257.

Demirbas, A. 1991. Fatty and resin acids recovered from spruce wood by supercritical acetone extraction. Holzforschung 45:337–339.

Demirbas, A.1997. Calculation of higher heating values of biomass fuels. Fuel 76:431–434.

Demirbas, A. 1998. Determination of combustion heat of fuels by using non-calorimetric experimental data. Energy Edu Sci Technol 1:7–12.

Demirbas, A. 2000. Mechanisms of liquefaction and pyrolysis reactions of biomass. Energy Convers Mgmt 41:633–646.

Demirbas, A. 2001. Biomass resource facilities and biomass conversion processing for fue and chemicals. Energy Convers Mgmt 42:1357–1378.

Demirbas, A. 2002. Fuel characteristics of olive husk and walnut, hazelnut, sunflower and almond shells. Energy Sources 24:213–219.

Demirbas, A. 2003. Fuelwood characteristics of six indigenous wood species from Eastern Black Sea Region. Energy Sources 25:309–316.

Demirbas A. 2004. Combustion characteristics of different biomass fuels. Prog Energy Combus Sci 30:219–230.

Demirbas A. 2006. Production and characterization of bio-chars from biomass *via* pyrolysis. Energy Sources Part A 28:413–422.

Demirbas, A. 2008a. Biodiesel: A Realistic fuel alternative for diesel engines. Springer, London.

Demirbas, A. 2008b. Heavy metal adsorption onto agro based waste materials: A review. J Hazard Mat 157:220–229.

Demirbas, A., Güllü, D., Caglar, A., Akdeniz, F. 1997. Determination of calorific values of fuel from lignocellulosics. Energy Sources 19:765–770.

Demirbas, A., Kucuk, M.M. 1993. Delignification of Ailanthus altissima and spruce orientalis with glycerol or alkaline glycerol at atmospheric pressure. Cellulose Chem Technol 27:679–686.

Demirbas, A., Ucan, H.I. 1991. Low temperature pyrolysis of black liquor and polymerization of products in alkali aqueous medium. Fuel Sci Technol Int 9:93–105.

Elliott, D. 1999. Prospects for renewable energy and green energy markets in the UK. Renewable Energy 16:1268–1271.

El-Saied, H., Nada, A.M.A., Ibrahem, A.A., Yousef, M.A. 1984. Waste liquers from cellulosic industries. III. Lignin from soda-spent liquor as a component in phenol-formaldehyde resin. Angew Makromol Chem 122:169–181.

Fengel, D., Wegener, G. 1983. In Wood chemistry, ultrastructure, reactions, Chap 7, p. 326.Walter de Gruyter, Berlin.

Freudenberg, K., Neish, A.C. 1968. Constitution and biosynthesis of lignin. Springer, New York.

Fridleifsson, I.B. 2003. Status of geothermal energy amongst the world's energy sources. Geothermics 32:379–388.

Garcia-Valls, R., Hatton, T.A. 2003. Metal ion complexation with lignin derivatives. Chem Eng J 94:99–105.

Gardziella, A., Pilato, L.A., Knop, A. 2000. Phenolic resins: Chemistry, applications, standardization, safety and ecology. Springer, New York.

Glasser, W.G., Sarkanen, S. (eds.) 1989. Lignin: Properties and materials, American Chemical Society, Washington, DC.

Goldemberg, J., Coelho, S.T. 2004. Renewable energy – Traditional biomass vs. modern biomass. Energy Policy 32:711–714.

Goldstein, I.S. 1981. Organic chemical from biomass, CRC Press, Boca Raton, FL.

Haag, A.P., Geesey, G.G., Mittleman, M.W. 2006. Bacterially derived wood adhesive. Int J Adhesion Adhesives.26: 177–183.

Haag, A.P., Maier, R.M. Combie, J., Geesey, G.G. 2004. Bacterially derived biopolymers as wood adhesives. Int J Adhesion Adhesives.24: 495–502.

Hamelinck, C., Faaij, A. 2006. Outlook for advanced biofuels. Energy Policy 34:3268–3283.

Hashem, A., Akasha, R.A.,. Ghith, A., Hussein, D.A. 2007. Adsorbent based on agricultural wastes for heavy metal and dye removal: A review. Energy Edu Sci Technol 19:69–86.

He, G., Yan, N. 2005. Effect of moisture content on curing kinetics of pMDI resin and wood mixtures. Int J Adhesion Adhesives 25: 450–455.

Hergert, H.L., Pye, E.K. 1992. Recent history of organosolv pulping, Tappi Notes-1992 Solvent Pulping Symposium, pp. 9–26.

Hoyt, C.H., Goheen, D.W. 1971. Lignins–Occurrence, formation, structure and reactions. Wiley-Interscience, New York.

IPCC. 1997. Greenhouse gas inventory reference manual: Revised 1996 IPCC guidelines for national greenhouse gas inventories. Report Vol. 3, p. 1.53, Intergovernmental Panel on Climate Change (IPCC), Paris, France (available from: www.ipcc.ch/pub/guide.htm).

IPCC. 2007. Intergovernmental Panel on Climate Change (IPCC) fourth assessment report, Working Group III (available from http://www.ipcc.ch).

Jain, R.K. 1992. Fuelwood characteristics of certain hardwood and softwood tree species of India.Biores Technol 41:129–133.

Jain, R.K., Singh, B. 1999. Fuelwood characteristics of selected indigenous tree species fromcentral India. Biores Technol 68:305–308.

Karaosmanoglu, F., Aksoy, H.A. 1988. The phase separation problem of gasoline-ethaol mixture as motor fuel alternatives. J Thermal Sci Technol 11:49–52.

Kartha, S., Larson, E.D. 2000. Bioenergy primer: Modernised biomass energy for sustainable Development. Technical Report UN Sales Number E.00.III.B.6, United Nations Development Programme, 1 United Nations Plaza, New York, NY 10017, USA.

Knop, A., Pilato, L.A.. 1985. Phenolic resin chemistry, application and performance, future directions. Springer, Heidelberg.

Knop, A., Scheibm W. 1979. Chemistry and application of phenolic resins. Springer, Berlin.

Kopf, P.W., Little, A.D. 1991. Phenolic resins. In Kirk-Othmer Encyclopedia of Chemical Technology, Vol. 18.

Kuznetsov, S.A., Kuznetsov, B.N., Aleksandrova, N.B., Danlov, V.G., Zhizhaev, A.M. 2005. Obtaining arabinigalactan, dihydrate quercetin and microcrystalline cellulose using machanochemical activation. Chem Sustain Develop 13:261–268.

Larson, E.D. 1993. Technology for fuels and electricity from biomass. Annual Rev. Energy Environ. 18:567–630.

Leite, J.L., Pires, A.T.N., Ulson de Souza, S.M.A.G., Ulson de Souza, A.A. 2004. Characterisation of a phenolic resin and sugar cane pulp composite. Brazilian J. Chem. Eng. 21:253– 260.

Li, K., Geng, X., Simonsen, J., Karchesy, J. 2004. Novel wood adhesives from condensed tannins and polyethylenimine. Int J Adhesion Adhesives 24:327–333.

Liu, Y., Li, K. 2007. Development and characterization of adhesives from soy protein for bonding wood. Int J Adhesion Adhesives 27:59–67.

Lorenz, L.F., Christiansen, A.W. 1995. Interactions of phenolic resin alkalinity, moisture content, and cure behavior. Ind Eng Chem Res 34: 4520–4523.

Lubin, G. 1969. Handbook of fiberglass and advanced plastics composites. Van Nostrand Reinhold, New York.

Mantanis, G.I., Young, R.A., Rowell, R.M. 1995. Swelling of compressed cellulose fiber webs in organic liquids. Cellulose 2:1–22.

Mohan, D., Pittman, C.U., Jr., Steele, P.H. 2006. Pyrolysis of wood/biomass for bio-oil: A critical review. Energy Fuels 20:848–889.

Morrison, R.T., Boyd, R.N. 1983. Organic chemistry, fourth ed., Chap. 24,. 960, Allyn and Bacon, New York.

Murphya, H., Niitsuma, H. 1999. Strategies for compensating for higher costs of geothermal electricity with environmental benefits. Geothermics 28:693–711.

Pérez, J.M., Rodríguez, F., Alonso, M.V., Oliet, M., Echeverría, J.M. 2007. Characterization of a novolac resin substituting phenol by ammonium lignosulfonate as filler or extender. BioRes 2:270–283.

Ragland, K.W., Aerts, D.J., Baker, A.J. 1991. Properties of wood for combustion analysis. Biores Technol 37:161–168.

Reddy, S.S., Kotaıah, B., Reddy, N.S.P., Velu, M. 2006. The removal of composite reactive dye from dyeing unit effluent using sewage sludge derived activated carbon. Turkish J Eng Env Sci 30:367–373.

Rydholm, S.A. 1965. Pulping Processes. Wiley-Interscience, New York.

Saga, K., Yokoyama, S., Imou, K., Kaizu, Y. 2008. A comparative study of the effect of CO_2 emission reduction by several bioenergy production systems. Int Energy J 9:53–60.

Sarkanen, K.V., Ludwig, C.H. (eds.) 971. Lignins: Occurrence, formation, structure and reactions, Wiley, New York.

Sellers, Jr., T. 1993. Modification of phenolic resin with organosolv lignins and evaluation of strandboards made by the resin as binder. PhD Thesis, The University of Tokyo, Japan.

Shafizadeh, F. 1982. Introduction to pyrolysis of biomass. J Anal Appl Pyrolysis 3:283–305.

Shafizadeh, F. 1985. In Fundamentals of thermochemicals biomass conversion. In: Overend, R.P.,Milne, T.A., Mudge, L.K. (eds.). Elsevier Applied Science, New York.

Sheehan, J., Dunahay, T., Benemann, J., Roessler, P. 1998. A look back at the U.S. Department of Energy's aquatic species program – Biodiesel from algae. National Renewable Energy Laboratory (NREL) Report: NREL/TP-580-24190. Golden, CO.

Tewfik, S.R. 2004. Biomass utilization facilities and biomass processing technologies. Energy Edu Sci Technol 14:1–19.

Theander, O. 1985. In: Fundamentals of thermochemical biomass conversion. In Overand, R.P., Mile, T.A., Mudge, L.K. (eds.). Elsevier Applied Science, New York.

Tillman, D.A. 1978. Wood as an Energy Resource, Academic Press, New York.

Timell, T.E. 1967. Recent progress in the chemistry of wood hemicelluloses. Wood Sci Technol 1:45–70.

UNDP (United Nations Development Programme). 2000. World Energy Assessment. Energy and the challange of sustainability.

Weimer, P.J., Koegel, R.G., Lorenz, L.F., Frihart, C.R., Kenealy, W.R. 2005. Wood adhesives prepared from lucerne fiber fermentation residues of Ruminococcus albus and Clostridium thermocellum. Appl Microbiol Biotechnol 66:635–640.

Wenzl, H.F.J., Brauns, F.E., Brauns, D.A. 1970. The chemical technology of wood. Academic Press, New York.

Wooten, A.L., Sellers, T., Tahir, P.M. 1988. Reaction of formaldehyde with lignin. Forest Products J 38(6): 45–46.

Yesil, Z.D., Karaoglanoglu, S., Akyil, M.S., Seven, N. 2007. Evaluation of the bond strength of different composite resins to porcelain and metal alloy. Int J Adhesion Adhesives 27:258–262.

Young, R.A. 1986. Structure, swelling and bonding of cellulose fibers. In Cellulose: Structure, modification, and hydrolysis, pp. 91–128. Wiley, New York.

Ysbrandy, R.E., Sanderson, R.D., Gerischer, G.F.R. 1992. Adhesives from autohydrolysis bagasse lignin. Part I. Holzforschung 46:249–252.

Zanzi, R. 2001. Pyrolysis of biomass. Dissertation, Royal Institute of Technology, Department ofChemical Engineering and Technology, Stockholm.

Zanzi, R,. Sjöström, K., Björnbom, E. 1996. Rapid high-temperature pyrolysis of biomass in afree-fall reactor. Fuel 75:545–550.

Chapter 3
Biofuels

3.1 Introduction to Biofuels

Petroleum-based fuels are limited reserves concentrated in certain regions of the world. The dramatic increase in the price of petroleum, the finite nature of fossil fuels, increasing concerns regarding environmental impact, especially related to greenhouse gas emissions, and health and safety considerations are forcing the search for new energy sources and alternative ways to power the world's motor vehicles.

There are two global biorenewable liquid transportation fuels that might replace gasoline and diesel fuel. These are bioethanol and biodiesel. Bioethanol is good alternate fuel that is produced almost entirely from food crops. Biodiesel has become more attractive recently because of its environmental benefits.

Transport is one of the main energy consuming sectors. It is assumed that biodiesel is used as a fossil diesel replacement and that bioethanol is used as a gasoline replacement. Biomass-based energy sources for heat, electricity and transportation fuels are potentially carbon dioxide neutral and recycle the same carbon atoms. Due to its widespread availability, biorenewable fuel technology will potentially employ more people than fossil-fuel-based technology (Demirbas, 2006).

The term biofuel is referred to as solid, liquid, or gaseous fuels that are predominantly produced from biorenewable or combustible renewable feedstocks (Demirbas, 2007a). Liquid biofuels are important for the future because they replace petroleum fuels. Biofuels are generally considered as offering many priorities, including sustainability, reduction of greenhouse gas emissions, regional development, social structure and agriculture, security of supply (Reijnders, 2006). The biggest difference between biofuels and petroleum feedstocks is oxygen content. Biofuels are non-polluting, locally available, accessible, sustainable, and are a reliable fuel obtained from renewable sources. Electricity generation from biofuels has been found to be a promising method for the near future. The future of biomass electricity generation lies in biomass integrated gasification/gas turbine technology, which offers high-energy conversion efficiencies.

A. Demirbas, *Biofuels,*
© Springer 2009

Table 3.1 Classification of renewable biofuels based on their production technologies

Generation	Feedstock	Example
First generation biofuels	Sugar, starch, vegetable oils, or animal fats	Bioalcohols, vegetable oil, Biodiesel, biosyngas, biogas
Second generation biofuels	Non-food crops, wheat straw, corn, wood, solid waste, energy crops	Bioalcohols, bio-oil, bio-DMF, Biohydrogen, bio-Fischer–Tropsch diesel, wood diesel
Third generation biofuels	Algae	Vegetable oil, biodiesel
Fourth generation biofuels	Vegetable oil, biodiesel	Biogasoline

First generation biofuels refer to biofuels made from sugar, starch, vegetable oils, or animal fats using conventional technology. The basic feedstocks for the production of first generation biofuels are often seeds or grains such as wheat, which yields starch that is fermented into bioethanol, or sunflower seeds, which are pressed to yield vegetable oil that can be used in biodiesel. Table 3.1 shows the classification of renewable biofuels based on their production technologies.

Second and third generation biofuels are also called advanced biofuels. Second generation biofuels are made from non-food crops, wheat straw, corn, wood, energy crop using advanced technology. Algae fuel, also called oilgae or third generation biofuel, is a biofuel from algae. Algae are low-input/high-yield (30 times more energy *per* acre than land) feedstocks to produce biofuels using more advanced technology. On the other hand, an appearing fourth generation is based in the conversion of vegoil and biodiesel into biogasoline using the most advanced technology.

Renewable liquid biofuels for transportation have recently attracted huge attention in different countries all over the world because of its renewability, sustainability, common availability, regional development, rural manufacturing jobs, reduction of greenhouse gas emissions, and its biodegradability. Table 3.2 shows the availability of modern transportation fuels.

There are several reasons for biofuels to be considered as relevant technologies by both developing and industrialized countries (Demirbas, 2006). They include energy security reasons, environmental concerns, foreign exchange savings, and

Table 3.2 Availability of modern transportation fuels

Fuel type	Availability	
	Current	Future
Gasoline	Excellent	Moderate-poor
Bioethanol	Moderate	Excellent
Biodiesel	Moderate	Excellent
Compressed natural gas (CNG)	Excellent	Moderate
Hydrogen for fuel cells	Poor	Excellent

socioeconomic issues related to the rural sector. Due to its environmental merits, the share of biofuel in the automotive fuel market will grow fast in the next decade (Kim and Dale, 2005; Demirbas and Balat, 2006). The advantages of biofuels are the following: (a) they are easily available from common biomass sources; (b) they represent a carbon dioxide-cycle in combustion, (c) they have a considerable environmentally friendly potential; (d) they have many benefits for the environment, economy, and consumers; and (e) they are biodegradable and contribute to sustainability (Puppan, 2002).

Various scenarios have resulted in high estimates of biofuel in the future energy system. The availability of resources is an important factor. The rationale is to facilitate the transition from the hydrocarbon economy to the carbohydrate economy by using biomass to produce bioethanol and biomethanol as replacements for traditional oil-based fuels and feedstocks. The biofuel scenario produced equivalent rates of growth in GDP and *per capita* affluence, reduced fossil energy intensities of GDP, reduced oil imports and gave an energy ratio. Each scenario has advantages whether it is rates of growth in GDP, reductions in carbon dioxide emissions, the energy ratio of the production process, the direct generation of jobs, or the area of plantation biomass required to make the production system feasible (Demirbas, 2006).

Ethanol is the most widely used liquid biofuel. It is an alcohol and is fermented from sugars, starches, or cellulosic biomass. Most commercial production of ethanol is from sugar cane or sugar beet, as starches and cellulosic biomass usually require expensive pretreatment. It is used as a renewable energy fuel source as well for manufacture of cosmetics, pharmaceuticals, and also for the production of alcoholic beverages. Ethyl alcohol is not only the oldest synthetic organic chemical used by man, but it is also one of the most important.

Carbohydrates (hemicelluloses and cellulose) in plant materials can be converted to sugars by the hydrolysis process. In an earlier study (Taherzadeh, 1999), the physiological effects of inhibitors on ethanol from lignocellulosic materials and fermentation strategies were comprehensively investigated. Fermentation is an anaerobic biological process in which sugars are converted to alcohol by the action of microorganisms, usually yeast. The resulting alcohol is bioethanol. The value of any particular type of biomass as feedstock for fermentation depends on the ease with which it can be converted to sugars. It is possible that wood, straw, and even household wastes may be economically converted to bioethanol. In 2004, 3.4 billion gallons of fuel ethanol were produced from over 10% of the corn crop. The ethanol demand is expected to more than double in the next ten years. For the supply to be available to meet this demand, new technologies must be moved from the laboratories to commercial reality (Bothast and Schlicher, 2005). The world ethanol production is about 60% from feedstock from sugar crops.

Vegetable oils from renewable oil seeds can be used when mixed with diesel fuels. Pure vegetable oil, however, cannot be used in direct-injection diesel engines, such as those regularly used in standard tractors, since engine coking occurs after several hours of use. Conversion of vegetable oils and animal fats into biodiesel has been undergoing further development over the past several years

(Prakash, 1998; Madras *et al.*, 2004; Haas *et al.*, 2006; Meher *et al.*, 2006). Bio-diesel represents an alternative to petroleum-based diesel fuel. Commonly accepted biodiesel raw materials include the oils from soy, canola, corn, rapeseed, and palm. New plant oils that are under consideration include mustard seed, peanut, sunflower, and cotton seed. The most commonly considered animal fats include those derived from poultry, beef, and pork (Usta *et al.*, 2005).

The Fischer–Tropsch Synthesis (FTS) produces hydrocarbons of different length from a gas mixture of H_2 and CO (syngas) from biomass gasification called biosyngas. The fundamental reactions of synthesis gas chemistry are methanol synthesis, FTS, oxosynthesis (hydroformylation), and methane synthesis (Prins *et al.*, 2004). The FTS process is capable of producing liquid hydrocarbon fuels from biosyngas. The process for producing liquid fuels from biomass, which integrates biomass gasification with FTS, converts a renewable feedstock into a clean fuel. The products from FTS are mainly aliphatic straight-chain hydrocarbons (C_xH_y). Besides C_xH_y, also branched hydrocarbons, unsaturated hydrocarbons, and primary alcohols are formed in minor quantities (Dry, 1999; Anderson, 1984; Bukur *et al.*, 1995; Schulz, 1999; Tijmensen *et al.*, 2002; May, 2003). The products from FTS are mainly aliphatic straight-chain hydrocarbons (C_xH_y). The product distribution obtained from FTS includes the light hydrocarbons methane (CH_4), ethene (C_2H_4) and ethane (C_2H_6), LPG (C_3–C_4, propane and butane), gasoline (C_5–C_{12}), diesel fuel (C_{13}–C_{22}), and light and waxes (C_{23}–C_{33}). The distribution of the products depends on the catalyst and the process parameters such as temperature, pressure, and residence time. FTS has been extensively investigated by many researchers (Rapagna *et al.*, 1998; Sie *et al.*, 1999; Ahón *et al.*, 2005; Mirzaei *et al.*, 2006)

3.1.1 Economic Impact of Biofuels

Policy drivers for renewable liquid biofuels have attracted particularly high levels of assistance in some countries given their promise of benefits in several areas of interest to governments, including agricultural production, greenhouse gas emissions, energy security, trade balances, rural development, and economic opportunities for developing countries. Total biofuel support estimates for US and EU in 2006 are given in Table 3.3 (Demirbas, 2008).

Table 3.3 Total biofuel support estimates for US and EU in 2006 (billions of US$)

	Biodiesel	Ethanol	Total biofuel
United States (US)	0.60	5.96	6.70
European Union (EU)	3.11	1.61	4.82
Total of world	3.65	7.85	11.79

Source: Demirbas, 2008

The continuous increasing of energy consumption in all economic sectors, the stringency of fossil resources and environmental pollution, are leading to the emerging of new sources of energy. Biomass in general and agricultural waste in particular seems to be a realistic alternative power generation leading to environmental, technical, and economical benefits (Bridgewater, 1995; Sami *et al.*, 2001).

Bioethanol can be used directly in cars designed to run on pure ethanol or blended with gasoline to make "gasohol". Anhydrous ethanol is required for blending with gasoline. No engine modification is typically needed to use the blend. Ethanol can be used as an octane-boosting, pollution-reducing additive in unleaded gasoline. World production of ethanol from sugar cane, maize, and sugar beet increased from less than 20 billion liters in 2000 to over 40 billion liters in 2005. This represents around 3% of global gasoline use. Production is forecasted to almost double again by 2010 (UN, 2006).

Biodiesel has become more attractive recently because of its environmental benefits. The cost of biodiesel, however, is the main obstacle to commercialization of the product. With cooking oils used as raw material, the viability of a continuous transesterification process and recovery of high quality glycerol as a biodiesel byproduct are primary options to be considered to lower the cost of biodiesel (Ma and Hanna, 1999; Zhang *et al.*, 2003). The possible impact of biodiesel on fuel economy is positive as given in Table 3.4 (EPA, 2002).

Green energy production is the principal contributor in economic development of a developing country. Its economical development is based on agricultural production and most people live in rural areas. Implementation of integrated community development programs is therefore very necessary.

Biorenewable alcohols are at present more expensive in the case of synthesis-ethanol from ethylene and methanol from natural gas. The simultaneous production of biomethanol (from sugar juice) parallel to the production of bioethanol, appears economically attractive in locations where hydroelectricity is available at very low cost.

The EU production of biofuels amounted to around 2.9 billion liters in 2004, with bioethanol totaling 620 million liters and biodiesel the remaining 2.3 billion liters. The feed stocks used for ethanol production are cereals and sugar beet, while biodiesel is manufactured mainly from rapeseed. In 2004, EU biodiesel production used 27% of EU rapeseed crop. In the same year, bioethanol production used 0.4% of EU cereals production and 0.8% of EU sugar beet production. The EU is by far the world's biggest producer of biodiesel with Germany producing over half of the EU's biodiesel. France and Italy are also important biodiesel producers, while Spain is the EU's leading bioethanol producer (UN, 2006).

Table 3.4 Fuel economy impacts of biodiesel use

Percent of biodiesel in diesel fuel	% reduction in miles/gallon
20	0.9–2.1
100	4.6–10.6

Fig. 3.1 World production
of ethanol and biodiesel,
1980–2007
Source: Demirbas, 2008

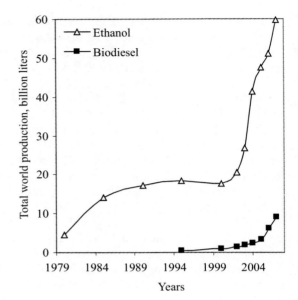

Figure 3.1 shows the world productions of ethanol and biodiesel between 1980 and 2007 (Demirbas, 2008). Between 1991 and 2001, world ethanol production rose from around 16 billion liters a year to 18.5 billion liters. From 2001 to 2007, production is expected to have tripled, to almost 60 billion liters a year. Brazil was the world's leading ethanol producer until 2005 when US production roughly equaled Brazil's. The United States became the world's leading ethanol producer in 2006. China holds a distant but important third place in world rankings, followed by India, France, Germany, and Spain. Figure 3.2 shows the top five bioethanol producers in 2006 (RFA, 2007).

Between 1991 and 2001, world biodiesel production grew steadily to approximately 1 billion liters. Most of this production was in OECD Europe and was based on virgin vegetable oils. Small plants using waste cooking oils started to be built in other OECD countries by the end of the 1990s, but the industry outside Europe remained insignificant until around 2004. Since then, governments around the world have instituted various policies to encourage development of the industry, and new capacities in North America, south-east Asia and Brazil has begun to come on stream at a brisk rate. As a result, between 2001 and 2007, biodiesel production will have grown almost tenfold, to 9 billion liters (Demirbas, 2008).

Biofuels production costs can vary widely depending on feedstock, conversion process, scale of production, and region. Only the ethanol produced in Brazil comes close to competing with gasoline. Ethanol produced from corn in the US is considerably more expensive than that produced from sugar cane in Brazil, and ethanol produced from grain and sugar beet in Europe is even more expensive. These differences reflect many factors, such as scale, process efficiency, feedstock costs, capital and labor costs, co-product accounting, and the nature of the estimates.

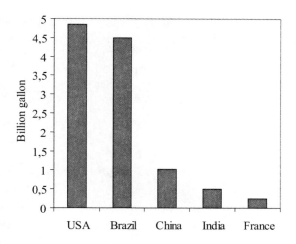

Fig. 3.2 The top five bio-ethanol producers (billion gallons) in 2006

Average international prices for common biocrude, fat, crops and oils used as feedstock for biofuel production in 2007 are given in Table 3.5. Agriculture ethanol is at present more expensive than synthesis-ethanol from ethylene. The simultaneous production of biomethanol in parallel to the production of bioethanol and appears economically attractive in locations where hydroelectricity is available at very low cost (~0.01 \$ Kwh) (Grassi, 1999).

Ethanol from sugar cane, produced mainly in developing countries with warm climates, is generally much cheaper to produce than that produced from grain or sugar beet in IEA countries. Estimates show that bioethanol in the EU becomes competitive when the oil price reaches US\$ 70 a barrel while in the United States it becomes competitive at US\$ 50–60 a barrel. For Brazil the threshold is much lower – between US\$ 25 and US\$ 30 a barrel. Other efficient sugar producing countries such as Pakistan, Swaziland, and Zimbabwe have production costs similar to Brazil's (Balat, 2007; Dufey, 2006).

The largest ethanol cost component is the plant feedstock. Operating costs, such as feedstock cost, co-product credit, chemicals, labor, maintenance, insurance, and taxes, represent about one third of the total cost *per* liter, of which the

Table 3.5 Average international prices for common biocrude, fat, crops, and oils used as feedstock for biofuel production in 2007 (US\$/ton)

Biocrude	167
Crude palm oil	703
Maize	179
Rapeseed oil	824
Soybean oil	771
Sugar	223
Wheat	215
Yellow grease	412

Source: Demirbas, 2008

energy needed to run the conversion facility is an important (and in some cases quite variable) component. Capital cost recovery represents about one-sixth of the total cost *per* liter. It has been shown that plant size has a major effect on cost (Whims, 2002). The plant size can reduce operating costs by 15% to 20%, saving another $ 0.02 to $ 0.03 *per* liter. Thus, a large plant with production costs of $ 0.29 *per* liter may be saving $ 0.05 to $ 0.06 *per* liter over a smaller plant (Whims, 2002).

3.1.2 Environmental Impact of Biofuels

Biodiesel is superior to conventional diesel in terms of its sulfur content, aromatic content, and flash point. It is essentially sulfur free and non-aromatic while conventional diesel can contain up to 500 ppm SO_2 and 20–40%wt aromatic compounds. These advantages may be a key solution to reducing the problem of urban pollution since gas emissions from the transportation sector contribute a significant amount to the total gas emissions. Diesel, in particular, is dominant for black smoke particulate together with SO_2 emissions and contributes to a one-third of the total transport generated greenhouse gas emissions (Nas and Berktay, 2007). There was an average of decreasing of 14% for CO_2, 17.1% for CO and 22.5% for smoke density when using biodiesel (Utlu, 2007).

Biofuels generally results lower emissions than those of fossil-based engine fuels. Many studies on the performances and emissions of compression ignition engines, fueled with pure biodiesel and blends with diesel oil, have been performed and are reported in the literature (Laforgia et al., 1994; Cardone et al., 1998).

Vegetable oils have become more attractive recently because of their environmental benefits and the fact that they are made from renewable resources. Dorado et al. (2003) describe experiments on the exhaust emissions of biodiesel from olive oil methyl ester as alternative diesel fuel in a diesel direct injection Perkins engine.

The methyl ester of vegetable oil has been evaluated as a fuel in CIE by researchers (Dunn, 2001). They concluded that the performance of the esters of vegetable oil does not differ greatly from that of diesel fuel. Carbon deposits inside the engine were normal, with the exception of intake valve deposits. The results showed the transesterification treatment decreased the injector coking to a level significantly lower than that observed with D2 (Shay, 1993). Although most researchers agree that vegetable oil ester fuels are suitable for use in CIE, a few contrary results have also been obtained.

Neat biodiesel and biodiesel blends reduce particulate matter (PM), hydrocarbons (HC) and carbon monoxide (CO) emissions and increase nitrogen oxides (NO_x) emissions compared with diesel fuel used in an unmodified diesel engine (EPA, 2002). The emission impacts of 20 vol% soybean-based biodiesel added to an average base petrodiesel is given in Table 3.6 (EPA, 2002).

The total net emission of carbon dioxide (CO_2) is considerably less than that of diesel oil, and the amount of energy required for the production of biodiesel is less

Table 3.6 Emission impacts of 20 vol% soybean-based biodiesel added to an average base petrodiesel

	Percent change in emissions
NO$_x$ (nitrogen oxides)	+2.0
PM (particulate matter)	−10.1
HC (hydrocarbons)	−21.1
CO (carbon monoxide)	−11.0

Table 3.7 Average biodiesel emissions (%) compared to conventional diesel

Emission type	Pure biodiesel B100	20% biodiesel + 80% petrodiesel B20
Total unburned hydrocarbons (HC)	−67	−20
Carbon monoxide	−48	−12
Particulate matter	−47	−12
NO$_x$	+10	+2
Sulfates	−100	−20
Polycyclic aromatic hydrocarbons[a]	−80	−13
Ozone potential of speciated HC	−50	−10

[a]Average reduction across all compounds measured.

Table 3.8 Average changes in mass emissions from diesel engines using the biodiesel mixtures relative to the standard diesel fuel (%)

Mixture	CO	NO$_x$	SO$_2$	Particular matter	Volatile organic compounds
B20	−13.1	+2.4	−20	−8.9	−17.9
B100	−42.7	+13.2	−100	−55.3	−63.2

than that obtained with the final product. Table 3.7 shows the average biodiesel emissions compared to conventional diesel, according to EPA (2002). Table 3.8 shows the average changes in mass emissions from diesel engines using the biodiesel mixtures relative to standard diesel fuel (Morris *et al.*, 2003).

Results indicate that the transformities of biofuels are greater than those of fossil fuels. This can be explained by the fact that natural processes are more efficient than industrial ones. On the other hand, the time involved in the formation of fossil fuels is considerably different from that required for the production of biomass (Carraretto *et al.*, 2004). Coconut BD can yield reductions of 80.8–109.3% in net CO_2 emissions relative to PD (Tan *et al.*, 2004). Unburned hydrocarbon emissions from biodiesel fuel combustion decrease compared to regular petroleum diesel. The use of blends of biodiesel and diesel oil are preferred in engines, in order to avoid some problems related to the decrease of power and torque, and to the increase of NO$_x$ emissions with increasing content of pure biodiesel in the blend (Schumacher *et al.*, 1992).

3.1.2.1   Emission Characteristics of Gasohol and Diesohol

Gasoline and ethanol mixtures are called as gasohol. Diesohol is a mixture of diesel fuel and hydrated ethanol blended using a chemical emulsifier. Diesohol is used in compression ignition engines as an alternative diesel fuel.

Biorenewable feedstocks can be converted into liquid and gaseous fuels through thermochemical and biological routes. Biofuel is a non-polluting, locally available, accessible, sustainable and reliable fuel obtained from renewable sources (Vasudevan *et al.*, 2005). Due to the negligible amounts of sulfur and nitrogen that biorenewables contain, the energy that is being utilized does not contribute to environmental pollution (Tsai *et al.*, 2007).

The increased utilization of biofuels for heat and power production has led to an increase in political support in European countries. This has resulted in a large number of biofuels being processed for energy conversion necessities and suitability for choosing the most appropriate method by using different materials to obtain valuable properties. In addition, it is necessary that new standard analytical methods be developed in order to apply new technologies for biofuel production from biomass materials.

Diesel-alcohol blends are known by a number of names – including E-diesel, M-diesel, oxy-diesel and diesohol. Diesohol is used in compression ignition engines as an alternative diesel fuel. Ethanol performs well as a fuel in cars, either in a neat form or in a mixture with gasoline. In addition to ethanol/gasoline blend markets, ethanol has other motor fuel applications including: (1) use as E85, 85% ethanol and 15%; gasoline, (2) use as E100, 100% ethanol with or without a fuel additive; and (3) use in oxy-diesel, typically a blend of 80% diesel fuel, 10% ethanol and 10% additives and blending agents.

Gasoline and ethanol mixtures are called as gasohol. E10, sometimes called gasohol, is a fuel mixture of 10% ethanol and 90% gasoline that can be used in the ICEs of most modern automobiles. Gasohol a gasoline extender made from a mixture of gasoline (90%) and ethanol (10%; often obtained by fermenting agricultural crops or crop wastes) or gasoline (97%) and methanol (3%).

Anhydrous ethanol will readily blend with petrol. Hydrated ethanol containing more than 2% volume of water is not completely miscible with petrol. Hydrated ethanol is not miscible with diesel but can form an emulsion using a suitable emulsifier. Diesohol is a mixture of diesel fuel and hydrated ethanol blended using a chemical emulsifier. Diesohol is a fuel containing alcohol that comprises a blend of diesel fuel (84.5%), hydrated ethanol (15%), and an emulsifier (0.5%). The emulsifier that allows the ethanol and the diesel to blend consists of a styrene-butadiene copolymer, which is dissolved in the diesel fuel, and a polyethyleneoxide-polystyrene copolymer, which is dissolved in the hydrated alcohol.

Hydrated (or azeotropic) ethanol is ethyl alcohol that contains approximately 5% water. Hydrated ethanol derived from sugar, or ethanol derived from wheat starch, may be used for production of diesohol. Hydrated ethanol production is a one-stage refining. Ethanol is produced by the fermentation of sugar solutions from sugar cane or grain crops. The action of yeast on the sugar produces a solu-

Fig. 3.3 A schematic diagram of fuel burner system for carbon monoxide, carbon dioxide, and particulate matter (PM) emissions measurements

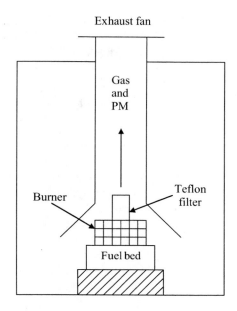

tion containing about 12% ethanol. The alcohol can be concentrated by distillation to produce up to 96% ethanol. Removal of the remaining 4% water requires special treatment.

Figure 3.3 shows the schematic diagram of fuel burner system for carbon monoxide, carbon dioxide, and particulate matter (PM) emissions measurements. Gas and particulate matter (PM) mixture emitted from burner is filtered with a Teflon filter indicating that particles do not interfere with the gas measurements. The PM on the Teflon filter is weight after passing of the gas and PM mixture in measured volume.

Gaseous products to obtain from pure gasoline and No. 2 diesel fuel, and gasohol and diesohol combustions can be analyzed in Orsat type gas analyzers. The gas samples are purified by passing on $MgSO_4$ pulverized glass wool. The soak time for the product gas was 20 min. The results obtained from pure gasoline and No. 2 diesel fuel, and gasohol and diesohol combustions under the same conditions were compared.

Table 3.9 shows the emission impacts of gasoline and diesel fuel blends with 15 vol% ethanol. The use of blends of ethanol with gasoline and No. 2 diesel fuel are preferred in internal combustion engines in order to decrease carbon monoxide and carbon dioxide emissions. The total net emissions of carbon dioxide from gasohol and diesohol blends are considerably less than those of gasoline and No. 2 diesel fuel. Ethanol reduces CO_2 emissions. The reductions of carbon monoxide and carbon dioxide were 32.0% and 29.0% for CO emissions and 43.0% and 45.0% for CO_2 emissions of E15 gasohol and diesohol, respectively.

Figure 3.4 shows the plots for reductions of unburned hydrocarbon (%) versus ethanol (% by volume) in the blend of gasoline (gasohol) and diesel fuel (diesohol). Unburned hydrocarbon emissions from ethanol blends (gasohol and dieso-

Table 3.9 Emission impacts of gasoline and diesel fuel blends with 15 vol% ethanol (percent change in emissions)

Emission type	Gasohol	Diesohol
Total unburned hydrocarbons	−27	−24
Carbon monoxide	−32	−29
Carbon dioxide	−43	−45
Particulate matter	−21	−18
Sulfates	−15	−15

hol) combustion decrease compared to regular gasoline and No. 2 diesel fuel. The reductions of unburned hydrocarbon are 27.0% and 24.0% for E15 blends and 74.0% and 69.8% for E60 blends of gasohol and diesohol, respectively.

Gasohol has higher octane, or antiknock, properties than gasoline and burns more slowly, coolly, and completely, resulting in reduced emissions of some pollutants, but it also vaporizes more readily, potentially aggravating ozone pollution in warm weather (Demirbas, 2007b).

The major effect of diesohol on engine performance is a significant reduction in visible smoke and particulate emissions. Engine thermal efficiency increases by up to 8% when operating on diesohol. There is also a significant overall reduction in emission of carbon dioxide (Demirbas, 2007b).

The cetane number of neat alcohols is very low (8 for ethanol and 3 for methanol), and as such they are extremely poor compression ignition engine fuels. The cetane number of diesohol is dependent on the ignition quality of the base diesel, the percentage of the alcohol in the blend, and the addition of cetane improver additives. Cetane will be important to manage for diesohol, especially given the potential for alcohols to reduce cetane. The cetane number appears to be the appropriate measure for diesohol as it reflects the addition of cetane improvers (Demirbas, 2007b).

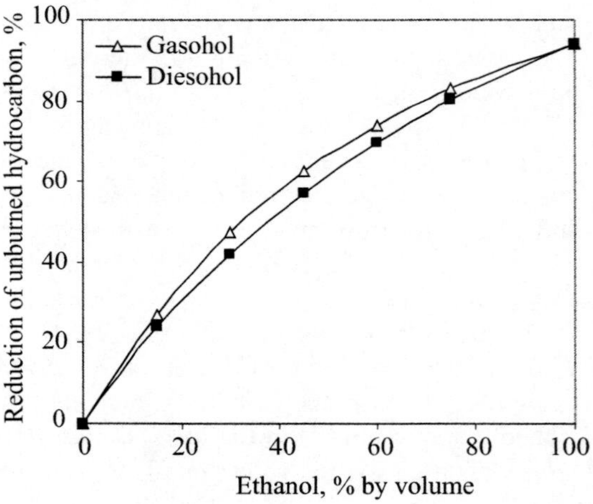

Fig. 3.4 Plots for reductions of unburned hydrocarbon (%) *versus* ethanol (% by volume) in the blend of gasoline (gasohol) and diesel fuel (diesohol)

Gasohol and diesohol are fuel mixtures that can be used in internal combustion engines. Gasohol has higher octane, or antiknock, properties than gasoline. The major effect of diesohol on engine performance is a significant reduction in visible smoke and particulate emissions. Engine thermal efficiency increases by up to 8% when operating on diesohol.

Biomethanol has such low emissions because the carbon content of the alcohol is primarily derived from carbon that was sequestered in the growing of the bio-feedstock and is only being re-released into the atmosphere (Difiglio, 1997).

References

Ahón, V.R., Costa, E.F. Jr., Monteagudo, J.E.P., Fontes, C.E., Biscaia, E.C. Jr., Lage, P.L.C. 2005. A comprehensive mathematical model for the Fischer–Tropsch synthesis in well-mixed slurry reactors. Chem Eng Sci 60:677–694.

Anderson, R.B. 1984. The Fischer–Tropsch synthesis, Academic Press, New York.

Balat, M. 2007. An overview of biofuels and policies in the European Union. Energy Sources Part B 2:167–181.

Bothast, R.J., Schlicher, M.A. 2005. Biotechnological processes for conversion of corn into ethanol. Appl Microbiol Biotechnol 67:19–25.

Bridgewater, A.V. 1995. The technical and economic feasibility of biomass gasification for power generation. Fuel 74:631–53.

Bukur, D.B., Nowicki, L., Manne, R.V., Lang, X. 1995. Activation studies with a precipitated iron catalysts for the Fischer–Tropsch synthesis. J Catalysis 155:366–375.

Cardone, M., Prati, M.V., Rocco, V., Senatore, A. 1998. Experimental analysis of performances and emissions of a diesel engines fuelled with biodiesel and diesel oil blends. Proceedings of MIS–MAC V, Rome, p. 211–25 [in Italian].

Carraretto, C., Macor, A., Mirandola, A., Stoppato, A., Tonon, S. 2004. Biodiesel as alternative fuel: Experimental analysis and energetic evaluations. Energy 29:2195–2211.

Dry, M.E. 1999. Fischer–Tropsch reactions and the environment. Appl Catal A: General 189:185–190.

Demirbas, A. 2006. Global biofuel strategies. Energy Edu Sci Technol 17:27–63.

Demirbas, A. 2007a. Progress and recent trends in biofuels. Prog Energy Combus Sci 33:1–18.

Demirbas, A. 2007b.Gasoline and diesel fuel blends with alcohols. Energy Edu Sci Technol 19:87–92.

Demirbas A. 2008. Economic and environmental impacts of the liquid biofuels. Energy Edu Sci Technol 22:37–58.

Demirbas, M.F., Balat, M. 2006. Recent advances on the production and utilization trends of biofuels: A global perspective. Energy Convers Mgmt 47:2371–2381.

Difiglio, C. 1997. Using advanced technologies to reduce motor vehicle greenhouse gas emissions. Energy Policy 25:1173–1178.

Dorado, M.P., Ballesteros, E.A., Arnal, J.M., Gomez, J., Lopez, F.J.. 2003. Exhaust emissions from a Diesel engine fueled with transesterified waste olive oil. Fuel 82:1311–1315.

Dry, M.E. 1999. Fischer–Tropsch reactions and the environment. Appl Catal A: General 189:185–190.

Dufey, A. 2006. Biofuels production, trade and sustainable development: emerging issuesEnvironmental Economics Programme, Sustainable Markets Discussion Paper No.2, International Institute for Environment and Development (IIED), London, September, 2006.

Dunn, R.O. 2001. Alternative jet fuels from vegetable-oils. Trans ASAE 44:1151–757.

EPA (US Environmental Protection Agency), 2002. A comprehensive analysis of biodiesel impacts on exhaust emissions. Draft Technical Report, EPA420-P-02-001, October 2002.

Grassi, G. 1999. Modern bioenergy in the European Union. Renewable Energy 6:985–990.

Haas, M.J., McAloon, A.J., Yee, W.C., Foglia, T.A. 2006. A process model to estimate biodiesel production costs. Biores Technol 97:671–678.

Kim, S., Dale, B.E. 2005. Life cycle assessment of various cropping systems utilized for producingbiofuels: Bioethanol and biodiesel. Biomass Bioenergy 29: 426–439.

Laforgia, D., Ardito, V. 1994. Biodiesel fuelled IDI e ngines: Performances, emissions and heat release investigation. Biores Technol 51:53–9.

Ma, F., Hanna, M.A. 1999. Biodiesel production: a review. Biores Technol 70:1–15.

May, M. 2003. Development and demonstration of Fischer-Tropsch fueled heavy-duty vehicles with control technologies for reduced diesel exhaust emissions. 9th Diesel Engine Emissions Reduction Conference. Newport, Rhode Island, 24–28 August 2003.

Meher, L.C., Sagar, D.V., Naik, S.N. 2006. Technical aspects of biodiesel production by transesterification—A review. Renew Sust Energy Rev 10:248–268.

Mirzaei, A.A., Habibpour, R., Faizi, M., Kashi, E. 2006. Characterization of iron-cobalt oxide catalysts: Effect of different supports and promoters upon the structure and morphology of precursors and catalysts. Applied Catalysis A: General 301:272–283.

Morris, R.E., Pollack, A.K., Mansell, G.E., Lindhjem, C., Jia, Y., Wilson, G. 2003. Impact of biodiesel fuels on air quality and human health. Summary report. Subconractor Report, National Renewable Energy Laboratory (NREL), Golden, CO.

Nas, B., Berktay, A. 2007. Energy potential of biodiesel generated from waste cooking oil: An environmental approach. Energy Sources Part B 2:63–71.

Prakash, C.B. 1998. A critical review of biodiesel as a transportation fuel in Canada. A Technical Report. GCSI – Global Change Strategies International Inc., Canada.

Prins, M.J., Ptasinski, K.J., Janssen, F.J.J.G. 2004. Exergetic optimisation of a production process of Fischer–Tropsch fuels from biomass. Fuel Proc Technol 86:375–389.

Puppan, D. 2002. Environmental evaluation of biofuels. Periodica Polytechnica Ser Soc Man Sci 2002;10:95–116.

Rapagna, S., Foscolo, P. U. 1998. Catalytic gasification of biomass to produce hydrogen rich gas. Int J Hydrogen Energy 23:551–557.

Reijnders, L. 2006. Conditions for the sustainability of biomass based fuel use. Energy Policy 34:863–876.

RFA. 2007. Renewable Fuels Association (RFA). Ethanol Industry Statistics, Washington, DC,USA.

Sami M, Annamalai K, Wooldridge M. 2001. Co-firing of coal and biomass fuel blends. Progress Energy Combus Sci 27:171–214.

Schulz, H. 1999. Short history and present trends of FT synthesis. Applied Catalysis A: General 186:1–16.

Schumacher, L.G., Hires, W.G., Borgelt, S.C. 1992. Fuelling diesel engines with methyl-ester soybean oil, liquid fuels from renewable resources. Proceedings of Alternative Energy Conference, Nashville, TN.

Sie, S.T., Krishna, R. 1999. Fundamentals and selection of advanced FT-reactors. Applied Catalysis A: General 186:55–70.

Shay, E.G. 1993. Diesel fuel from vegetable oils: Satus and opportunities. Biomass Bioenergy 4:227–42.

Taherzadeh, M.J. 1999. Ethanol from Lignocellulose: Physiological effects of inhibitors and fermentation strategies. Thesis for the Degree of Doctor of Philosophy, Department of Chemical Reaction Engineering, Chalmers University of Technology, Göteborg, Sweden.

Tan, R.R., Culaba, A.B., Purvis, M.R.I. 2004. Carbon balance implications of coconut biodiesel utilization in the Philippine automotive transport sector. Biomass Bioenergy 26:579–585.

Tijmensen, M.J.A., Faaij, A.P.C., Hamelinck, C.N., van Hardeveld, R.M.R. 2002. Exploration of the possibilities for production of Fischer–Tropsch liquids and power *via* biomass gasification. Biomass Bioenergy 23:129–152.

Tsai, W.T., Lee, M.K., Chang, Y.M. 2007. Fast pyrolysis of rice husk: Product yields and compositions. Biores Technol 98:22–28.

UN (United Nations). 2006. The emerging biofuels market: Regulatory, trade and development implications. United Nations Conference on Trade and Development, New York and Geneva..Available from: www.verasun. com/pdf/RFA.

Usta, N., Ozturk, E., Can, O., Conkur, E.S., Nas, S., Con, A.H., Can, A.C., Topcu, M. 2005. Combustion of biodiesel fuel produced from hazelnut soapstock/waste sunflower oil mixture in a Diesel engine. Energy Convers Mgmt 46:741–755.

Utlu, Z. 2007. Evaluation of biodiesel fuel obtained from waste cooking oil. Energy Sources Part A 29:1295–1304.

Vasudevan, P., Sharma, S., Kumar, A. 2005. Liquid fuel from biomass: An overview. J Sci Ind Res 64:822–831.

Whims, J. 2002. Corn based ethanol costs and margins. Attachment 1, AGMRC, Kansas State University. Available from: http://www.agmrc.org/corn/info/ksueth1.pdf.

Zhang, Y., Dube, M.A., McLean, D.D., Kates, M. 2003. Biodiesel production from waste cooking oil: 2. Economic assessment and sensitivity analysis. Bioresour Technol 90:229–240.

Chapter 4
Biorenewable Liquid Fuels

4.1 Introduction to Biorenewable Liquid Fuels

The aim of this chapter is to provide a global approach on liquid biofuels such as bioethanol, biodiesel, vegetable oils, bio-oils, and other biorenewable liquid fuels for the future of energy for transportation. It is possible that sugar cane, corn, wood, straw, and even household wastes may be economically converted to bioethanol. Bioethanol is derived from alcoholic fermentation of sucrose or simple sugars, which are produced from biomass by a hydrolysis process. Biodiesel is an environmentally friendly alternative liquid fuel that can be used in any diesel engine without modification. There has been renewed interest in the use of vegetable oils for making biodiesel due to its less polluting and renewable nature as against conventional petroleum diesel fuel (Demirbas, 2008).

Due to its environmental merits, the share of biofuel such as bioethanol and biodiesel in the automotive fuel market will grow fast in the next decade. An important reason for interest in renewable energy sources is the concern for the greenhouse effect. Biorenewable liquid fuels have gained a lot of attention due to their environmental and technological advantages. Development of ethanol as a motor fuel can work to fulfill this commitment. Greenhouse gas emission reductions should be estimated on an annual basis. Where the levels from year to year vary significantly these should be specified on an annual basis. If ethanol from biomass is used to drive a light-duty vehicle, the net CO_2 emission is less than 7% of that from the same car using reformulated gasoline (Bergeron, 1996).

The most important biorenewable liquid fuels are bioethanol and biodiesel made from plant material and recycled elements of the food chain. Biodiesel is a diesel alternative. Bioethanol is a petrol additive/substitute. Biorenewable fuels are safely and easily biodegradable and so are particularly well suited to the environment. Biodiesel, a biofuel that can replace directly petroleum-derived diesel without engine modifications, has gained a lot of attention due to its environmental and technological advantages. These advantages include its being from

A. Demirbas, *Biofuels*,
© Springer 2009

a renewable source, completely biodegradable, cleaner burning than petroleum-based diesel, low in sulfur, and resulting in CO_2 recycling.

4.1.1 Evaluation of Gasoline-Alcohol Mixtures as Motor Fuel Alternatives

Production of motor fuel alternatives from biomass materials is an important application area of biotechnological methods. Table 4.1 shows potential and available motor fuels. Biorenewable sourced motor fuel alternatives are:

1. Gasoline-alcohol mixtures
2. Alcohol substituting gasoline
3. Gasoline-vegetable oil mixtures
4. Diesel fuel-vegetable oil mixtures
5. Vegetable oil substituting diesel fuel.

The application of alcohol and gasoline-alcohol mixtures in gasoline (Otto) engines began in the first half of the 20th century. It is possible to find information about various studies on the change in octane numbers of gasoline-alcohol mixtures, composition of the exhaust gases, motor tests, and about the materials used in constructing the engines. Between 1980 and 1985 studies intensified on solving the phase separation problem of gasoline-alcohol mixtures.

In gasoline-alcohol mixtures ethanol and methanol are generally used, and in gasoline engine mixtures containing 20% or less alcohol by volume can be used without altering the construction of the engine. Because of the hygroscopic properties of ethanol and methanol the gasoline-alcohol mixtures are in fact ternary mixtures composed of gasoline-alcohol and water. In the evaluation of such mixtures as motor fuel there is the phase separation problem, which depends on several factors. It can be seen in the literature that there have been numerous attempts to over come this problem (Mislavskaya et al., 1982; Osten and Sell, 1983).

In gasoline-methanol mixtures containing 0.1% water i-propanol is added to the environment (medium) in order to decrease the phase separation temperature, and

Table 4.1 Potential and available motor fuels

Fuel type	Available motor fuel
Traditional fuels	Diesel and gasoline
Oxygenated fuels	Ethanol 10% (E10), methanol, methyl tertiary butyl ether (MTBE), ethyl tertiary butyl ether (ETBE), tertiary butyl alcohol (TBA), and tertiary amyl methyl ether (TAME)
Alternative fuels	Liquefied petroleum gases (LPG), ethanol, 85% (E85), ethanol, 95% (E95), methanol, 85% (M85), methanol, neat (M100), compressed natural gas (CNG), liquefied natural gas (LNG), biodiesel (BD), hydrogen, electricity

fuels containing different ratios of gasoline-methanol-i-propanol and water are composed, which are proved to be stable in the climatic conditions. An increase in the aromatic character of the gasoline, a decrease in the water content of the mixture, an increase in the amount of the additive used results in a decrease in the phase separation temperature of the mixture. In gasoline-ethanol mixtures the additive used is also i-propanol. In gasoline-alcohol mixtures various additives like i-propanol, n-butanol, i-butanol, and i-amylalcohol are used.

4.1.2 Evaluation of Vegetable Oils and Diesel Fuel Mixtures as Motor Fuel Alternatives

Since the 1980s important progress has been made on evaluating some low grade oils, oil production wastes, and residues as motor fuel (Pryor *et al.*, 1983). However, direct usage of vegetable oils causes a number of problems concerning the engine because of their high viscosity and the excessive carbonaceous deposits left in the cylinders and on the injector nozzles. Therefore, chemical conversion of vegetable oils was suggested. In order to lower the viscosities and flash points of vegetable oils the transesterification method has been applied and it is reported that the alcoholysis products of soybean, sunflower, rapeseed, and used frying oils were proposed as diesel fuel alternatives.

The deregulation of domestic crude oil prices and the formation of OEC have been largely responsible for high fuel prices. The farmer is highly dependent on diesel fuel for crop production. Alternative fuels such as vegetable oils may help easy the petroleum dependence of farmers. Recently, the demand for crude oil has decreased because of conservation practices, but ultimately a liquid fuel resource problem exists (Pryor *et al.*, 1983).

4.2 Bioalcohols

The alcohols that can be used for motor fuels are methanol (CH_3OH), ethanol (C_2H_5OH), propanol (C_3H_7OH), and butanol (C_4H_9OH). However, only first two are technically and economically suitable as fuels for internal combustion engines (ICEs). Main commercial bioalcohols from renewable feedstocks are bioethanol and biomethanol in the world's energy marketing. Bioethanol currently accounts for more than 94% of global biofuel production, with the majority coming from sugarcane. About 60% of global bioethanol production comes from sugarcane and 40% from other crops. Brazil and the United States are the world leaders, which together accounted for about 70% of the world bioethanol production exploiting sugarcane and corn respectively. Ethanol was used in Germany and France as early as 1894 by the then incipient industry of internal combustion engines. Brazil has utilized ethanol as a fuel since 1925. Currently, ethanol is produced from sugar beets and from molasses. A typical yield is 72.5 liters of ethanol *per* ton of sugar cane. Modern crops yield 60 tons of sugar cane *per* hectare of land. Production of ethanol from biomass is one way to reduce both the consumption of crude oil and environmental pollution. The use of gasohol (an ethanol and gasoline mixture) as an alternative motor fuel has been steadily increasing in the world for a number of reasons. Domestic production and use of ethanol for fuel can decrease dependence on foreign oil, reduce trade deficits, create jobs in rural areas, reduce air pollution, and reduce global climate change carbon dioxide build-up (Bala, 2005).

4.2.1   Alternate Fuels to Gasoline

Gasoline is a blend of hydrocarbons with some contaminants, including sulfur, nitrogen, oxygen, and certain metals. The four major constituent groups of gasoline are olefins, aromatics, paraffins, and naphthenes. The important characteristics of gasoline are density, vapor pressure, distillation range, octane, and chemical composition. To be attractive, a motor gasoline must have (a) desirable volatility, (b) antiknock resistance (related to octane rating), (c) good fuel economy, (d) minimal deposition on engine component surfaces, and (e) complete combustion and low pollutant emissions (Chigier, 1981).

Alternative fuels for Otto engines or light-duty vehicles (LDVs; cars and light trucks) contain (1) reformulated gasoline, (2) compressed natural gas, (3) methanol and ethanol, (4) liquid petroleum gas, (5) liquefied natural gas, (6) Fischer–Tropsch liquids from natural gas, (7) hydrogen, and (8) electricity. The electricity can be obtained from (a) spark ignition port injection engines, (b) spark ignition direct injection engines, (c) compression ignition engines, (d) electric motors with battery power, (e) hybrid electric propulsion options, and (f) fuel cells (Demirbas, 2005a).

Figure 4.1 shows the whole sale prices of a number of possible alternative fuels on an energy equivalent basis compared to conventional gasoline (AICHE, 1997). Only compressed natural gas (CNG) and liquid petroleum gas (LPG) appear to have some economic advantage relative to gasoline while ethanol, methanol, and electricity are at a severe economic disadvantage (Piel, 2001).

The most commonly used measure of a gasoline's ability to burn without knocking is its octane number. Table 4.2 shows hydrocarbon octane numbers. By 1922 a number of compounds had been discovered that were able to increase the octane number of gasoline. Adding as little as 6 mL of tetraethyl lead (Fig. 4.2) to a gallon of gasoline, for example, can increase the octane number by 15 to 20 units. This discovery gave rise to the first "ethyl" gasoline, and enabled the petroleum industry to produce aviation gasolines with octane numbers greater than 100. The octane numbers and some properties of common fuels are given in Table 4.3. As seen in Table 4.3, the octane numbers of ethanol and methanol are 107 and 112, respectively.

About 10% of the product of the distillation of crude oil is a fraction known as straight-run gasoline. Straight-run gasoline burns unevenly in high-compression

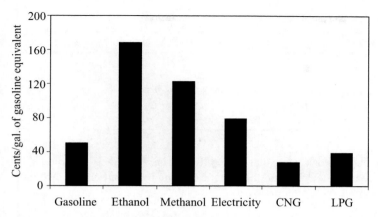

Fig. 4.1 Economics of gasoline alternative fuels

Table 4.2 Hydrocarbon octane numbers

Hydrocarbon	Octane Number (ON)
Heptane	0
2-Methylheptane	23
Hexane	25
2-Methylhexane	44
1-Heptene	60
Pentane	62
1-Pentene	84
Butane	91
Cyclohexane	97
2,2,4-Trimethylpentane (isooctane)	100
Benzene	101
Toluene	112

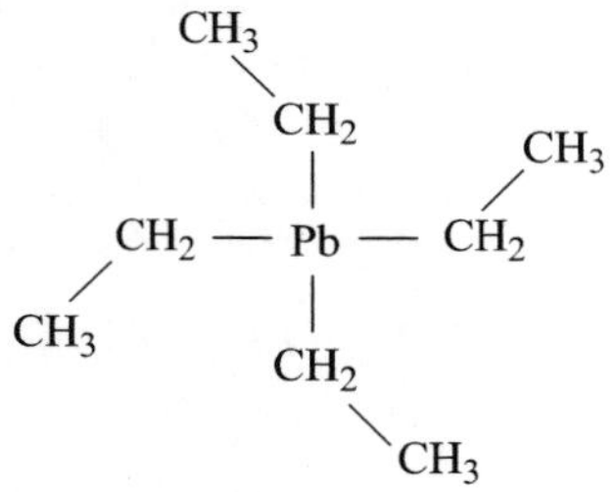

$$
\begin{array}{c}
CH_3 \\
| \\
CH_2 \qquad\qquad CH_3 \\
| \qquad\qquad\quad | \\
CH_2 - Pb - CH_2 \\
/ \qquad | \\
CH_3 \qquad CH_2 \\
\qquad\qquad \backslash \\
\qquad\qquad CH_3
\end{array}
$$

Fig. 4.2 Chemical structure of tetraethyl lead

Table 4.3 Octane number and some properties of common fuels

Fuel property Ethanol	Gasoline	Diesel No. 2	Isooctane	Methanol
Cetane number	–	50	–	5	8
Octane number	96	–	100	112	107
Auto-ignition temperature (K)	644	588	530	737	606
Latent heat of vaporization (MJ/Kg)	0.35	0.22	0.26	1.18	0.91
Lower heating value (MJ/Kg)	44.0	42.6	45.0	19.9	26.7

engines, producing a shock wave that causes the engine to "knock" or "ping". The relationship between knocking and the structure of the hydrocarbons in gasoline is summarized in the following general rules.

- Branched alkanes and cycloalkanes burn more evenly than straight-chain alkanes.
- Short alkanes (C_4H_{10}) burn more evenly than long alkanes (C_7H_{16}).
- Alkenes burn more evenly than alkanes.
- Aromatic hydrocarbons burn more evenly than cycloalkanes.

4.3 Bioethanol

Nowadays ethanol is the most popular fuel. Biothanol is ethyl alcohol, grain alcohol, or chemically C_2H_5OH or EtOH. The use of ethanol as a motor fuel has as long a history as the car itself. Ethanol can be produced from cellulose feedstocks such as corn stalks, rice straw, sugar cane, bagasse, pulpwood, switchgrass, and municipal solid waste is called bioethanol. Bioethanol is a renewable green fuel. Utilization of ethyl alcohol as engine fuel is one way to reduce both the consumption of crude oil and environmental pollution. Primary consideration involves the production of ethyl alcohol from renewable resources and determination of the economic and technical feasibility of using alcohol as an automotive fuel blended with gasoline. An important reason for interest in renewable energy sources is the concern for the greenhouse effect. Development of ethanol as a motor fuel can work to fulfill this commitment. Greenhouse gas emission reductions should be estimated on an annual basis. Where the levels from year to year vary significantly these should be specified on an annual basis. If ethanol from biomass is used to drive a light-duty vehicle, the net CO_2 emission is less than 7% of that from the same car using reformulated gasoline (Bergeron, 1996).

4.3.1 Synthetic Ethanol Production Processes

The hydration of ethylene is the oldest process among the two major ethanol production methods from ethylene ($CH_2 = CH_2$), and started more than 100 years ago.

The ethanol is prepared from ethylene in a three-step process using sulfuric acid (H_2SO_4). In the first step, the hydrocarbon feedstock containing 35–95% ethylene is exposed to 95–98% sulfuric acid in a column reactor to form monosulfate:

$$CH_2 = CH_2 + H_2SO_4 \rightarrow CH_3CH_2OSO_3H \tag{4.1}$$

It is subsequently hydrolyzed with enough water to give 50–60% aqueous sulfuric acid solution:

$$CH_3CH_2OSO_3H + H_2O \rightarrow CH_3CH_2OH + H_2SO_4 \tag{4.2}$$

The ethanol is then separated from the dilute sulfuric acid in a stripper column. The last step of this process is to concentrate the sulfuric acid and recycle to the process.

In the direct hydration process, an ethylene-rich gas is combined with water and passes through a fixed-bed catalyst reactor, in which ethanol is formed according to the following reaction (Nelson and Courter, 1954):

$$CH_2 = CH_2 + H_2O \rightarrow CH_3CH_2OH \tag{4.3}$$

The ethanol is then recovered in a distillation system.

Ethanol can be obtained from acetylene process in the presence of a proper catalyst such as H_2SO_4 and $HgSO_4$; acetylene/ethyne reacts with water to yield acetaldehyde:

$$C_2H_2 + H_2O \rightarrow CH_3CHO \tag{4.4}$$

Acetaldehyde can be readily reduced by catalytic hydrogenation to ethyl alcohol:

$$CH_3CHO + H_2 \rightarrow CH_3CH_2OH \tag{4.5}$$

The classical catalyst is octacarbonyldicobalt, $Co_2(CO)_8$, formed by reaction of metallic cobalt with carbon monoxide.

4.3.2 Production of Ethanol from Biomass

Ethanol from biorenewable feedstocks has the potential to contribute substantially to bioethanol for transportation. In the process evaluated, prehydrolysis with dilute sulfuric acid is employed to hydrolyze hemicellulose and make the cellulose more accessible to hydrolysis by enzymes. Residual biomass from hydrolysis and extraction of carbohydrates can be burned in a power plant to generate electricity and process steam.

Carbohydrates in plant materials can be converted to sugars by a hydrolysis process. Fermentation is an anaerobic biological process in which sugars are converted to alcohol by the action of microorganisms, usually yeast. The resulting alcohol is bioethanol. The value of any particular type of biomass as feedstock for fermentation depends on the ease with which it can be converted to sugars. Bioethanol and the biorefinery concept are closely linked. It is possible that wood, straw, and even household wastes may be economically converted to bioethanol.

The corn-starch-to-fuel ethanol industry has matured during the past 30 years by bioethanol research. Most bioethanol researchers are focusing on the challenge of producing bioethanol from lignocellulosic biomass instead of corn starch. Toward this end, researchers have already developed an effective technology to thermochemically pretreat biomass; to hydrolyze hemicellulose to break it down into its component sugars and open up the cellulose to treatment; to enzymatically hydrolyze cellulose to break it down to sugars; and to ferment both five-carbon sugars from hemicellulose and six-carbon sugars from cellulose. Figure 4.3 shows the flow chart for the production of bioethanol from cereal grain or straw.

Bioethanol can be produced from a large variety of carbohydrates with a general formula of $(CH_2O)_n$. The chemical reaction is composed of enzymatic hydrolysis of sucrose followed by fermentation of simple sugars. Fermentation of sucrose is performed using commercial yeast such as *Saccharomyces ceveresiae*. First, the invertase enzyme in the yeast catalyzes the hydrolysis of sucrose to convert it into glucose and fructose.

$$C_{12}H_{22}O_{11} \rightarrow C_6H_{12}O_6 + C_6H_{12}O_6 \qquad (4.6)$$
Sucrose   Glucose   Fructose

Second, zymase, another enzyme also present in the yeast, converts the glucose and the fructose into ethanol.

$$C_6H_{12}O_6 \rightarrow 2\,C_2H_5OH + 2CO_2 \qquad (4.7)$$

The gluco-amylase enzyme converts the starch into D-glucose. The enzymatic hydrolysis is then followed by fermentation, distillation, and dehydration to yield

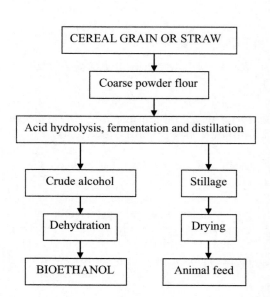

Fig. 4.3 Flow chart for the production bioethanol from cereal grain or straw

anhydrous bioethanol. Corn (60–70% starch) is the dominant feedstock in the starch-to-bioethanol industry worldwide.

Cellulose hydrolysis produces glucose, which is readily fermented with existing organisms in much the same way as has been done for centuries. Hemicelluloses hydrolysis produces both hexose and pentose sugars: mannose, galactose, xylose, and arabinose that are not all fermented with existing strains.

The hemicelluloses fraction typically produces a mixture of sugars including xylose, arabinose, galactose, and mannose. These are both pentosans: xylose and arabinose, and hexosans: galactose and mannose. The quantities are dependent on the material and also the growing environment and storage history of the material.

The amount of product formed *per* unit of substrate consumed by the organism is a useful way to refer to yields. The yields of fermentation are expressed on either molar or weight basis. In this case the primary stoichiometric equations for the bioethanol production are as follows.

Pentosan to pentose:

$$\text{n } C_5H_8O_4 + \text{n } H_2O \rightarrow \text{n } C_5H_{10}O_5 \qquad\qquad (4.8)$$
$$1\,\text{g} \qquad\quad 0.136\,\text{g} \qquad 1.136\,\text{g}$$

Hexosan to hexose:

$$\text{n } C_6H_{10}O_5 + \text{n } H_2O \rightarrow \text{n } C_6H_{12}O_6 \qquad\qquad (4.9)$$
$$1\,\text{g} \qquad\quad 0.111\,\text{g} \qquad 1.111\,\text{g}$$

Pentose and hexose to bioethanol, 0.511 grams *per* gram hexose or pentose:

$$3\,C_5H_{10}O_5 \rightarrow 5\,C_2H_5OH + 5\,CO_2 \qquad\qquad (4.10)$$
$$1\,\text{g} \qquad\qquad 0.511\,\text{g} \qquad 0.489\,\text{g}$$

$$C_6H_{12}O_6 \rightarrow 2\,C_2H_5OH + 2\,CO_2 \qquad\qquad (4.11)$$
$$1\,\text{g} \qquad\qquad 0.511\,\text{g} \qquad 0.489\,\text{g}$$

A reduction in yield below the theoretical value always occurs since the microorganism requires a portion of the substrate for cell growth and maintenance. For *E. coli* and *S. cerevisiae* these values are approximately 0.054 and 0.018 grams of glucose/g dry cell weight-hour, respectively (Roels and Kossen, 1978).

Microorganisms for ethanol fermentation can best be described in terms of their performance parameters and other requirements such as compatibility with existing products, processes, and equipment. The performance parameters of fermentation are: temperature range, pH range, alcohol tolerance, growth rate, productivity, osmotic tolerance, specificity, yield, genetic stability, and inhibitor tolerance.

4.3.3   Sugars from Biomass by Hydrolysis

Hydrolysis breaks down the hydrogen bonds in the hemicelluloses and cellulose fractions into their sugar components: pentoses and hexoses. These sugars can then be fermented into ethanol. Pretreatment methods refer to the solubilization and separation of one or more of the four major components of biomass to make the remaining solid biomass more accessible to further chemical or biological treatment.

After the pretreatment process, there are two types of processes to hydrolyze the cellulosic biomass for fermentation into ethanol. The most commonly applied methods can be classified in two groups: chemical hydrolysis and enzymatic hydrolysis.

Both enzymatic and chemical hydrolyses require a pretreatment to increase the susceptibility of cellulosic materials. In the chemical hydrolysis, the pretreatment and the hydrolysis may be carried out in a single step. There are two basic types of acid hydrolysis processes commonly used: dilute acid and concentrated acid, each with variations.

Dilute acid processes are conducted under high temperature and pressure, and have reaction times in the range of seconds or minutes, which facilitates continuous processing. Most dilute acid processes are limited to a sugar recovery efficiency of around 50%. The reason for this is that at least two reactions are part of this process. Not only does sugar degradation reduce sugar yield, but the furfural and other degradation products can be poisonous to the fermentation microorganisms. The biggest advantage of dilute acid processes is their fast rate of reaction, which facilitates continuous processing.

Since 5-carbon sugars degrade more rapidly than 6-carbon sugars, one way to decrease sugar degradation is to have a two-stage process. The first stage is conducted under mild process conditions to recover the 5-carbon sugars while the second stage is conducted under harsher conditions to recover the 6-carbon sugars. The primary advantage of the concentrated process is the high sugar recovery efficiency (Demirbas, 2004a).

4.3.3.1 Dilute Acid Hydrolysis

The dilute acid process is conducted under high temperature and pressure, and has a reaction time in the range of seconds or minutes, which facilitates continuous processing. As an example, using a dilute acid process with 1% sulfuric acid in a continuous flow reactor at a residence time of 0.22 minutes and a temperature of 510 K with pure cellulose provided a yield over 50% sugars. The combination of acid and high temperature and pressure dictate special reactor materials, which can make the reactor expensive. The first reaction converts the cellulosic materials to sugar and the second reaction converts the sugars to other chemicals.

The biggest advantage of dilute acid processes is their fast rate of reaction, which facilitates continuous processing. Since 5-carbon sugars degrade more rapidly than 6-carbon sugars, one way to decrease sugar degradation is to have a two-stage process. The first stage is conducted under mild process conditions to recover the 5-carbon sugars while the second stage is conducted under harsher conditions to recover the 6-carbon sugars.

4.3.3.2 Concentrated Acid Hydrolysis

Concentrated acid hydrolysis uses relatively mild temperatures, and the only pressures involved are those created by pumping materials from vessel to vessel. Reaction times are typically much longer than for dilute acid. This method generally uses concentrated sulfuric acid followed by a dilution with water to dissolve and hydrolyze or convert the substrate into sugar. This process provides a complete and rapid conversion of cellulose to glucose and hemicelluloses to 5-carbon sugars with little degradation. The critical factors needed to make this process economically viable are to optimize sugar recovery and cost effectively recovers the acid for recycling.

The primary advantage of the concentrated acid process is the potential for high sugar recovery efficiency. Table 4.4 shows the yields of bioethanol by concentrated sulfuric acid hydrolysis from cornstalks. The acid and sugar are separated *via* ion exchange and then acid is re-concentrated *via* multiple effect evaporators. The low temperatures and pressures employed allow the use of relatively low cost materials such as fiberglass tanks and piping. The low temperatures and pressures also minimize the degradation of sugars. Unfortunately, it is a relatively slow process and cost effective acid recovery systems have been difficult to develop. Without acid recovery, large quantities of lime must be used to neutralize the acid in the sugar solution. This neutralization forms large quantities of calcium sulfate, which requires disposal and creates additional expense.

Table 4.4 Yields of bioethanol by concentrated sulfuric acid hydrolysis from cornstalks (% dry weight)

Amount of cornstalk (kg)	1000
Cellulose content (kg)	430
Cellulose conversion and recovery efficiency	0.76
Ethanol stoichiometric yield	0.51
Glucose fermentation efficiency	0.75
Ethanol yield from glucose (kg)	130
Amount of cornstalk (kg)	1000
Hemicelluloses content (kg)	290
Hemicelluloses conversion and recovery efficiency	0.90
Ethanol stoichiometric yield	0.51
Xylose fermentation efficiency	0.50
Ethanol yield from xylose (kg)	66
Total ethanol yield from 1000 kg of cornstalk	196 kg (225.7 L = 59 gal)

Source: Demirbas, 2005b

4.3.3.3 Enzymatic Hydrolysis

Another basic method of hydrolysis is enzymatic hydrolysis. Enzymes are naturally occurring plant proteins that cause certain chemical reactions to occur. However, for enzymes to work, they must obtain access to the molecules to be hydrolyzed. There are two technological developments: enzymatic and direct microbial conversion methods. Another basic method of hydrolysis is enzymatic hydrolysis.

The chemical pretreatment of the cellulosic biomass is necessary before enzymatic hydrolysis. The first application of enzymatic hydrolysis was used in separate hydrolysis and fermentation steps. Enzymatic hydrolysis is accomplished by cellulolytic enzymes. Different kinds of "cellulases" may be used to cleave the cellulose and hemicelluloses. A mixture of endoglucanases, exoglucanases, ß-glucosidases, and cellobiohydrolases is commonly used (Ingram and Doran, 1995; Laymon *et al.*, 1996). The endoglucanases randomly attack cellulose chains to produce polysaccharides of shorter length, whereas exoglucanases attach to the non-reducing ends of these shorter chains and remove cellobiose moieties. ß-glucosidases hydrolyze cellobiose and other oligosaccharides to glucose (Philippidis and Smith, 1995).

For enzymes to work efficiently, they must obtain access to the molecules to be hydrolyzed. This requires some kind of pretreatment process to remove hemicelluloses and break down the crystalline structure of the cellulose or removal of the lignin to expose hemicelluloses and cellulose molecules.

Hydrolysis is often used to pretreat lignocellulosic feedstocks to break down the cellulose and hemicelluloses from the lignocellulose and break down the compounds into simple sugars. Hydrolysis can be catalyzed by use of acids (either strong or weak), enzymes, and/or hydrothermal means, the latter including hot water and supercritical methods. Figure 4.4 shows a schematic flow diagram of enzymatic hydrolysis process.

Cellulosic materials are comprised of lignin, hemicelluloses, and cellulose and are thus sometimes called lignocellulosic materials. Cellulose molecules consist of long chains of glucose molecules (6-carbon sugars) as do starch molecules, but have a different structural configuration. Since pentose molecules (5-carbon sugars) comprise a high percentage of the available sugars, the ability to recover and ferment them into ethanol is important for the efficiency and economics of the process.

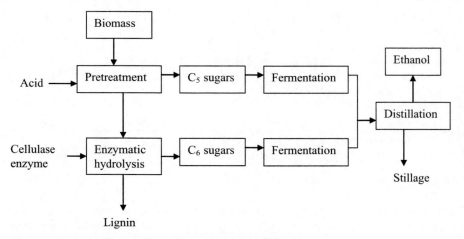

Fig. 4.4 Schematic flow diagram of the enzymatic hydrolysis process

Recently, special microorganisms have been genetically engineered that can ferment 5-carbon sugars into ethanol with relatively high efficiency. Bacteria have drawn special attention from researchers because of their speed of fermentation.

The biggest advantage of dilute acid processes is their fast rate of reaction, which facilitates continuous processing. Since 5-carbon sugars degrade more rapidly than 6-carbon sugars, one way to decrease sugar degradation is to have a two-stage process. The first stage is conducted under mild process conditions to recover the 5-carbon sugars while the second stage is conducted under harsher conditions to recover the 6-carbon sugars.

A new generation of enzymes and enzymes production technology is needed to cost-effectively hydrolyze cellulose to glucose. Technical barriers for enzymatic hydrolysis include: low specific activity of current commercial enzymes, high cost of enzyme production, and lack of understanding of enzyme biochemistry and mechanistic fundamentals.

Biomass can be hydrolyzed to create fermentable sugars for producing ethanol. The lignin component is also unconvertible in a hydrolysis and fermentation system. Sugars can also be converted to levulinic acid and citric acid. Levulinic acid is a versatile chemical that is a precursor to other specialty chemicals, fuels and fuels additives, herbicides, and pesticides. The largest application for citric acid is in the beverage industry, which accounts for about 45% of the market for this product. Citric acid is also used in a wide variety of candies, frozen foods, and processed cheeses, and as a preservative in canned goods, meats, jellies, and preserves. Residual acids in the sugar stream can be neutralized using lime. The use of lime as a neutralizing agent yields gypsum, which can be sold as a soil amendment or to wallboard manufacturers.

The first application of enzymatic hydrolysis was used in separate hydrolysis and fermentation steps. Enzymatic hydrolysis is accomplished by cellulolytic enzymes. Different kinds of cellulases may be used to cleave the cellulose and hemicelluloses. A mixture of endoglucanases, exoglucanases, ß-glucosidases, and cellobiohydrolases is commonly used. The endoglucanases randomly attack cellulose chains to produce polysaccharides of shorter length, whereas exoglucanases attach to the non-reducing ends of these shorter chains and remove cellobiose moieties. ß-glucosidases hydrolyze cellobiose and other oligosaccharides to glucose.

For enzymes to work efficiently, they must obtain access to the molecules to be hydrolyzed. This requires some kind of pretreatment process to remove hemicelluloses and break down the crystalline structure of the cellulose or removal of the lignin to expose the hemicelluloses and cellulose molecules.

4.3.4 Bioethanol Production by Fermentation of Carbohydrates

Fermentation using genetically engineered yeast or bacteria will utilize all five of the major biomass sugars: glucose, xylose, mannose, galactose, and arabinose. Ethanol may be produced by direct fermentation of sugars, or from other carbohydrates that can be converted to sugar, such as starch and cellulose. Fermentation of sugars by yeast, the oldest synthetic chemical process used by man, is still of enormous importance for the preparation of ethyl alcohol. Sugars come from a variety of sources, mostly molasses from sugar cane, or starch obtained from various grains; the name "grain alcohols" has been given to ethyl alcohol for this reason. The raw materials are classified into three categories of agricultural raw materials: (1) simple sugars from sugar cane, sugar beet, molasses, and fruit; (2) starch from grains, potatoes, and root crops; and (3) cellulose from wood, agricultural residue, municipal solid wastes, waste papers, and crop residues.

Some sugars can be converted directly to ethanol, whereas starch and cellulose must first be hydrolyzed to sugar before conversion to ethanol. Most of the polymeric raw materials are available at prices lower than refined sugars. However, transportation costs of the raw materials make it necessary to use locally available raw material. Consequently, each country may preferably develop ethanol production based on the available raw material in that country (Taherzadeh, 1999).

The fermentation method generally uses three steps: (a) the formation of a solution of fermentable sugars, (b) the fermentation of these sugars to ethanol, and (c) the separation and purification of the ethanol, usually by distillation.

Fermentation involves microorganisms that use the fermentable sugars for food and in the process produces ethyl alcohol and other byproducts. These microorganisms can typically use the 6-carbon sugars, one of the most common being glucose. Therefore, cellulosic biomass materials containing high levels of glucose or precursors to glucose are the easiest to convert to ethanol. Microorganisms, termed ethanologens, presently convert an inadequate portion of the sugars from biomass to ethanol. Although fungi, bacteria, and yeast microorganisms can be used for fermentation, specific yeast (*Saccharomyces cerevisiae* also known as Bakers' yeast) is frequently used to ferment glucose to ethanol.

Operating temperatures are less than desired and the organism performance can be inhibited by components inherent in the process. Three recombinant fermentation strains were considered to be candidates for short term improvement by the participants: *Saccharomyces, E. Coli and Zymomonas*. There is a clear preference for yeast by the existing grain ethanol producers, particularly *Saccharomyces*. This yeast is widely used, known and fits existing equipment.

The hemicelluloses fraction typically produces a mixture of sugars including xylose, arabinose, galactose, and mannose. These are both pentosans: xylose and arabinose, and hexosans: galactose, and mannose. The quantities are dependent on the material and also the growing environment and storage history of the material.

A reduction in yield below the theoretical value always occurs since the microorganism requires a portion of the substrate for cell growth and maintenance. For *E. coli* and *S. cerevisiae* these values are approximately 0.054 and 0.018 grams of glucose/g dry cell weight-hour respectively (Demirbas, 2004a).

4.3.4.1 Fungal Enzymes to Convert Biomass to Bioethanol

Bioethanol production from biomass is one such already proven industrial process for renewable energy production. Bioethanol can be directly mixed with petrol and used in today's cars, or converted to electricity. Currently there are two type blends of ethanol/gasoline on the market: blends of 10% ethanol and 90% gasoline (E10) and blends of 85% ethanol and 15% gasoline (E85). In the US, many states currently mandate E10 in gasoline. It is generally accepted that bioethanol gives a 70% carbon dioxide reduction, which means 7% in an E10 blend or 50% in an E85 blend. Increased bioethanol usage may reduce U.S. greenhouse-gas emissions to 1.7 billion tons/year. Recent researches have established that polysaccharides in biomass can be hydrolyzed enzymatically into glucose sugar that is fermented to bioethanol.

When manufactured from agricultural sources, like corn or wood, ethanol is commonly referred to as bioethanol. Theoretically, bioethanol production should be able to yield 0.5 g of ethanol *per* gram of raw biomass, which translates into an energy recovery of approximately 90%. Corn is a common substrate for bioethanol manufacture because the process is relatively free of technical obstacles. Microorganisms are involved in the transformation of corn to bioethanol in two ways: they catalyze the hydrolysis of starches using amylases and amyloglucosidases and they ferment the resulting sugars to bioethanol. The fermentation step is generally carried out by yeast, but certain strains of bacteria, including *Zymomonas mobilis* and recombinant strains of *Escherichia coli* and *Klebsiella oxytoca*, are also capable of producing high yields of bioethanol.

Because bioethanol is a liquid, bioethanol fits into the current fuel infrastructure, although transport requires special handling to prevent water accumulation. Bioethanol can be cost-competitive with petroleum. However, using solid substrates in converting lignocellulose to sugars poses a dilemma in bioethanol production. Cellulose and lignocellulose are in much greater supply than starch and sugars and are, therefore, preferred substrates for ethanol production. However, producing ethanol from cellulose and lignocellulose is comparatively difficult and expensive.

Lignocellulose includes such diverse sources as switchgrass, cornstalks, and wood chips. Fungal enzymes and fermentative yeasts are then used to transform lignocellulose first to sugars and then to bioethanol. The low lignocellulose reactivity limits the production of bioethanol. The expense of enzyme production is the biggest economic barrier to lignocellulose conversion to bioethanol.

Biological pretreatments use fungi to solubilize the lignin. Biodelignification is the biological degradation of lignin by microorganisms. Biodelignification will be a useful delignification method in the future, although at that time it was inadequate and expensive, required a long process time and the microorganisms were poisoned by lignin derivatives. These technologies could greatly simplify pretreatment, but the rates are slow, yields are low, and there is little experience with such approaches.

Recent research has focused on enzyme catalysts called *cellulases* that can attack these chains more efficiently, leading to very high yields of fermentable sugars. Fungal *cellulases* and *beta-glucosidases* produced in separate aerobic reactors can be extracted in very high yields, but because these enzymes have low specific activities, they must be used in large quantities to achieve lignocellulose conversion.

Fungi produce a plethora of enzymes that are used to degrade complex polysaccharides and proteins into simpler sugars and amino acids, and are long established as a key source of a wide variety of industrially important enzymes. Some of these enzymes have already been harnessed in releasing fermentable sugars from a variety of biomass feedstocks, including waste paper, food stuffs, cereals, sugar crops, grains, and woods.

Both bacteria and fungi can produce cellulases for the hydrolysis of lignocellulosic materials. These microorganisms can be aerobic or anaerobic, mesophilic or thermophilic. Bacteria belonging to *Clostridium, Cellulomonas, Bacillus, Thermomonospora, Ruminococcus, Bacteriodes, Erwinia, Acetovibrio, Microbispora,* and *Streptomyces* can produce cellulases (Sun, 2002).

The widely accepted mechanism for enzymatic cellulose hydrolysis involves synergistic actions by endoglucanses or endo-1,4-β-glucanases (EG), exoglucanases or cellobiohydrolases (CBH), and β-glucosidases (BGL). EG play an important role in cellulose hydrolysis by cleaving cellulose chains randomly and thus encouraging strong degradation. EG hydrolyze accessible intramolecular β-1,4-glucosidic bonds of cellulose chains randomly to produce new chain ends; exoglucanases processively cleave cellulose chains at the ends to release soluble cellobiose or glucose; and BGL hydrolyze cellobiose to glucose in order to eliminate cellobiose inhibition (Zhang *et al.*, 2006). BGL complete the hydrolysis process by catalyzing the hydrolysis of cellobiose to glucose.

Filamentous fungi are the major source of *cellulases* and *hemicellulases*. Mutant strains of *Trichoderma sp.* (*T. viride, T. reesei, T. longibrachiatum*) have long been considered to be the most productive and powerful destroyers of crystalline cellulose (Gusakov *et al.*, 2005). CBH I and CBH II are the major *T. reesei* enzymes; the content of CBH I comprises up to 60% of the total cellulolytic protein; whereas, the content of CBH II is about 20% (Gusakov *et al.*, 2005). Similarly, EG I and EG II are the dominant endoglucanases in *T. reesei*, and presumably acting as important partners to CBH I in nature (Väljamäe *et al.*, 2001). Such protein yields are comparable or exceed the respective parameters for the best *Trichoderma sp.* strains (35–40 g/L) (Gusakov *et al.*, 2006). Yeast and fungi tolerate a range of 3.5 to 5.0 pH. The ability to lower pH below 4.0 offers a method for present operators using yeast in less than aseptic equipment to minimize loss due to bacterial contaminants.

Fungal lignocellulolytic enzymes for conversion of lignocellulosic biomass to fermentable sugars for the production of bioethanol were used (Tabka *et al.*, 2006). Wheat straw was pretreated by acid treatment with diluted sulfuric acid followed by steam explosion. Several enzymatic treatments implementing hydrolases (cellulases and xylanases from *Trichoderma reesei*, recombinant feruloyl esterase (FAE) from *Aspergillus niger* and oxidoreductases (laccases from

Pycnoporus cinnabarinus) were investigated to the saccharification of exploded wheat straw. A synergistic effect between cellulases, FAE and xylanase was proven under a critical enzymatic concentration (10 U/g of cellulases, 3 U/g of xylanase and 10 U/g of FAE). The yield of enzymatic hydrolysis was enhanced by increasing the temperature from 310 K to 323 K and addition of a non-ionic surfactant, Tween 20 (Tabka *et al.*, 2006).

4.3.5 Bioethanol Feedstocks

Biological feedstocks that contain appreciable amounts of sugar – or materials that can be converted into sugar, such as starch or cellulose – can be fermented to produce ethanol to be used in gasoline engines. Ethanol feedstocks can be conveniently classified into three types: (1) sucrose-containing feedstocks, (2) starchy materials, and (3) lignocellulosic biomass.

Today the production cost of bioethanol from lignocellulosic materials is still too high, which is the major reason why bioethanol has not made its breakthrough yet. When producing bioethanol from maize or sugar cane the raw material constitutes about 40–70% of the production cost.

Feedstock for bioethanol is essentially comprised of sugar cane and sugar beet. The two are produced in geographically distinct regions. Sugar cane is grown in tropical and subtropical countries, while sugar beet is only grown in temperate climate countries.

The conversion of sucrose into ethanol is easier compared to starchy materials and lignocellulosic biomass because previous hydrolysis of the feedstock is not required since this disaccharide can be broken down by the yeast cells; in addition, the conditioning of the cane juice or molasses favors the hydrolysis of sucrose.

Another type of feedstock, which can be used for bioethanol production, is starch-based materials. Starch is a biopolymer and defines as a homopolymer consisting of only one monomer, D-glucose. To produce bioethanol from starch it is necessary to break down the chains of this carbohydrate to obtain glucose syrup, which can be converted into ethanol by yeasts. The single greatest cost in the production of bioethanol from corn, and the cost with the greatest variability, is the cost of the corn.

Lignocellulosic biomass, such as agricultural residues (corn stover and wheat straw), wood and energy crops, is attractive materials for ethanol fuel production since it is the most abundant reproducible resources on Earth. Lignocellulosic perennial crops (*e.g.*, short rotation coppices and grasses) are promising feedstock because of high yields, low costs, good suitability for low quality land (which is more easily available for energy crops), and low environmental impacts.

Cellulosic biomass is a complex mixture of carbohydrate polymers from plant cell walls known as cellulose and hemicelluloses. The structural composition of various types of cellulosic biomass materials are given in Table 4.5. A cellulosic biomass differs from corn kernels for the methanol conversion process. Corn kernel is easier to process into ethanol than is cellulosic biomass. However, cellulosic

Table 4.5 Composition of various types of cellulosic biomass materials (% dry weight)

Material	Cellulose	Hemicelluloses	Lignin	Ash	Extractives
Algae (green)	20–40	20–50	–	–	–
Grasses	25–40	25–50	10–30	–	–
Hardwoods	45±2	30±5	20±4	0.6±0.2	5±3
Softwoods	42±2	27±2	28±3	0.5±0.1	3±2
Cornstalks	39–47	26–31	3–5	12–16	1–3
Wheat straw	37–41	27–32	13–15	11–14	7±2
Newspapers	40–55	25–40	18–30	–	–
Chemical pulps	60–80	20–30	2–10	–	–

Source: Demirbas, 2005b

Table 4.6 Relative abundance of individual sugars in carbohydrate fraction of wood (% by weight)

Sugar	In softwoods	In hardwoods
Glucose	61–65	55–73
Xylose	9–13	20–39
Mannose	7–16	0.4–4
Galactose	6–17	1–4
Arabinose	<3.5	<1

biomass is less expensive to produce than corn kernels by a factor of roughly 2 on a *per* ton basis, and the amount of ethanol that can be produced *per* acre of land of a given quality is higher for cellulosic biomass than for to corn kernels. Relative to corn kernel, production of a perennial cellulosic biomass crop such as switchgrass requires lower inputs of energy, fertilizer, pesticide, and herbicide, and is accompanied by less erosion and improved soil fertility.

Cellulose molecules consist of long chains of glucose molecules (6-carbon sugars) as do starch molecules, but they have a different structural configuration. These structural characteristics plus the encapsulation by lignin makes cellulosic materials more difficult to hydrolyze than starchy materials. Hemicelluloses are also comprised of long chains of sugar molecules; but contain pentoses, in addition to glucose. The relative abundance of individual sugars in carbohydrate fraction of wood is shown in Table 4.6.

4.3.6   Fuel Properties of Ethanol

Ethanol is a liquid biofuel which can be produced from several different biomass feedstocks and conversion technologies. Bioethanol is an attractive alternative fuel because it is a renewable bio-based resource and it is oxygenated, thereby providing the potential to reduce particulate emissions in compression-ignition engines.

Ethanol can be made synthetically from petroleum or by microbial conversion of biomass materials through fermentation. In 1995, about 93% of the ethanol in the world was produced by the fermentation method and about 7% by the synthetic method. The fermentation method generally uses three steps: (1) the formation of a solution of fermentable sugars, (2) the fermentation of these sugars to ethanol, and (3) the separation and purification of the ethanol, usually by distillation.

Ethanol, unlike gasoline, is an oxygenated fuel that contains 35% oxygen, which reduces particulate and NO_x emissions from combustion. Ethanol is produced a more environmentally benign fuel. The systematic effect of ethyl alcohol differs from that of methyl alcohol. Ethyl alcohol is rapidly oxidized in the body to carbon dioxide and water, and in contrast to methyl alcohol no cumulative effect occurs. Ethanol is also a preferred alcohol in the transesterification process compared to methanol because it is derived from agricultural products and is renewable and biologically less objectionable in the environment.

Bioethanol has a higher octane number, broader flammability limits, higher flame speeds, and higher heats of vaporization than gasoline. These properties allow for a higher compression ratio, shorter burn time, and leaner burn engine, which leads to theoretical efficiency advantages over gasoline in an internal combustion engine. Disadvantages of bioethanol include its lower energy density than gasoline (but about 35% higher than that of methanol), its corrosiveness, low flame luminosity, lower vapor pressure (making cold starts difficult), miscibility with water, and toxicity to ecosystems.

Because bioethanol contains oxygen (35% of oxygen content) in the fuel, it can effectively reduce particulate matter emission in the diesel engine. Bioethanol is appropriate for the mixed fuel in the gasoline engine because of its high octane number, and its low cetane number and high heat of vaporization impede self-ignition in the diesel engine. The most popular blend for light-duty vehicles is known as E85, and contains 85% bioethanol and 15% gasoline. In Brazil, bioethanol for fuel is derived from sugarcane and is used pure or blended with gasoline in a mixture called gasohol (24% bioethanol, 76% gasoline). In several states of the United States, a small amount of bioethanol (10% by vol.) is added to gasoline, known as gasohol or E10. Blends having higher concentrations of bioethanol in gasoline are also used, *e.g.*, in flexible-fuel vehicles that can operate on blends of up to 85% bioethanol – E85. Some countries have exercised a biofuel program form the bioethanol-gasoline blend program, for example, the United States (E10 and for Flexible Fuel Vehicle (FFV) E85), Canada (E10 and for FFV E85), Sweden (E5 and for FFV E85), India (E5), Australia (E10), Thailand (E10), China (E10), Columbia (E10), Peru (E10), Paraguay (E7), and Brazil (E20, E25 and FFV any blend) (Kadiman, 2005).

As biomass hydrolysis and sugar fermentation technologies approach commercial viability, advancements in product recovery technologies will be required. For cases in which fermentation products are more volatile than water, recovery by distillation is often the technology of choice. Distillation technologies that allow the economic recovery of dilute volatile products from streams containing a variety of impurities have been developed and commercially demonstrated. A distillation system separates the bioethanol from water in the liquid mixture.

The first step is to recover the bioethanol in a distillation or beer column, where most of the water remains with the solids part. The product (37% bioethanol) is then concentrated in a rectifying column to a concentration just below the azeotrope (95%). The remaining bottoms product is fed to the stripping column to remove additional water, with the bioethanol distillate from stripping being recombined with the feed to the rectifier. The recovery of bioethanol in the distillation columns in the plant is fixed to be 99.6% to reduce bioethanol losses.

After the first effect, solids are separated using a centrifuge and dried in a rotary dryer. A portion (25%) of the centrifuge effluent is recycled to fermentation and the rest is sent to the second and third evaporator effects. Most of the evaporator condensate is returned to the process as fairly clean condensate (a small portion, 10%, is split off to waste water treatment to prevent build-up of low-boiling compounds) and the concentrated syrup contains 15–20% by weight total solids (Balat *et al.*, 2008).

4.4 Biomethanol

Methanol (CH_3OH or MeOH) is also known as "wood alcohol", Generally, methanol is easier to find than ethanol. Sustainable methods of methanol production are currently not economically viable. Methanol is produced from synthetic gas or biogas and evaluated as a fuel for internal combustion engines. The production of methanol is a cost intensive chemical process. Therefore, in current conditions, only waste biomass such as old wood or biowaste is used to produce methanol (Vasudevan *et al.*, 2005).

The use of methanol as a motor fuel received attention during the oil crises of the 1970s due to its availability and low cost. Problems occurred early in the development of gasoline-methanol blends. As a result of its low price some gasoline marketers over blended. Many tests have shown promising results using 85–100% by volume methanol as a transportation fuel in automobiles, trucks, and buses.

Methanol can be used as one possible replacement for conventional motor fuels. Methanol has been seen as a possible large volume motor fuel substitute at various times during gasoline shortages. It was often used in the early part of 20th century to power automobiles before inexpensive gasoline was widely introduced. In the early 1920s, some viewed it as a source of fuel before new techniques were developed to discover and extract oil. Synthetically produced methanol was widely used as a motor fuel in Germany during the World War.

Before modern production technologies were developed in the 1920s, methanol was obtained from wood as a co-product of charcoal production and, for this reason, was commonly known as wood alcohol. However, the yield from this method of production was very low. One ton of hardwood yielded only 1% or 2% methanol. This led to its eventual replacement by less expensive alternatives. Biomass resources can be used to produce methanol. The pyroligneous acid obtained from wood pyrolysis consists of about 50% methanol, acetone, phenols, and water. Biomass resources include crop residues, forage, grass, crops, wood resources, forest residues, short-rotation wood energy crops, and the lignocellulosic components of municipal solid waste. As a renewable resource, biomass represents a potentially inexhaustible supply of feedstock for methanol production. The product yield for the conversion process is estimated to be 185 kg of methanol *per* metric ton of solid waste. Methanol is currently made from natural gas but can be made using wood waste or garbage *via* partial oxidation reaction into syngas, followed by catalytic conversion into methanol called biomethanol. Adding sufficient hydrogen to the syngas to convert all of the biomass carbon into methanol carbon would more than double the methanol produced from the same biomass base. Current natural gas feedstocks are so inexpensive that even with tax incentives renewable methanol has not been able to compete economically. Technologies are being developed that may eventually result in commercial viability of renewable methanol.

Biomass and coal can be considered as a potential fuel for gasification and further syngas production and methanol synthesis. The feasibility of achieving the conversion has been demonstrated in a large scale system in which a product gas is initially produced by pyrolysis and gasification of a carbonaceous matter. The synthetic process developed in 1927 replaced this traditional method. Methanol from biochar was about 1–2% of volume or 6 gallons *per* ton. Syngas from biomass is altered by catalyst under high pressure and temperature to form methanol. This method will produce 100 gallons of methanol *per* ton of feed material. Figure 4.5 shows biomethanol obtained from carbohydrates by gasification and partial oxidation with O_2 and H_2O.

Methanol is poisonous and burns with an invisible flame. Just like ethyl alcohol methanol has a high octane rating and hence an Otto engine is preferable. Most processes require supplemental oxygen for the intermediate conversion of the biomass into a synthesis gas (H_2 + CO). A readily available supply of hydrogen and oxygen, therefore, should improve the overall productivity of biomass derived methanol (Ouellette *et al.*, 1997).

Before modern production technologies were developed in the 1920s, methanol was obtained from wood as a co-product of charcoal production and, for this reason, was commonly known as wood alcohol. Methanol is currently manufactured worldwide by conversion or derived from syngas, natural gas, refinery off-gas, coal or petroleum ($2H_2$ + CO $\rightarrow$ CH_3OH). The chemical composition of syngas from coal and then from natural gas can be identical with the same H_2/CO ratio. A variety of catalysts are capable of causing the conversion, including reduced NiO-based preparations, reduced Cu/ZnO shift preparations, Cu/SiO$_2$ and Pd/SiO$_2$, and Pd/ZnO (Takezawa *et al.*, 1987; Iwasa *et al.*, 1993).

Methanol is currently made from natural gas but can also be made using biomass *via* partial oxidation reactions (Demirbas and Güllü, 1998). Biomass and coal can be considered as a potential fuel for gasification and further syngas production and methanol synthesis (Takezawa *et al.*, 1987). Adding sufficient hydrogen to the synthesis gas to convert all of the biomass into methanol carbon then

Fig. 4.5 Biomethanol from carbohydrates by gasification and partial oxidation with oxygen and water
Source: Demirbas, 2008b

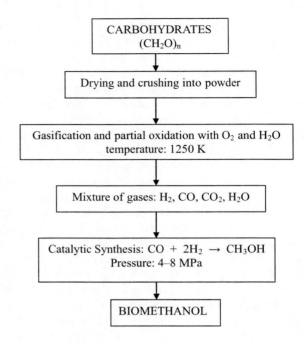

double the methanol produced from the same biomass base (Phillips *et al.*, 1990). Waste material can be partially converted to methanol, for which the product yield for the conversion process is estimated to be 185 kg of methanol *per* metric ton of solid waste (Brown *et al.*, 1952; Sorensen, 1983). Agriculture-(m)ethanol is at present more expensive than synthesis-ethanol from ethylene and of methanol from natural gas (Grassi, 1999).

Biomass resources can be used to produce methanol. The pyroligneous acid obtained from wood pyrolysis consists of about 50% methanol, acetone, phenols, and water (Demirbas and Güllü, 1998; Güllü and Demirbas, 2001). As a renewable resource, biomass represents a potentially inexhaustible supply of feedstock for methanol production. The composition of biosyngas from biomass for producing methanol is presented in Table 4.7. Current natural gas feedstocks are so inexpensive that even with tax incentives renewable methanol has not been able to compete economically. Technologies are being developed that may eventually result in thr commercial viability of renewable methanol.

Methanol from coal may be a very important source of liquid fuel in the future. The coal is first pulverized and cleaned, then fed to a gasifier bed where it is reacted with oxygen and steam to produce syngas. Once these steps have been taken, the production process is much the same as with the other feedstocks with some variations in the catalyst used and the design of the converter vessel in which the reaction is carried out. Methanol is made using synthesis gas (syngas) with hydrogen and carbon monoxide in a 2:1 ratio (Table 4.7). The syngas is transformed to methanol in a fixed catalyst bed reactor. Coal-derived methanol has many preferable properties

Table 4.7 Composition of biosyngas from biomass gasification

Constituents	% by volume (dry and nitrogen free)
Carbon monoxide (CO)	28–36
Hydrogen (H_2)	22–32
Carbon dioxide (CO_2)	21–30
Methane (CH_4)	8–11
Ethene (C_2H_4)	2–4

such as being free of sulfur and other impurities, and may replace petroleum in transportation, or be used as a peaking fuel in combustion turbines, or supply a source of hydrogen for fuel cells. The technology for making methanol from natural gas is already in place and requires only efficiency improvements and scale-up to make methanol an economically viable alternative transportation fuel (Demirbas, 2000).

In recent years, a growing interest has been observed in the application of methanol as an alternative liquid fuel, which can be used directly for powering Otto engines or fuel cells (Chmielniak and Sciazko, 2003). The feasibility of achieving the conversion has been demonstrated in a large scale system in which a product gas is initially produced by pyrolysis and gasification of a carbonaceous matter. Syngas from biomass is altered by a catalyst under high pressure and temperature to form methanol. This method produces 100 gallons of methanol *per* ton of feed material (Rowell and Hokanson, 1979).

The gases produced can be steam reformed to produce hydrogen and followed by a water–gas shift reaction to further enhance hydrogen production. When the moisture content of biomass is higher than 35%, it can be gasified in a supercritical water condition (Hao and Guo, 2002). Supercritical water gasification is a promising process to gasify biomass with high moisture contents due to the high gasification ratio (100% achievable) and high hydrogen volumetric ratio (50% achievable) (Yoshida *et al.*, 2004; Matsumura and Minowa, 2004). Hydrogen produced by biomass gasification has been reported to be comparable to that by natural gas reforming (Bowen *et al.*, 2003). The process is more advantageous than fossil fuel reforming in consideration of environmental benefits. It is expected that biomass thermochemical conversion will become one of the most economical large-scale renewable hydrogen technologies.

The strategy is based on producing hydrogen from biomass pyrolysis using a co-product strategy to reduce the cost of hydrogen and it has been concluded that only this strategy may compete with the cost of the commercial hydrocarbon-based technologies (Wang *et al.*, 1998). This strategy will demonstrate how hydrogen and biofuel are economically feasible and can foster the development of rural areas when practiced on a larger scale. The process of biomass to activated carbon is an alternative route to hydrogen with a valuable co-product that is practiced commercially (Demirbas, 1998).

The simultaneous production of bio-methanol (obtained by the hydrogenation of CO_2 developed during the fermentation of sugar juice), in parallel to the production of bio-ethanol, appears to be economically attractive in locations where

Table 4.8 Main production facilities of methanol and biomethanol

Methanol	Biomethanol
Catalytic synthesis from CO and H_2	Catalytic synthesis from CO and H_2
Natural gas	Distillation of liquid from wood pyrolysis
Petroleum gas	Gaseous products from biomass gasification
Distillation of liquid from coal pyrolysis	Synthetic gas from biomass and coal

hydroelectricity is available at very low cost (~0.01 $ Kwh) and where lignocellulosic residues are available as surpluses.

The gas is converted to methanol in a conventional steam-reforming/water–gas shift reaction followed by high-pressure catalytic methanol synthesis:

$$CH_4 + H_2O \rightarrow CO + 3H_2 \tag{4.12}$$

$$CO + H_2O \rightarrow CO_2 + H_2 \tag{4.13}$$

Equations 4.12 and 4.13 are called as gasification/shift reactions.

$$CO + 2H_2 \rightarrow CH_3OH \tag{4.14}$$

or

$$CO_2 + 3H_2 \rightarrow CH_3OH + H_2O \tag{4.15}$$

Equations 4.14 or 4.15 are methanol synthesis reactions.

The energy value of residues generated worldwide in agriculture and the forest products industry amounts to more than one-third of the total commercial primary energy used at present (Hall *et al.*, 1993). Bioenergy supplies can be divided into two broad categories: (a) organic municipal waste and residues from the food and materials sectors, and (b) dedicated energy crops plantations. Bioenergy from biomass, both residues and energy crops, can be converted into modern energy carriers such as hydrogen, methanol, ethanol, or electricity (Azar *et al.*, 2003).

Methanol can be produced from biomass essentially any primary energy source. Thus, the choice of fuel in the transportation sector is to some extent determined by the availability of biomass. As regards the difference between hydrogen and methanol production costs, conversion of natural gas, biomass, and coal into hydrogen is generally more energy efficient and less expensive than conversion into methanol (Azar *et al.*, 2003). The main production facilities of methanol and biomethanol are given in Table 4.8.

4.5 Vegetable Oils

Vegetable oils and animal fats are chemically triglyceride molecules in which the three fatty acids groups are esters attached to one glycerol molecule (Gunstone and Hamilton, 2001). Fats and oils are primarily water-insoluble, hydrophobic substances in the plant and animal kingdom that are made up of one mole of glycerol and three moles of fatty acids and are commonly referred to as triglycerides (Sonntag, 1979). Triglycerides are esters of glycerin with different carboxylic acids. The triglyceride molecules differ by the nature of the alkyl chain bound to glycerol. The proportions of the various acids vary from fat to fat; each fat has its characteristic composition. Although thought of as esters of glycerin and a varying blend of fatty acids, these oils in fact contain free fatty acids and diglycerides as well. Triglyceride vegetable oils and fats include not only edible, but also inedible vegetable oils and fats such as linseed oil, castor oil, and tung oil, used in lubricants, paints, cosmetics, pharmaceuticals, and for other industrial purposes. Some physical properties of the most common fatty acids occurring in vegetable oils and animal fats as well as their methyl esters are listed in Tables 4.9 and 4.10.

Vegetable oils have become more attractive recently because of their environmental benefits and the fact that they are made from renewable resources. More than 100 years ago, Rudolf Diesel tested vegetable oil as the fuel for his engine (Shay, 1993). Vegetable oils have the potential to substitute for a fraction of the petroleum distillates and petroleum-based petrochemicals in the near future. Vegetable oil fuels are not yet petroleum competitive fuels because they are more expensive than petroleum fuels. However, with the recent increases in petroleum prices and the uncertainties concerning petroleum availability, there is renewed interest in using vegetable oils in diesel engines (Giannelos *et al.*, 2002).

Table 4.9 Selected properties of some common fatty acids in vegetable oils and animal fats

Common name (systematic name)	Carbon number: double bond number	Molecular weight	Melting point (K)
Caprylic acid (octanoic acid)	8:0	144.22	289.7
Capric acid (decanoic acid)	10:0	172.27	304.7
Lauric acid (dodecanoic acid)	12:0	200.32	317.2
Myristic acid (tetradecanoic acid)	14:0	228.38	331.2
Palmitic acid (hexadecanoic acid)	16:0	256.43	336.2
Stearic acid (octadecanoic acid)	18:0	284.48	344.2
Oleic acid (9Z-octadecenoic acid)	18:1	282.47	289.2
Linoleic acid (9Z,12Z-octadecadienoic acid)	18:2	280.45	268.2
Linolenic acid (9Z,12Z,15Z-octadecatrienoic acid)	18:3	278.44	262.2
Erucic acid (13Z-docosenoic acid)	22:1	338.58	305–307

Table 4.10 Selected properties of some common fatty acid methyl esters

Common name (systematic name)	Carbon number: double bond number	Molecular weight	Melting point (K)	Cetane number
Methyl caprylate (methyl octanoate)	8:0	158.24	–	33.6
Methyl caprate (methyl decanoate)	10:0	186.30	–	47.7
Methyl laurate (methyl dodecanoate)	12:0	214.35	278.2	61.4
Methyl myristate (methyl tetradecanoate)	14:0	242.41	291.7	66.2
Methyl palmitate (methyl hexadecanoate)	16:0	270.46	303.7	74.5
Methyl stearate (methyl octadecanoate)	18:0	298.51	312.3	86.9
Methyl oleate (methyl 9Zocta-decenoate)	18:1	296.49	253.2	47.2
Methyl linoleate (methyl 9Z, 12Z–octadecadienoate)	18:2	294.48	238.2	28.5
Methyl linolenate (methyl 9Z, 12Z, 15Z–octadecatrienoate)	18:3	292.46	216.2	20.6
Methyl erucate (methyl 13Zdocosenoate)	22:1	352.60	–	76.0

Vegetable oils and fats are substances derived from plants that are composed of triglycerides. Oils extracted from plants have been used in many cultures since ancient times. The oily seed and nut kernels contain 20–60% oil. Fatty acid compositions of vegetable oils and fats are listed in Table 4.11.

As can be seen from Table 4.11, palmitic (16:0) and stearic (18:0) are two most common saturated fatty acids, with every vegetable oil containing at least a small amount of each one. Similarly, oleic (18:1) and linoleic (18:2) are the most common unsaturated fatty acids. Many of the oils also contain some linolenic acid (18:3).

Today, the world's largest producer of soybeans is the US with the majority of cultivation located in the midwest and south. Soybeans must be carefully cleaned, dried, and dehulled prior to oil extraction. There are three main methods for extracting oil from soybeans. These procedures are hydraulic pressing, expeller pressing, and solvent extraction.

There are more than 350 oil-bearing crops identified, among which only soybean, palm, sunflower, safflower, cottonseed, rapeseed, and peanut oils are considered as potential alternative fuels for diesel engines (Goering *et al.*, 1982; Pryor *et al.*, 1982). Table 4.12 shows the oil species that can be used in biodiesel production. Worldwide consumption of soybean oil is the highest in 2003 (27.9 million

Table 4.11 Fatty acid compositions of vegetable oils and fats*

Sample	16:0	16:1	18:0	18:1	18:2	18:3	Others
Cottonseed	28.7	0	0.9	13.0	57.4	0	0
Poppyseed	12.6	0.1	4.0	22.3	60.2	0.5	0
Rapeseed	3.8	0	2.0	62.2	22.0	9.0	0
Safflower seed	7.3	0	1.9	13.6	77.2	0	0
Sunflower seed	6.4	0.1	2.9	17.7	72.9	0	0
Sesame seed	13.1	0	3.9	52.8	30.2	0	0
Linseed	5.1	0.3	2.5	18.9	18.1	55.1	0
Wheat grain[a]	20.6	1.0	1.1	16.6	56.0	2.9	1.8
Palm	42.6	0.3	4.4	40.5	10.1	0.2	1.1
Corn marrow	11.8	0	2.0	24.8	61.3	0	0.3
Castor[b]	1.1	0	3.1	4.9	1.3	0	89.6
Tallow	23.3	0.1	19.3	42.4	2.9	0.9	2.9
Soybean	11.9	0.3	4.1	23.2	54.2	6.3	0
Bay laurel leaf[c]	25.9	0.3	3.1	10.8	11.3	17.6	31.0
Peanut kernel[d]	11.4	0	2.4	48.3	32.0	0.9	4.0
Hazelnut kernel	4.9	0.2	2.6	83.6	8.5	0.2	0
Walnut kernel	7.2	0.2	1.9	18.5	56.0	16.2	0
Almond kernel	6.5	0.5	1.4	70.7	20.0	0	0.9
Olive kernel	5.0	0.3	1.6	74.7	17.6	0	0.8
Coconut[e]	7.8	0.1	3.0	4.4	0.8	0	65.7

*xx:y: xx number of carbon atoms; y number of double bonds.
[a]Wheat grain oil contains 11.4% of 8:0 and 0.4% of 14:0 fatty acids.
[b]Castor oil contains 89.6% ricinoleic acid.
[c]Bay laurel oil contains 26.5% of 12:0 and 4.5% of 14:0 fatty acids.
[d]Peanut kernel oil contains about 2.7% of 22:0 and 1.3% of 24:0 fatty acids.
[e]Coconut oil contains about 8.9% of 8:0, 6.2% 10:0, 48.8% of 12:0 and 19.9% of 14:0 fatty acids.

Table 4.12 Oil species for biofuel production

Group	Source of oil
Major oils	Coconut (copra), corn (maize), cottonseed, canola (a variety of rapeseed), olive, peanut (ground nut), safflower, sesame, soybean, sunflower
Nut oils	Almond, cashew, hazelnut, macadamia, pecan, pistachio, walnut
Other edible oils	Amaranth, apricot, argan, artichoke, avocado, babassu, bay laurel, beech nut, ben, Borneo tallow nut, carob pod (algaroba), cohune, coriander seed, false flax, grape seed, hemp, kapok seed, lallemantia, lemon seed, macauba fruit (*Acrocomia sclerocarpa*), meadowfoam seed, mustard, okra seed (hibiscus seed), perilla seed, pine nut, poppyseed, prune kernel, quinoa, ramtil, niger pea), rice bran, tallow, tea (camellia), thistle seed, and wheat germ
Inedible oils	Algae, babassu tree, copaiba, honge, jatropha or ratanjyote, jojoba, karanja or honge, mahua, milk bush, nagchampa, neem, petroleum nut, rubber seed tree, silk cotton tree, tall
Other oils	Castor, radish, tung

Table 4.13 World vegetable and marine oil consumption (million metric tons)

Oil	1998	1999	2000	2001	2002	2003
Soybean	23.5	24.5	26.0	26.6	27.2	27.9
Palm	18.5	21.2	23.5	24.8	26.3	27.8
Rapeseed	12.5	13.3	13.1	12.8	12.5	12.1
Sunflower seed	9.2	9.5	8.6	8.4	8.2	8.0
Peanut	4.5	4.3	4.2	4.7	5.3	5.8
Cottonseed	3.7	3.7	3.6	4.0	4.4	4.9
Coconut	3.2	3.2	3.3	3.5	3.7	3.9
Palm kernel	2.3	2.6	2.7	3.1	3.5	3.7
Olive	2.2	2.4	2.5	2.6	2.7	2.8
Fish	1.2	1.2	1.2	1.3	1.3	1.4
Total	80.7	85.7	88.4	91.8	95.1	98.3

metric tons). Table 4.13 shows the world vegetable and marine oil consumption between 1998 and 2003.

Vegetable oil is a biorenewable fuel and is a potentially inexhaustible source of energy having an energy content close to diesel fuel. On the other hand, extensive use of vegetable oils may cause other significant problems such as starvation in developing countries. Vegetable oil fuels have not been acceptable because they are more expensive than petroleum fuels. However, with recent increases in petroleum prices and uncertainties concerning petroleum availability, there is renewed interest in vegetable oil fuels for diesel engines (Demirbas, 2003b; Giannelos et al., 2002).

The first use of vegetable oils as a fuel was in 1900. Vegetable oils have become more attractive recently because of their environmental benefits and the fact that they are made from renewable resources. The advantages of vegetable oils as diesel fuel are liquidity, ready availability, renewability, lower sulfur and aromatic content, and biodegradability (Goering et al., 1982). The main disadvantages of vegetable oils as diesel fuel are higher viscosity, lower volatility, and the reactivity of unsaturated hydrocarbon chains. The problems met in long-term engine tests, according to results obtained by earlier researchers (Komers et al., 2001; Darnoko and Cheryan, 2000) may be classified as follows: injector coking on the injectors, more carbon deposits, oil ring sticking and thickening, and gelling of the engine lubricant oil. Vegetable oils were all extremely viscous, with viscosities 8–15 times greater than No. 2 diesel fuel (No. 2 diesel fuel is a diesel engine fuel with 10–20 carbon number hydrocarbons) (Srivastava and Prasad, 2000).

A variety of biolipids can be used to produce biodiesel. These are: (a) virgin vegetable oil feedstock; rapeseed and soybean oils are most commonly used, though other crops such as mustard, palm oil, sunflower, hemp and even algae show promise; (b) waste vegetable oil; (c) animal fats including tallow, lard, and yellow grease; and (d) non-edible oils such as jatropha oil, neem oil, mahua oil, castor oil, tall oil, etc.

Soybeans are common use in the United States for food products this has led to soybean biodiesel becoming the primary source for biodiesel in that country. In

Malaysia and Indonesia palm oil is used as a significant biodiesel source. In Europe, rapeseed is the most common base oil used in biodiesel production. In India and Southeast Asia, the jatropha tree is used as a significant fuel source.

Algae can grow practically in every place where there is enough sunshine. Some algae can grow in saline water. The most significant different of algal oil is in the yield and hence its biodiesel yield. According to some estimates, the yield (*per* acre) of oil from algae is over 200 times the yield from the best-performing plant/ vegetable oils (Sheehan *et al.*, 1998). Microalgae are the fastest growing photo-synthesizing organisms. They can complete an entire growing cycle every few days. Approximately 46 tons of oil/hectare/year can be produced from diatom algae. Different algae species produce different amounts of oil. Some algae produce up to 50% oil by weight. The production of algae to harvest oil for biodiesel has not been undertaken on a commercial scale, but working feasibility studies have been conducted to arrive at the above number.

4.5.1 *Alternatives to Diesel Fuel*

The diesel engine is used mainly for heavy vehicles. Diesel fuel consists of hydro-carbons with between 9 and 27 carbon atoms in a chain as well as a smaller amount of sulfur, nitrogen, oxygen, and metal compounds. It is a general property of hydrocarbons that the auto-ignition temperature is higher for more volatile hydrocarbons. The hydrocarbons present in the diesel fuels include alkanes, naphthenes, olefins, and aromatics.

The main advantage of the diesel engine is that the level of efficiency is greater than in the Otto cycle engine. This means that a greater part of the energy content of the fuel is exploited. The efficiency of a diesel engine is at best 45%, compared with 30% for the Otto engine.

Diesel emissions contain low concentrations of carbon monoxide and hydro-carbons. The major problem with diesel emissions are nitric oxides and particles, as these are the most difficult to reduce.

Alternate fuels for diesel engines or heavy-duty vehicles (HDVs) contain Fischer–Tropsch (F–T) diesel, dimethyl ether (DME), vegetable oil, and biodiesel. Both F–T and DME can be manufactured from natural gas and are therefore not limited by feedstock availability. Biodiesel on the other hand, is produced from vegetable oils whose supply for non-nutritional uses is presently quite limited.

Dimethyl ether (DME or CH_3-O-CH_3) is a new fuel which has attracted much attention recently. Today DME is made from natural gas, but DME can also be produced by gasifying biomass. DME can be stored in liquid form at 5–10 bar pressure at normal temperature. A major advantage of DME is its naturally high cetane number, which means that self-ignition is easier. The high cetane rating makes DME most suitable for use in diesel engines, which implies that the high level of efficiency of the diesel engine is retained when using DME. The energy content of DME is lower than in diesel.

DME can be produced effectively from syngas in a single-stage, liquid phase (LPDME) process. The origin of syngas includes a wide spectrum of feedstocks such as coal, natural gas, biomass, and others. Non-toxic, high density, liquid DME fuel can be easily stored at modest pressures. The production of DME is very similar to that of methanol. DME conversion to hydrocarbons, lower olefins in particular, is studied using ZSM-5 catalysts with varying SiO_2/Al_2O_3 ratios, whereas the DME carbonization reaction to produce methyl acetate is studied over a variety of group VIII metal substituted heteropolyacid catalysts.

The single-stage, LPDME process is very significant from both scientific and commercial perspectives. DME can be effectively converted to gasoline-range hydrocarbons, lower olefins, and other oxygenates. It may be used directly as a transportation fuel in admixtures with methanol or as a fuel additive. In particular, dimethyl ether is shaping up as an ultra-clean alternative fuel for diesel engines. The advantages of using DME are ultra-low emissions of nitrogen oxides (NO_x), reduced engine noise or quiet combustion, practically soot-free or smokeless operation and hence no exhaust after treatment, and high diesel thermal efficiency.

A sustainable biofuel has two favorable properties which are availability from renewable raw material and its lower negative environmental impact compared to fossil fuels. As an alternative fuel vegetable oil is one of the renewable fuels. Vegetable oil is a potentially inexhaustible source of energy with an energetic content close to diesel fuel.

Vegetable oils, such as palm, soybean, sunflower, peanut, and olive oils as alternative fuels can be used for diesel engines. Due to the rapid decline in crude oil reserves, the use of vegetable oils as diesel fuels is again being promoted in many countries. The effect of coconut oil as a diesel fuel alternative or as a direct fuel blend has been investigated using a single-cylinder, direct-injection diesel engine (Machacon *et al.*, 2001).

Vegetable oils have the potential to substitute a fraction of petroleum distillates and petroleum-based petrochemicals in the near future. Vegetable oil fuels are not petroleum-competitive fuels because they are more expensive than petroleum fuels. However, with recent increases in petroleum prices and uncertainties concerning petroleum availability, there is renewed interest in using vegetable oils in diesel engines.

Short term tests are showing promising results by using neat vegetable oil. Problems appear only after the engine has been operating on vegetable oil for longer periods of time. Problems met in long-term engine tests according to results obtained by earlier researchers may be classified as follows: injector coking and trumpet formation on the injectors, more carbon deposits, oil ring sticking, and thickening and gelling of the engine lubricant oil (Demirbas, 2003b).

Biodiesel is a fuel consisting of long-chain fatty acid alkyl esters made from renewable vegetable oils, recycled cooking greases, or animal fats. Vegetable oil (m)ethyl esters, commonly referred to as "biodiesel", are prominent candidates as alternate diesel fuels (Vicente *et al.*, 2004). Biodiesel is technically competitive with or offers technical advantages compared to conventional petroleum diesel fuel. Compared to No. 2 diesel fuel, all vegetable oils are much more viscous, while

Table 4.14 Fuel properties of methyl ester biodiesels

Source	Viscosity cSt at 313.2 K	Density g/mL at 288.7 K	Cetane number	References
Sunflower	4.6	0.880	49	(Pischinger *et al.*, 1982)
Soybean	4.1	0.884	46	(Schwab *et al.*, 1987)
Palm	5.7	0.880	62	(Nelson and Courter, 1954)
Peanut	4.9	0.876	54	(Srivastava and Prasad, 2000)
Babassu	3.6	–	63	(Srivastava and Prasad, 2000)
Tallow	4.1	0.877	58	(Ali *et al.*, 1995)

methyl esters of vegetable oils are the slightly more viscous. The methyl esters are more volatile than those of the vegetable oils. The soaps obtained from the vegetable oils can be pyrolyzed into hydrocarbon-rich products (Demirbas, 2002).

The properties of biodiesel are close to diesel fuels. The biodiesel was characterized by determining its viscosity, density, cetane number, cloud and pour points, characteristics of distillation, flash and combustion points, and higher heating value (HHV) according to ISO norms. Some fuel properties of methyl ester biodiesels are presented in Table 4.14.

The viscosity of the distillate was $10.2\,mm^2/s$ at 311 K, which is higher than the ASTM specification for No. 2 diesel fuel ($1.9–4.1\,mm^2/s$) but considerably below that of soybean oil ($32.6\,mm^2/s$). Used cottonseed oil from the cooking process was decomposed with Na_2CO_3 as catalyst at 725 K to give a pyrolyzate containing mainly C_{8-20} alkanes (69.6%) besides alkenes and aromatics. The pyrolyzate had lower viscosity, pour point, and flash point than No. 2 diesel fuel and equivalent heating values. The cetane number of the pyrolyzate was lower than that of No. 2 diesel fuel (Bala, 2005).

Engine tests have demonstrated that methyl esters produce slightly higher power and torque than ethyl esters (Encinar *et al.*, 2002). The methyl ester of vegetable oil has been evaluated as a fuel in compression ignition engines (CIE) by researchers (Isigigur *et al.*, 1994; Kusdiana and Saka, 2001).

Biodiesel fuels have generally been found to be non-toxic and are biodegradable, which may promote their use in applications where biodegradability is desired. Neat biodiesel and biodiesel blends reduce particulate matter (PM), hydrocarbons (HC) and carbon monoxide (CO) emissions and increase nitrogen oxides (NO_x) emissions compared with petroleum-based diesel fuel used in an unmodified diesel engine (EPA, 2002).

4.5.2   Vegetable Oil Resources

World annual petroleum consumption and vegetable oil production is about 4.018 and 0.107 billion tons, respectively. Global vegetable oil production increased 56 million tons in 1990 to 88 million tons in 2000, following a below-normal in-

Table 4.15 World vegetable and marine oil consumption (million metric tons)

Oil	1998	1999	2000	2001	2002	2003
Soybean	23.5	24.5	26.0	26.6	27.2	27.9
Palm	18.5	21.2	23.5	24.8	26.3	27.8
Rapeseed	12.5	13.3	13.1	12.8	12.5	12.1
Sunflower seed	9.2	9.5	8.6	8.4	8.2	8.0
Peanut	4.5	4.3	4.2	4.7	5.3	5.8
Cottonseed	3.7	3.7	3.6	4.0	4.4	4.9
Coconut	3.2	3.2	3.3	3.5	3.7	3.9
Palm kernel	2.3	2.6	2.7	3.1	3.5	3.7
Olive	2.2	2.4	2.5	2.6	2.7	2.8
Fish	1.2	1.2	1.2	1.3	1.3	1.4
Total	80.7	85.7	88.4	91.8	95.1	98.3

crease. World vegetable and marine oil consumption is tabulated in Table 4.15. Figure 4.6 shows the plots of percentages the world oil consumption according to years. Leading the gains in vegetable oil production was a recovery in world palm oil output, from 18.5 million tons in 1998 to 27.8 million in 2003.

The major exporters of vegetable oils are Malaysia, Argentina, Indonesia, the Philippines, and Brazil. The major importers of vegetable oils are China, Pakistan, Italy, and the United Kingdom. A few countries such as the Netherlands, Germany, the United States, and Singapore are both large exporters as well as importers of vegetable oils (Bala, 2005).

Global vegetable oil consumptions rose modestly from 79.5 million tons in 1998 to 96.9 million in 2003. A large portion of the gain went to India, where even

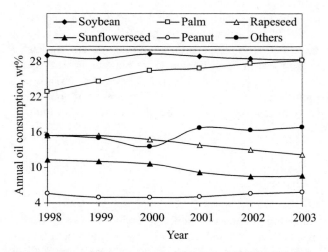

Fig. 4.6 Plots of percentages of world oil consumption according to years

Table 4.16 Oils and fats feedstock distribution top ten developed economy countries with self-sufficiency potential in 2006

Feedstock	%
Animal fats	52
Soybean oil	20
Rapeseed oil	11
Palm oil	6
Sunflower oil	5
Other vegetable oils	5

small price shifts can cause a substantial change in consumption. Indian palm oil imports climbed to a record 2.5 million tons. Similarly, Pakistan, Iran, Egypt, and Bangladesh sharply increased their vegetable oil imports. In 1999, Pakistan reacted to falling vegetable oil prices with a series of increases that doubled the import duties on soybean oil and palm oil, while eliminating duties on oilseeds. Pakistan also raised the import duty on soybean meal from 10% to 35% to stem the influx of Indian exports (Erickson *et al.*, 1980).

Table 4.16 shows the oils and fats feedstock distribution top ten developed economy countries with self-sufficiency potential in 2006.

4.5.2.1 Inedible Oil Resources

The main commodity sources for biodiesel production from inedible oils can be obtained from plant species such as jatropha or ratanjyote or seemaikattamankku (*Jatropha curcas*), karanja or honge (*Pongamia pinnata*), nagchampa (*Calophyllum inophyllum*), rubber seed tree (*Hevca brasiliensis*), neem (*Azadirachta indica*), mahua (*Madhuca indica and Madhuca longifolia*), silk cotton tree (*Ceiba pentandra*), jojoba (*Simmondsia chinensis*), babassu tree, *Euphorbia tirucalli*, microalgae, *etc.* They are easily available in many parts of the world and are very cheap compared to edible oils in India. (Karmee and Chadha, 2005).

Two major species of the genus *Madhuca indica* and *Madhuca longifolia* are found in India. Oil of the rubber seed tree *(Hevea brasiliensis)* is a inedible source of biodiesel production. It is found mainly in Indonesia, Malaysia, Liberia, India, Srilanka, Sarawak, and Thailand. Rubber seed kernels (50–60% of seed) contain 40–50% of brown color oil (Ramadhas *et al.*, 2004). Two major species of the genus the oil palms, *Elaeis guineensis and Elaeis oleifera*, can be found in Africa and Central/South America, respectively. Among vegetable oils, the price of palm oil is the cheapest in palm-producing countries such as Malaysia, Indonesia, Thailand, and Korea. Neem oil is a vegetable oil pressed from the fruits and seeds of neem (*Azadirachta indica*), an evergreen tree that is endemic to the Indian subcontinent and has been introduced to many other areas native to India and Burma, growing in tropical and semitropical regions. Jojoba oil is produced in the seed of the jojoba (*Simmondsia chinensis*) plant, a shrub native to southern Arizona, southern California and northwestern Mexico (Wikipedia, 2008). Oil of the silk cotton tree (*Ceiba pentandra*) is a inedible source of biodiesel production. The tree belongs to the *Bornbacacea* family. The silk cotton tree has great economic importance for both domestic and industrial uses in Nigeria. The seeds are also used as food/feed for humans and livestock in many part of the world such as India, Tanzania, and Mozambique. *Ceiba pentendra* crude oil was extracted for 24 hours using a Soxhlet extractor with n-hexane as a solvent (Das *et al.*, 2002). The babassu tree is a species of palm tree that is resource of light yellow clear oil. There are both edible and non-edible species of babassu oils. Non-edible species of the oil is obtained from the babassu tree that is widely grown in Brazil. Viscosity at 313.2 K and cetane number values of babassu oil are 3.6 cSt and 63, respectively (Srivastava and Prasad, 2000).

Fatty acid profiles of seed oils of 75 plant species having 30% or more fixed oil in their seed/kernel were examined and *Azadirachta indica, Calophyllum inophyllum, Jatropha curcas*, and *Pongamia pinnata* were found to be the most suitable for use as biodiesel (Shah *et al.*, 2005; Azam *et al.*, 2005). The seed oil of *Jatropha* was used as a diesel fuel substitute during World War II and as blends with diesel (Foidl *et al.*, 1996; Gubitz *et al.*, 1999; Shah *et al.*, 2004; Shah and Gupta, 2007). Thus *Jatropha curcas* and *Pongamia pinnata (Karanja)* are the most suitable for the purpose of production of renewable fuel as biodiesel (Meher *et al.*, 2006a; 2006b). Jatropha and Karanja have a high oil content (25–30%) (Foidl *et al.*, 1996; Shah *et al.*, 2004).

From 1978 to 1996, the US Department of Energy's Office of Fuels Development funded a program to develop renewable transportation fuels from algae (Sheehan *et al.*, 1998). Most current research on oil extraction is focused on microalgae to produce biodiesel from algal-oil. Algal-oil processes into biodiesel as easily as oil derived from land-based crops. The lipid and fatty acid contents of microalgae vary in accordance with culture conditions. All algae contain proteins, carbohydrates, lipids and nucleic acids in varying proportions. Algal-oil contains saturated and monounsaturated fatty acids. The fatty acids were determined in the algal oil in the following proportions: 36% oleic (18:1), 15% palmitic (16:0), 11% stearic (18:0), 8.4% iso-17:0, and 7.4% linoleic (18:2). The high proportion of saturated and monounsaturated fatty acids in this alga is considered optimal from a fuel quality standpoint, in that fuel polymerization during combustion would be substantially less than what would occur with polyunsaturated fatty acid-derived fuel (Sheehan *et al.*, 1998).

Algae can grow practically in every place where there is enough sunshine. Some algae can grow in saline water. The most significant different of algal oil is in the yield and hence its biodiesel yield. According to some estimates, the yield (*per* acre) of oil from algae is over 200 times the yield from the best-performing plant/vegetable oils (Sheehan *et al.*, 1998). Microalgae are the fastest growing photosynthesizing organisms. They can complete an entire growing cycle every few days. Approximately 46 tons of oil/hectare/year can be produced from diatom algae. Different algae species produce different amounts of oil. Some algae produce up to 50% oil by weight. Microalgae have much faster growth rates than terrestrial crops. The *per* unit area yield of oil from algae is estimated to be from between 5,000 to 20,000 gallons *per* acre, *per* year; this is 7 to 31 times greater than the next best crop, palm oil.

4.5.3   The Use of Vegetable Oils as Diesel Fuel

The first engine like diesel engine was developed in the 1800s for fossil fuels. The famous German inventor Rudolf Diesel designed the original diesel engine to run on vegetable oil. Dr. Rudolf Diesel used peanut oil to fuel one of this his engines at the Paris Exposition of 1900 (Nitschke and Wilson, 1965). Because of the high temperatures created, the engine was able to run a variety of vegetable oils including hemp and peanut oil. Life for the diesel engine began in 1893 when the Dr. Diesel published a paper entitled "The theory and construction of a rational heat engine". At the 1911 World Fair in Paris, Dr. Diesel ran his engine on peanut oil and declared that "the diesel engine can be fed with vegetable oils and will help considerably in the development of the agriculture of the countries which use it". One of the first uses of transesterified vegetable oil (biodiesel) was to power heavy-duty vehicles in South Africa before World War II.

The first International Conference on Plant and Vegetable Oils as fuels was held in Fargo, North Dakota in August 1982. The primary concerns discussed were the cost of the fuel, the effects of vegetable oil fuels on engine performance, durability and fuel preparation, specifications and additives (Ma and Hanna, 1999). Oil production, oilseed processing and extraction also were considered at this meeting (ASAE, 1982).

Diesel fuel can also be replaced by biodiesel made from vegetable oils. Biodiesel is now mainly being produced from soybean and rapeseed oils. Soybean oil is of primary interest as biodiesel source in the United States while many European countries are concerned with rapeseed oil, and countries with tropical climate prefer to utilize coconut oil or palm oil.

Palm oil is widely grown in South East Asia and 90% of the palm oil produced is used for food and the remaining 10% for non-food consumption, such as production of oleo-chemicals (Leng *et al.*, 1999). An alternative use could be its conversion to liquid fuels and chemicals. Conversion of palm oil to biodiesel using methanol has been reported (Yarmo *et al.*, 1992). There are great differences between palm oil and palm kernel oil in physical and chemical characteristics. Palm oil contains mainly palmitic (16:0) and oleic (18:1) acids, the two common fatty acids and about 50% saturated acids, while palm kernel oil contains mainly lauric acid (12:0) and more than 89% saturated acids (Demirbas, 2003c).

Rapeseed had been grown in Canada since 1936. Hundreds of years ago, Asians and Europeans used rapeseed oil in lamps. Cottonseed oil is used almost entirely as a food material. Sesame, olive, and peanut oils can be used to add flavor to a dish. Walnut oil is high quality edible oil refined by purely physical means from quality walnuts. Poppy seeds are tiny seeds contained within the bulb of the poppy flower, also known as the opium plant (*Papaver somniferum*). Poppy seed oil is high in linoleic acid (typically 60–65%) and oleic acid (typically 18–20%) (Bajpai *et al.*, 1999).

Table 4.17 shows the comparisons of some fuel properties of vegetable oils with No. 2 diesel fuel. The heat contents of vegetable oils are approximately 88% of that of No. 2 diesel. There is little difference between the gross heat content of any of the vegetable oils (Demirbas, 1998). The density values of vegetable oils are between 912 and 921 kg/m^3 while that of No. 2 diesel fuel is 815 kg/m^3. The kinematic viscosity values of vegetable oils vary between 39.2 and 65.4 mm^2/s at 300 K. The vegetable oils were all extremely viscous, with viscosities 8–15 times greater than No. 2 diesel fuel (Table 4.17).

12 (cottonseed, poppyseed, rapeseed, safflower seed, sunflower seed, soybean, corn marrow, sesame seed, linseed, castor, wheat grain, and bay laurel leaf) and 5 kernels (peanut, hazelnut, walnut, almond, and olive) samples used in this study were supplied from different Turkish agricultural sources. Physical analyses of the samples were carried out according to standard test methods: ASTM D445, ASTM D613, and ASTM D524 for kinematic viscosity (KV), cetane number (CN), and carbon residue (CR), respectively. However, as too low amounts of wheat grain, bay laurel leaf, and corn marrow oils were available for determination of cetane numbers using the standard method a calculated cetane number was established as according to Goering *et al.* (1982). Chemical analyses of the samples were carried out according to the standard test methods: ASTM D 2015-85, ASTM D5453-93, ASTM D482-91, AOCS CD1-25, and AOCS CD3 for higher heating value (HHV), sulfur content (SC), ash content (AC), iodine value (IV), and saponification value (SV), respectively. The other standard test methods for fuel properties are presented in Table 4.18. Fatty acid compositions of oil samples were determined by GC analysis. The oil samples were saponified for 3.5 h and 338 K with 0.5 N methanolic KOH to liberate the fatty acids present as their esters. After acidification of the saponified solutions with 1.5 N HCl acid, the acids were weighed and methylated with diazomethane according to the method of Schelenk and Gellerman (1960). The methyl esters of the fatty acids were analyzed by GC (Hewlett–Packard 5790) on a 12 m long and 0.2 mm inside diameter capillary

Table 4.17 Comparisons of some fuel properties of vegetable oils with No. 2 diesel fuel

Fuel type	Heating value (MJ/kg)	Density (kg/m^3)	Viscosity at 300 K (mm^2/s)	Cetane number[a]
No. 2 diesel fuel	43.4	815	4.3	47.0
Sunflower oil	39.5	918	58.5	37.1
Cottonseed oil	39.6	912	50.1	48.1
Soybean oil	39.6	914	65.4	38.0
Corn oil	37.8	915	46.3	37.6
Opium poppy oil	38.9	921	56.1	–
Rapeseed oil	37.6	914	39.2	37.6

[a]Cetane number (CN) is a measure of ignition quality of diesel fuel.

column coated with Carbowax PEG 20. The detector was a FID. Helium was used as a carrier gas. The flame ionization detector temperature was 500 K. The oven temperature was kept at 450 K for 25 min. After that, the oven was heated with heat ratio 5 K/min to 495 K. The spectra of methyl esters were recorded with a VG 70-250-SE mass spectrometer with double focusing. Ionization was carried out at 70 eV. The mass spectrometer was fitted to the gas chromatograph by means of a capillary glass jet separator.

Table 4.19 lists the physical and chemical properties of the oil samples. Viscosity values (KVs) of the oil samples range from 23.2–42.4 mm^2/s at 311 K. Vegeta-

Table 4.18 Determination of physical and chemical properties using standard test methods

Property	Symbol	Standard method	Unit
Density	d	ASTM D4052-91	g/ml
Iodine value	IV	AOCS CD1-25 1993	centigram I/g oil
Saponification value	SV	AOCS CD3 1993	mg KOH/g oil
Higher heating value	HHV	ASTM D2015-85	MJ/kg
Cloud point	CP	ASTM D2500-91	K
Pour point	PP	ASTM D97-93	K
Flash point	FP	ASTM D93-94	K
Cetane number	CN	ASTM D613	–
Kinematic viscosity	KV	ASTM D445	mm^2/s at 311 K
Sulfur content	SC	ASTM D5453-93	wt%
Carbon residue	CR	ASTM D 524	wt%
Ash content	AC	ASTM D482-91	wt%

Table 4.19 Physical and chemical properties of oil samples

Vegetable oil	KV	CR	CN	HHV	AC	SC	IV	SV
Cottonseed	33.7	0.25	33.7	39.4	0.02	0.01	113.20	207.71
Poppyseed	42.4	0.25	36.7	39.6	0.02	0.01	116.83	196.82
Rapeseed	37.3	0.31	37.5	39.7	0.006	0.01	108.05	197.07
Safflower seed	31.6	0.26	42.0	39.5	0.007	0.01	139.83	190.23
Sunflower seed	34.4	0.28	36.7	39.6	0.01	0.01	132.32	191.70
Sesame seed	36.0	0.25	40.4	39.4	0.002	0.01	91.76	210.34
Linseed	28.0	0.24	27.6	39.3	0.01	0.01	156.74	188.71
Wheat grain	32.6	0.23	35.2	39.3	0.02	0.02	120.96	205.68
Corn marrow	35.1	0.22	37.5	39.6	0.01	0.01	119.41	194.14
Castor	29.7	0.21	42.3	37.4	0.01	0.01	88.72	202.71
Soybean	33.1	0.24	38.1	39.6	0.006	0.01	69.82	220.78
Bay laurel leaf	23.2	0.20	33.6	39.3	0.03	0.02	105.15	220.62
Peanut kernel	40.0	0.22	34.6	39.5	0.02	0.01	119.55	199.80
Hazelnut kernel	24.0	0.21	52.9	39.8	0.01	0.02	98.62	197.63
Walnut kernel	36.8	0.24	33.6	39.6	0.02	0.02	135.24	190.82
Almond kernel	34.2	0.22	34.5	39.8	0.01	0.01	102.35	197.56
Olive kernel	29.4	0.23	49.3	39.7	0.008	0.02	100.16	196.83

ble oils are all extremely viscous, with viscosities ranging from 10–20 times that of the ASTM upper limit given for diesel fuels ($2.7 \, mm^2/s$).

Higher heating values (HHVs) of the oil samples range from 39.3 to 39.8 MJ/kg. Castor oil has exceptional HHV (37.4 MJ/kg). The oxygen content of castor oil is higher than those of vegetable oils due to ricinoleic acid in its structure. Ricinoleic is the only 18:1 fatty acid that contains a hydroxyl group. Because castor oil contains a hydroxyl group, its HHV is lower than that of other oils.

Cetane numbers (CNs) of the oil samples range from 27.6–52.9. Iodine values (IVs) and saponification values (SVs) of the oil samples range from 69.82–156.74 and from 188.71–220.78, respectively.

The saponification value (SV) of an oil decreases with increase of its molecular weight. The percentages of C and H in an oil decrease with increase in molecular weight. The increase in SV results in a decrease in the heat content of an oil. The increase in iodine value (IV) (*i.e.*, carbon–carbon double bond, –C=C–, content of oil) results in a decrease in the heat content of an oil. The heat contents of the oil depend on the saponification and iodine values. Therefore, for calculation of the HHVs (MJ/kg) of oil samples, Eq. 4.16 was suggested (Demirbas, 1998).

$$HHV = 49.43 - 0.041 \, (SV) - 0.015 \, (IV) \tag{4.16}$$

Thus the heating values of the vegetable oils can be calculated by using their SVs and IVs obtained from simple chemical analyses without using a calorimeter.

The carbon residues (CRs), sulfur contents (SCs) and ash contents (ACs) of the oil samples range from 0.20–0.31 wt.%, from 0.01–0.02 wt.%, from 0.002–0.02 wt.%, respectively (Table 4.19).

The increase in heat content results from a high increase in the number of carbons and hydrogens, as well as an increase in the ratio of these elements relative to oxygen. A decrease in heat content is the result of fewer hydrogen atoms (*i.e.*, greater unsaturation) in the fuel molecule. Determination of data obtained for a great many compounds has shown that the HHV of an aliphatic hydrocarbon agrees rather closely with that calculated by assuming a certain characteristic contribution from each structural unit (Morrison and Boyd, 1983). For open-chain alkanes, each methylene group, $-CH_2-$, contributes very close to 46,956 kJ/kg.

4.5.3.1   Direct Use of Vegetable Oils in Diesel Engines

The use of vegetable oils as alternative renewable fuel competing with petroleum was proposed in the beginning of 1980s. The advantages of vegetable oils as diesel fuel are (Demirbas, 2003b): (a) liquid nature-portability, (b) ready availability, (c) renewability, (d) higher heat content (about 88% of No. 2 diesel fuel), (e) lower sulfur content, (f) lower aromatic content, and (g) biodegradability,

Full combustion of a fuel requires in existence the amount of stoichiometric oxygen. However, the amount of stoichiometric oxygen generally is not enough for full combustion so as not to oxygenate the fuel. The structural oxygen content of fuel increases combustion efficiency of the fuel due to increased mixing of oxygen with the fuel during combustion. For these reasons, the combustion efficiency and cetane number of vegetable oil is higher than that of diesel fuel and the combustion efficiency methanol and ethanol is higher than that of gasoline.

The main disadvantages of vegetable oils as diesel fuel are (Pryde, 1983): (a) higher viscosity, (b) lower volatility, and (c) the reactivity of unsaturated hydrocarbon chains.

Problems appear only after the engine has been operating on vegetable oils for longer periods of time, especially with direct-injection engines. The problems include: (a) coking and trumpet formation on the injectors to such an extent that fuel atomization does not occur properly or is even prevented as a result of plugged orifices, (b) carbon deposits, (c) oil ring sticking and (d) thickening and gelling of the lubricating oil as a result of contamination by the vegetable oils (Ma and Hanna, 1999).

Among the renewable resources for the production of alternative fuels, triglycerides have attracted much attention as alternative diesel engine fuels (Shay, 1993). However, the direct use of vegetable oils and/or oil blends is generally considered to be unsatisfactory and impractical for both direct injection and indirect type diesel engines because of their high viscosity and low volatility, injector coking and trumpet formation on the injectors, more carbon deposits, oil ring sticking, and thickening and gelling of the engine lubricant oil, acid composition (the reactivity of unsaturated hydrocarbon chains), and free fatty acid content (Ma and Hanna, 1999; Darnoko and Cheryan, 2000; Srivastava and Prasad, 2000; Komers et al. 2001). Consequently, different ways have been considered to reduce the viscosity of vegetable oils such as dilution, microemulsification, pyrolysis, catalytic cracking, and transesterification. Methods based on pyrolysis (Dandik and Aksoy, 1998; Bhatia et al., 1998; Lima et al., 2004) and microemulsification (Billaud et al., 1995) have been studied but are not entirely satisfactory.

Vegetable oils can be used as fuels for diesel engines, but their viscosities are much higher than usual diesel fuel and require modifications of the engines (Kerschbaum and Rinke, 2004). Vegetable oils can possibly only replace a very small fraction of transport fuel. Different ways have been considered to reduce the high viscosity of vegetable oils:

1. Dilution of 25 parts of vegetable oil with 75 parts of diesel fuel.
2. Microemulsions with short chain alcohols such as ethanol or methanol.
3. Transesterification with ethanol or methanol, which produces biodiesel.
4. Pyrolysis and catalytic cracking, which produces alkanes, cycloalkanes, alkenes, and alkylbenzenes.

Dilution of oils with solvents and microemulsions of vegetable oils lowers the viscosity, some engine performance problems, such as injector coking and more carbon deposits, etc. To dilute vegetable oils the addition of 4% ethanol increases the brake thermal efficiency, brake torque, and brake power, while decreasing brake specific fuel consumption. Since the boiling point of ethanol is less than those of vegetable oils the development of the combustion process may be assisted through unburned blend spray (Bilgin et al., 2002).

The viscosity of oil can be lowered by blending with pure ethanol. 25 parts sunflower oil and 75 parts diesel have been blended as diesel fuel (Ziejewski *et al.*, 1986). The viscosity was 4.88 cSt at 313 K, while the maximum specified ASTM value is 4.0c St at 313 K. This mixture was not suitable for long-term use in a direct injection engine. Another study was conducted by using the dilution technique on the same frying oil (Karaosmonoglu, 1999).

The addition of 4% ethanol to No. 2 diesel fuel increases brake thermal efficiency, brake torque, and brake power, while decreasing the brake specific fuel consumption. Since the boiling point of ethanol is less than that of D2 fuel, it may assist the development of the combustion process through an unburned blend spray (Bilgin *et al.*, 2002).

To reduce of the high viscosity of vegetable oils, microemulsions with immiscible liquids such as methanol, ethanol, and ionic or non-ionic amphiphiles have been studied (Billaud *et al.*, 1995). Short engine performances of both ionic and non-ionic microemulsions of ethanol in soybean oil were nearly as good as that of No. 2 diesel fuel (Goering *et al.*, 1982). To solve the problem of the high viscosity of vegetable oils, microemulsions with solvents such as methanol, ethanol, and 1-butanol have been studied. All microemulsions with butanol, hexanol, and octanol met the maximum viscosity requirement for No. 2 diesel fuel. The 2-octanol was found to be an effective amphiphile in the micellar solubilization of methanol in triolein and soybean oil (Schwab *et al.*, 1987; Ma and Hanna, 1999).

Ziejewski *et al.* (1986) prepared an emulsion of 53% (vol) alkali-refined and winterized sunflower oil, 13.3% (vol) 190-proof ethanol and 33.4% (vol) 1-butanol. This non-ionic emulsion had a viscosity of 6.31 cSt at 313 K, a cetane number of 25, and an ash content of less than 0.01%. Lower viscosities and better spray patterns (more even) were observed with an increase of 1-butanol. In a 200 h laboratory screening endurance test, no significant deteriorations in performance were observed, but irregular injector needle sticking, heavy carbon deposits, incomplete combustion, and an increase of lubricating oil viscosity were reported (Ma and Hanna, 1999).

A microemulsion prepared by blending soybean oil, methanol, 2-octanol, and cetane improver in the ratio of 52.7:13.3:33.3:1.0 also passed the 200 h EMA test (Goering, 1984). Schwab *et al.* (1987) used the ternary phase equilibrium diagram and the plot of viscosity *versus* solvent fraction to determine the emulsified fuel formulations. All microemulsions with butanol, hexanol, and octanol met the maximum viscosity requirement for No. 2 diesel. The 2-octanol was an effective amphiphile in the micellar solubilization of methanol in triolein and soybean oil. Methanol was often used due to its economic advantage over ethanol.

Among all these alternatives, transesterification seems to be the best choice, as the physical characteristics of fatty acid esters (biodiesel) are very close to those of diesel fuel, and the process is relatively simple. In the esterification of an acid, an alcohol acts as a nucleophilic reagent; in hydrolysis of an ester, an alcohol is displaced by a nucleophilic reagent. This alcoholysis (cleavage by an alcohol) of an ester is called transesterification (Gunstone and Hamilton, 2001).

Transesterified vegetable oils have proven to be a viable alternative diesel engine fuel with characteristics similar to those of diesel fuel. The transesterification reaction proceeds with a catalyst or any unused catalyst by using primary or secondary monohydric aliphatic alcohols having 1–8 carbon atoms as follows. Transesterification is catalyzed by a base (usually alkoxide ion) or acid (H_2SO_4 or dry HCl). The transesterification is an equilibrium reaction. To shift the equilibrium to the right, it is necessary to use a large excess of the alcohol or else to remove one of the products from the reaction mixture. Furthermore, the methyl or ethyl esters of fatty acids can be burned directly in unmodified diesel engines, with very low deposit formation. Although short-term tests using neat vegetable oil showed promising results, longer tests led to injector coking, more engine deposits, ring sticking, and thickening of the engine lubricant. These experiences led to the use of modified vegetable oil as a fuel.

Technical properties of biodiesel, such as the physical and chemical characteristics of methyl esters related are close to, such as physical and chemical characteristics of methyl esters related to its performance in compression ignition engines are close to petroleum diesel fuel (Saucedo, 2001). Compared with transesterification, the pyrolysis process has more advantages. The liquid fuel produced from pyrolysis has similar chemical components to conventional petroleum diesel fuel (Zhenyi et al., 2004).

4.5.4 New Biorenewable Fuels from Vegetable Oils

World annual petroleum consumption and vegetable oil production is about 4.018 and 0.107 billion tons, respectively. Vegetable oils from bio renewable oil seeds can be used when mixed with diesel fuels. Vegetable oils can be used as fuels for diesel engines, but their viscosities are much higher than usual diesel fuel and require modifications of the engines. Different ways have been considered to reduce the viscosity of vegetable oils such as dilution, microemulsification, pyrolysis, catalytic cracking, and transesterification. Compared with transesterification, the pyrolysis process has more advantages. The liquid fuel produced from pyrolysis has similar chemical components to conventional petroleum diesel fuel. Vegetable oils can be converted to a maximum of liquid and gaseous hydrocarbons by pyrolysis, decarboxylation, deoxygenation, and catalytic cracking processes.

Pyrolysis utilizes biomass to produce a product that is used both as an energy source and a feedstock for chemical production (Demirbas, 2000). A current comprehensive review focuses on the recent developments in the biomass pyrolysis and reports the characteristics of the resulting bio-oils, which are the main products of pyrolysis (Mohan et al., 2006).

Triglycerides are esters of glycerin with different fatty acids. The proportions of the various acids vary from oil to oil; each oil has its own characteristic composition. Triglyceride vegetable oils and fats include not only edible, but also inedible vegetable oils and fats such as linseed oil, castor oil, and tung oil.

Vegetable oils have the potential to substitute a fraction of petroleum distillates and petroleum-based petrochemicals in the near future. Possible acceptable converting processes of vegetable oils into reusable products are transesterification, solvent extraction, cracking, and pyrolysis (Bhatia *et al.*, 1998). Pyrolysis has received a significant amount of interest as this gives products of better quality compared to any other thermochemical process. The liquid fuel produced from vegetable oil pyrolysis has similar chemical components to conventional petroleum diesel fuel. Pyrolysis of triglycerides has been investigated for more than 100 years, especially in areas of the world that lack of deposits of petroleum (Zhenyi *et al.*, 2004).

Vegetable oils are biorenewable and potentially inexhaustible sources of energy with energetic contents close to diesel fuel. There are more than 350 identified oil-bearing crops, among which only sunflower, safflower, soybean, cottonseed, rapeseed, and peanut oils are considered as potential alternative fuels for diesel engines (Goering *et al.*, 1982; Pryor *et al.*, 1982). The major problem associated with the use of pure vegetable oils as fuels, for diesel engines are caused by high fuel viscosity in compression ignition.

Because of the possibility of production in wide variety of products by the high temperature pyrolysis reactions, many investigators have studied the pyrolysis of plant oils to obtain products suitable for fuel under different reaction conditions with and without catalysts (Dandik and Aksoy, 1998; Bhatia *et al.*, 1999; Lima *et al.*, 2004).

4.5.4.1 New Transportation Fuels from Vegetable Oils *via* Pyrolysis

Limitations to vegetable oil use are costs and potential production. Production of vegetable oil is limited by the land area available. Vegetable oil fuels are not petroleum-competitive fuels because they are more expensive than petroleum fuels. However, with recent increases in petroleum prices and uncertainties concerning petroleum availability, there is renewed interest in using vegetable oils in diesel engines.

Biodiesel is a fuel consisting of long-chain fatty acid alkyl esters made from renewable vegetable oils, recycled cooking greases, or animal fats. Vegetable oils are generally converted to their methyl esters by transesterification reaction in the presence catalyst. Methyl esters of vegetable oils (biodiesels) have several outstanding advantages among other new-renewable and clean engine fuel alternatives. Compared to No. 2 diesel fuel, all vegetable oils are much more viscous, while methyl esters of vegetable oils are the slightly more viscous. The methyl esters are more volatile than those of the vegetable oils.

The soaps obtained from the vegetable oils can be pyrolyzed into hydrocarbon-rich products (Demirbas, 2002). Pyrolysis of Na-soaps may be obtained from vegetable oil as follows:

$$2RCOONa \rightarrow R\text{-}R + Na_2CO_3 + CO \tag{4.17}$$

Table 4.20 Yields of pyrolysis products from used sunflower oil sodium soaps at different temperatures (% by weight)

400 K	450 K	500 K	520 K	550 K	570 K	590 K	610 K
2.8	8.4	29.0	45.4	62.4	84.6	92.7	97.5

Source: Demirbas, 2002

The soaps obtained from the vegetable oils can be pyrolyzed into hydrocarbon-rich products according to Eq. 4.17 with higher yields at lower temperatures (Demirbas, 2002). Table 4.20 shows the yields of pyrolysis products from used sunflower oil sodium soaps at different temperatures. These findings are in general agreement with results given in the literature (Barsic and Humke, 1981).

Pyrolysis of used sunflower oil was carried out in a reactor equipped with a fractionating packed column at 673 and 693 K in the presence of sodium carbonate (1, 5, 10, and 20% based on oil weight) as a catalyst. The conversion of oil was high (42–83 wt%) and the product distribution depended strongly on the reaction temperature, residence times, and catalyst content. The pyrolysis products consisted of gas and liquid hydrocarbons, carboxylic acids, CO, CO_2, H_2 and water (Dandik and Aksoy, 1998).

The three vegetable oils (soybean, palm, and castor oils) were pyrolyzed to obtain light fuel products at 503–673 K (Lima *et al.*, 2004). These results show that the soybean, palm, and castor oils present a similar behavior considering the pyrolysis temperature range. On the other hand, palm oil reacts in a lower temperature range with a higher yield in the heavy fraction (Lima *et al.*, 2004). Short pyrolysis time (less than 10 s) leads to a high amount of alkanes, alkenes and aldehydes instead of carboxylic acids. On the other hand, higher temperature and long pyrolysis time did not favor an extensive pyrolysis of this material. In this case, desorption like process becomes more likely than the pyrolytic process (Lima *et al.*, 2004; Fortes and Baugh, 1999). The parameters of the pyrolysis systematically affected the pyrolytic process.

An increase of Na_2CO_3 content and the temperature increased the formation of liquid hydrocarbon and gas products and decreased the formation of aqueous phase, acid phase and coke–residual oil. The highest C_5–C_{11} yield (36.4%) was obtained by using 10% Na_2CO_3 and a packed column of 180 mm at 693 K. The use of packed column increased the residence times of the primer pyrolysis products in the reactor, which caused catalytic and thermal degradation reactions (Dandik and Aksoy, 1998).

The properties of pyrolysis oil from rapeseed are given in Table 4.21 The higher heating value of pyrolysis oil from vegetable oils is quite high. The higher heating value of pyrolysis oil from rapeseed (38.4 MJ/kg) is slightly lower than that of gasoline (47 MJ/kg), diesel fuel (43 MJ/kg) or petroleum (42 MJ/kg), but higher than coal (32–37 MJ/kg) (Sensoz *et al.*, 2000).

Table 4.22 shows the yields of $ZnCl_2$ catalytic pyrolysis from sunflower oil at different temperatures (Demirbas, 2003b). Yield of conversion to products of

$ZnCl_2$ catalytic pyrolysis from the sunflower oil increased with the increase in reaction temperature. The yield of conversion to the products from the sunflower oil reached the maximum 78.3% at 660 K. The decrease in yield of conversion could probably be due to higher coke and gas formation at the pyrolysis temperatures higher than 660 K. As more coke might be deposited on the catalyst's surface the effect of pyrolysis reduced. The gasoline content reached a maximum (35.8% of the conversion products) at 660 K. The aromatic and gas oil contents of conversion products showed a similar trend (Demirbas, 2003b).

Thermal degradation of aliphatic long chain compounds is known as cracking. Higher molecules generally convert into smaller molecules by the cracking process. Large alkane molecules are converted into smaller alkanes and some hydrogen in the cracking process. The smaller hydrocarbons can be obtained by hydrocracking process. Hydrocracking is carried out in the presence of a catalyst and hydrogen, at high pressure and at much lower temperatures (525–725 K).

Table 4.21 Properties of pyrolysis oil from rapeseed

Properties	Methods	Pyrolysis oil
Density, 303 K (kg/m^3)	ASTM D 1298	918
Water content (w/w%)	ASTM D 1744	None
Viscosity, 323 K (cSt)	ASTM D 88	43
Flash point (K)	ASTM D 93	359
Conradson residue (w/w%)	ASTM D 189	0.05
Heating value (MJ/kg)	ASTM D 3286	38.4
Elemental analysis (w/w%)		
Carbon	ASTM D 482	74.04
Hydrogen	ASTM D 3177	10.29
Oxygen (by diff.)		11.70
Nitrogen		3.97
H/C Molar ratio	Calculation	1.67
O/C Molar ratio	Calculation	0.12
Empirical formula	Calculation	$CH_{1:67}O_{0:12}N_{0:046}$

Table 4.22 Yields of $ZnCl_2$ catalytic pyrolysis from sunflower oil (SFO) at different temperatures

Temperature (K)	610	630	650	660	670	690
Conversion, wt% of SFO	35.6	60.7	71.5	78.3	74.9	68.4
Gaseous product, wt%	3.4	5.1	6.4	7.0	8.9	10.6
Aromatic content, wt%	8.5	9.3	9.0	9.6	8.2	8.8
Gasoline content, wt%	28.6	30.4	29.4	35.8	32.7	29.3
Gas oil content, wt%	6.6	7.3	8.4	10.7	8.6	7.9
Coke residue, wt%	0.2	0.3	0.4	0.5	2.4	6.8
Water formation, wt%	3.4	3.7	4.1	4.5	4.1	3.8
Unidentified, wt%	49.3	43.9	42.3	31.9	35.1	32.8

Source: Demirbas, 2003b

Higher boiling petroleum fractions (typically, gas oil) are obtained from silica-alumina catalytic cracking at 725–825 K and under lover pressure. The catalytic cracking not only increases the yield of gasoline by breaking large molecules into smaller ones, but also improves the quality of the gasoline: this process involves carbocations (a group of atoms that contains a carbon atom bearing only six electrons)

Palm oil stearin and copra oil was subjected to conversion over different catalysts, like silica-alumina and zeolite (Pioch *et al.*, 1993). It was found that the conversion of palm and copra oil was 84 wt% and 74 wt%, respectively. The silica-alumina catalyst was highly selective for obtaining aliphatic hydrocarbons, mainly in the kerosene boiling point range (Katikaneni, 1995). The organic liquid products obtaining with the silica-alumina catalyst contained between 4–31 wt% aliphatic hydrocarbons and 14–58 wt% aromatic hydrocarbons. The conversion was high and ranged between 81 wt% and 99 wt%. Silica-alumina catalysts are suitable for converting vegetable oils to aliphatic hydrocarbons. The zinc chloride catalyst, as a Lewis acid, contributes to hydrogen transfer reactions and formation of hydrocarbons in the liquid phase. Palm oil has been converted to hydrocarbons using a shape selective zeolite catalyst (Leng *et al.*, 1999). Palm oil can be converted to aromatics and hydrocarbons in the gasoline, diesel and kerosene range, light gases, and coke and water with the yield of 70 wt%. The maximum yield of gasoline range hydrocarbons was 40 wt% of the total product.

Palm oil has been cracked at atmospheric pressure, at a reaction temperature of 723 K to produce biofuel in a fixed-bed microreactor. The reaction was carried out over microporous HZSM-5 zeolite, mesoporous MCM-41, and composite micro-mesoporous zeolite as catalysts. The products obtained were gas, organic liquid, water, and coke. The organic liquid product was composed of hydrocarbons corresponding to gasoline, kerosene, and diesel boiling point range. The maximum conversion of palm oil, 99 wt%, and gasoline yield of 48 wt% was obtained with composite micromesoporous zeolite. Table 4.23 presents the conversion of palm oil over HZSM-5 with different Si/Al ratios of catalysts by catalytic cracking (Sang *et al.*, 2003). The gasoline yield increased with the increase in the Si/Al ratio due to the decrease in the secondary cracking reactions and the drop in the yield of gaseous products. The vegetable oils were converted to liquid products

Table 4.23 Catalytic cracking of palm oil over HZSM-5 with different Si/Al ratios

Catalyst ID	HZSM-5(50)	HZSM-5(240)	HZSM-5(400)
Conversion (wt%)	96.9	96.0	94.0
Gas yield (wt%)	17.5	14.0	8.2
Water yield (wt%)	6.8	4.6	6.1
OLP (wt%)			
Total organic liquid yield	70.9	76.0	78.0
Gasoline (wt%)	44.6	45.9	49.3
Kerosene (wt%)	19.6	24.6	26.1
Diesel (wt%)	6.7	5.5	2.6
Coke (wt%)	1.7	1.4	1.7

containing gasoline boiling range hydrocarbons. The results show that the product compositions are affected by catalyst content and temperature.

4.5.4.2 Pyrolysis Mechanisms of Vegetable Oils

Soybean, rapeseed, sunflower, and palm oils have been the most studied for the preparation of bio-oil. The viscosity of the distillate was $10.2\,mm^2/s$ at $311\,K$, which is higher than the ASTM specification for No. 2 diesel fuel (1.9–$4.1\,mm^2/s$) but considerably below that of soybean oil ($32.6\,mm^2/s$). Used cottonseed oil from the cooking process was decomposed with Na_2CO_3 as catalyst at $725\,K$ to give a pyrolyzate containing mainly $C_{8\text{-}20}$ alkanes (69.6%) besides alkenes and aromatics. The pyrolyzate had lower viscosity, pour point, and flash point than No. 2 diesel fuel and equivalent heating values (Bala, 2005).

A mechanism for catalytic decarboxylation of vegetable oils is proposed in Fig. 4.7. Vegetable oils contain mainly palmitic, stearic, oleic, and linoleic acids. These fatty acids underwent various reactions, resulting in the formation of different types of hydrocarbons.

The variety of reaction paths and intermediates makes it difficult to describe the reaction mechanism. Besides, the multiplicity of possible reactions of mixed triglycerides make the pyrolysis reaction more complicated (Zhenyi et al., 2004). Generally, thermal decomposition of triglycerides proceed through either a free-radical or carbonium ion mechanism (Srivastava and Prasad, 2000).

Vegetable oil is converted to lower molecular products by two simultaneous reactions: cracking and condensation. In a first step of pyrolysis the triglycerides are converted into carboxylic acids, which are further decomposed by decarboxylation (leading to alkanes and carbon dioxide) or decarbonylations (forming alkenes, water and carbon monoxide) (Gusmao, et al., 1989).

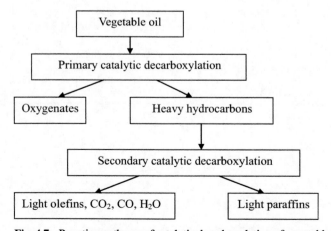

Fig. 4.7   Reaction pathway of catalytic decarboxylation of vegetable oils

The distributions of pyrolysis products depend on the dynamics and kinetics control of different reactions. Maximum of gasoline fraction can be obtained under appropriate reaction conditions. Thermodynamic calculation shows that the initial decomposition of vegetable oils occurs with the breaking of the C–O bond at lower temperature, and fatty acids are the main product. The pyrolysis temperature should be higher than 675 K; at this temperature, the maximum of diesel yield with high content of oxygen can be obtained (Zhenyi *et al.*, 2004). The effect of temperature, the use of catalysts, and the characterization of the products have been investigated (Srivastava and Prasad, 2000). In pyrolysis, the high molecular materials are heated to high temperatures, so their macromolecular structures are broken down into smaller molecules, and a wide range of hydrocarbons are formed. These pyrolytic products can be divided into a gas fraction, a liquid fraction consisting of paraffins, olefins and naphthenes, and solid residue. The cracking process yields a highly unstable low-grade fuel oil, which can be acid-corrosive, tarry, and discolored along with a characteristically foul odor (Demirbas, 2004b).

It has been proposed that thermal and catalytic cracking of triglyceride molecules occurs at the external surface of the catalysts to produce small molecular size components, comprising of mainly heavy liquid hydrocarbons and oxygenates (Demirbas, 2004). In general, it is assumed that the reactions predominantly occur within the internal pore structure of zeolite catalyst.

The catalyst acidity and pore size affect the formation of aromatic and aliphatic hydrocarbons. Hydrogen transfer reactions, which are essential for hydrocarbon formation, are known to increase with catalyst acidity. The high acid density of $ZnCl_2$ catalyst contributes greatly to high amounts of hydrocarbons in the liquid product.

Pyrolysis is the conversion of one substance into another by means of heat or by heat with the aid of a catalyst (Sonntag, 1979). It involves heating in the absence of air or oxygen and cleavage of chemical bonds to yield small molecules. The pyrolyzed material can be vegetable oils, animal fats, natural fatty acids, and methyl esters of fatty acids.

Catalytic cracking of vegetable oils to produce biofuels has been studied (Pioch *et al.*, 1993). Copra oil and palm oil stearin were cracked over a standard petroleum catalyst SiO_2/Al_2O_3 at 723 K to produce gases, liquids and solids with lower molecular weights. The condensed organic phase was fractionated to produce biogasoline and biodiesel fuels.

Pyrolysis liquid products of vegetable oils can be used as alternative engine fuel. Vegetable oils may be converted to liquid product containing gasoline boiling range hydrocarbons. The product compositions are affected by catalyst content and temperature. In pyrolysis, the high molecular materials are heated to high temperatures, so their macromolecular structures are broken down into smaller molecules and a wide range of hydrocarbons are formed.

A single-step direct process for the production of gasoline-like fuel catalytic pyrolysis of vegetable oil is a promising alternative route for environmentally friendly liquid fuels. Vegetable oil has been cracked to obtain liquid fuels at atmospheric pressure, with a reaction temperature of 700–750 K in the presence zeolite as catalysts. The products obtained were gas, organic liquid product, water, and coke. The organic liquid product was composed of hydrocarbons corresponding to gasoline, kerosene, and diesel boiling point range.

4.5.4.3   Gasoline-rich Liquid from Sunflower Oil by Alumina Catalytic Pyrolysis

Recycling and re-refining are the applicable processes for upgrading of vegetable oils by converting them into reusable products such as gasoline and diesel fuel. Possible acceptable processes are transesterification, cracking, and pyrolysis.

The samples of sunflower seed oil were used in the experiments. The sunflower oils were obtained from commercial sources and used without further purification. Aluminum oxide (Al_2O_3, also known as alumina) was obtained from bauxite by the caustic leach method. The catalyst was treated with 10% sodium hydroxide solution before using in the pyrolysis. 2.5 g of NaOH and 25 g alumina (Al_2O_3) were added to 250 ml of deionized water with stirring in a water bath for 45 minutes. 0.5 g of $AlCl_3$ was slowly added to the mixture and stirred vigorously for 30 minutes. The solid material was thoroughly washed, filtered, dried at room temperature overnight, and then calcined at 850 K for 6 h. The pyrolysis experiments were performed on a laboratory scale apparatus. The main element of this device was a vertical cylindrical reactor of stainless-steel, 127.0 mm height, 17.0 mm inner diameter, and 25.0 mm outer diameter inserted vertically into an electrically heated furnace and provided with an electrical heating system power source.

Heat to the vertical cylindrical reactor was supplied from external heater and the power was adjusted to give an appropriate heat up time. The simple thermocouple (NiCr–Constantan) or a 360° degree thermometer with mercury was placed directly in the pyrolysis medium. For each run, the heater was started at 298 K and terminated when the desired temperature was reached. The sunflower seed oil samples were treated with 3% sodium hydroxide solutions in a separatory funnel and then washed with water before pyrolysis. The catalyst (1%, 3%, 4%, and 5% by weight of used sample) was used in the pyrolysis experiments. In addition, the sunflower seed oil samples were pyrolyzed in catalytic runs with 2% and 5% potassium hydroxide. The pyrolysis products were collected within three different groups as condensable liquid products, non-condensable gaseous products, and solid residue. The liquid product was collected in two glass traps with cooled ice–salt mixture and ice, respectively. The gas products were trapped over a saturated solution of NaCl in a gas holder.

Figure 4.8 shows the plots for yield of liquid products from pyrolysis of the sunflower oil at different temperatures in the presence of KOH. The nominal pyrolysis time was 30 min. The yields of liquid products increase temperature and the amounts of KOH. The yield sharply increases between 580 K and 610 K and then it reaches a plateau value with 2% KOH run. Qualitative observations show that the pyrolytic liquid products from the runs with KOH are highly viscous as compared with waste cooking sunflower oil. The repolymerization degree of the pyrolytic liquid products increases with increasing temperature.

Figure 4.9 shows the plots for the yields of liquid products from pyrolysis of the sunflower oil at different temperatures in the presence of an aluminum oxide

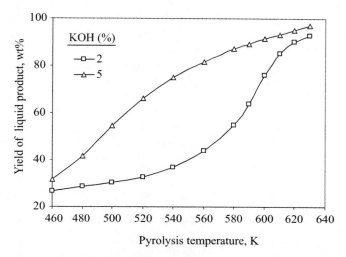

Fig. 4.8 Plots for yield of liquid products from pyrolysis of sunflower oil at different temperatures in the presence of KOH. Pyrolysis time: 30 min

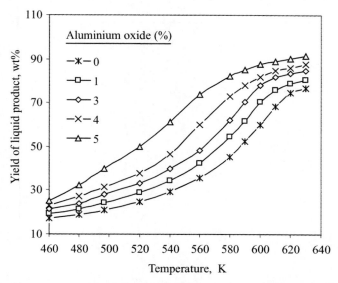

Fig. 4.9 Plots for yield of liquid products from pyrolysis of sunflower oil at different temperatures in the presence of aluminum oxide. Pyrolysis time: 30 min. Particle size: 80–120 mesh

catalyst, respectively. The catalyst was treated with 10% sodium hydroxide solution before being used in the pyrolysis. The particle size of the catalyst was between 80 mesh and 120 mesh. The yields of liquid products generally increase temperature and the percent of the catalyst.

From Fig. 4.9, the yield of liquid product from the sunflower oil sharply increases between 500 K and 680 K in 5% catalytic runs. The yields from noncatalytic runs were 22.1 and 76.8% at 500 and 630 K, respectively. The yields from 5% catalytic runs were 39.8 and 91.4% at 500 and 630 K, respectively. The yields of liquid products reach plateau values between 600 K and 630 K.

The liquid products from pyrolysis of used samples have gasoline-like fractions. Table 4.24 shows the average gasoline percentages of liquid products from pyrolysis of sunflower seed oil at different temperatures in the presence of sodium hydroxide treated aluminum oxide. As seen from Table 4.24, the properties of liquid products obtained from catalytic pyrolysis are similar to gasoline. The highest yields of gasoline were 53.8% for the gasoline from sunflower oil, which can be obtained from the pyrolysis with 5% catalytic runs.

Table 4.24 Average gasoline percentages of liquid products from pyrolysis of sunflower seed oil at different temperatures in the presence of aluminum oxide

Al_2O_3 (%)	560 K	580 K	600 K	620 K	630 K
0	5.7	9.6	11.8	13.7	16.8
1	12.5	19.3	23.7	27.9	32.5
3	17.3	23.5	28.4	32.8	38.2
4	24.9	29.4	33.9	38.7	47.4
5	33.5	39.6	42.5	48.1	53.8

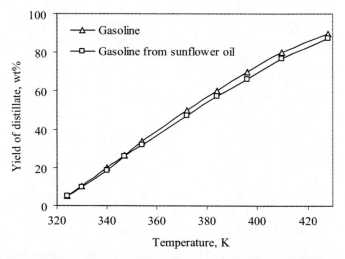

Fig. 4.10 Distillation curves of gasoline and sunflower oil gasoline

Figure 4.10 shows the curves of distillation of the petroleum-based gasoline and the gasoline from sunflower oil. The distillation curve of the gasoline from used lubricant oil by the catalytic pyrolysis is similar to that of gasoline. The petroleum-based gasoline is slightly more volatile than the sunflower oil gasoline.

4.5.5   Properties of Triglycerides

4.5.5.1   Emissions of Neat Vegetable Oil Fuel

While neat vegetable oils are competitive with conventional DF in some emission categories, problems were identified for other kinds of emissions. For example, it was shown that PAH emissions were lower for neat vegetable oils, especially very little amounts of alkylated PAHs, which are common in the emissions of conventional diesel fuel (Mills and Howard, 1983). Besides higher NO_x levels (Geyer et al., 1984), aldehydes are reported to present problems with neat vegetable oils. Total aldehydes increased dramatically with vegetable oils (Geyer et al., 1984).

4.5.5.2   Viscosity

Viscosity is a measure of the internal fluid friction or resistance of oil to flow, which tends to oppose any dynamic change in the fluid motion (Song, 2000). As the temperature of oil is increased its viscosity decreases and it is therefore able to flow more readily. It is also important for flow of oil through pipelines, injector nozzles and orifices (Radovanovic et al., 2000). The lower the viscosity of the oil, the easier it is to pump and atomize and achieve finer droplets (Islam and Beg, 2004).

Viscosity is measured on several different scales, including Redwood No. 1 at 100F, Engler Degrees, Saybolt Seconds, etc. The Redwood viscosity value is the number of seconds required for 50 ml of an oil to flow out of a standard Redwood viscosimeter at a definite temperature. Viscosity is the most important property of biofuel since it affects the operation of fuel injection equipment, particularly at low temperatures when the increase in viscosity affects the fluidity of the fuel. Biodiesel has viscosity close to diesel fuels. High viscosity leads to poorer atomization of the fuel spray and less accurate operation of the fuel injectors.

The vegetable oils were all extremely viscous with viscosities of 10–20 times greater than No. 2 diesel fuel. Castor oil is in a class by itself with a viscosity more than 100 times that of No. 2 diesel fuel. The viscosity of oil can be lowered by blending with pure ethanol. To reduce of the high viscosity of vegetable oils, microemulsions with immiscible liquids such as methanol and ethanol, and ionic or non-ionic amphiphiles have been studied (Ramadhas et al., 2004; Mittelbach and Gangl, 2001). Short engine performances of both ionic and non-ionic microemulsions of ethanol in soybean oil were nearly as good as that of No. 2 diesel (D2) fuel. All microemulsions with butanol, hexanol, and octanol met the maximum viscosity requirement for D2 fuel. 2-octanol was found to be an effective amphiphile in the micellar solubilization of methanol in triolein and soybean oil.

4.5.5.3   Density

Density is another important property of biofuel. Density is the mass *per* unit volume of any liquid at a given temperature. Specific gravity is the ratio of the density of a liquid to the density of water. Density is important in diesel engine performance, since fuel injection operates on a volume metering system (Song, 2000). Also, the density of the liquid product is required for the estimation of the cetane index (Srivastava and Prasad, 2000). The densities were determined using a density meter at 298 K according to ASTM D5002-94. The density meter was calibrated using reverse osmosis water at room temperature.

4.5.5.4   Cetane Number

The cetane number (CN) is a measure of ignition quality or ignition delay, and is related to the time required for a liquid fuel to ignite after injection into a compression ignition engine. CN is based on two compounds, namely hexadecane with a cetane of 100 and heptamethylnonane with a cetane of 15. The CN scale also shows that straight-chain, saturated hydrocarbons have higher CN compared to branched-chain or aromatic compounds of similar molecular weight and number of carbon atoms. It relates to the ignition delay time of a fuel upon injection into the combustion chamber. The CN is a measure of ignition quality of diesel fuels and high CN implies short ignition delay. The longer the fatty acid carbon chains and the more saturated the molecules, the higher the CN. The CN of biofuel from animal fats is higher than that of vegetable oils. CN is determined from real engine tests. The cetane index (CI) is a calculated value derived from the density and volatility obtained from the boiling characteristics of the fuel. CI usually gives a reasonably close approximation to the real cetane number (Song, 2000).

4.5.5.5   Cloud and Pour Points

Two important parameters for low temperature applications of a fuel are the cloud point (CP) and the pour point (PP). CP is the temperature at which a cloud of crystals first appears in a liquid when cooled under conditions as described in ASTM D2500-91. PP is the temperature at which the amount of wax out of the solution is sufficient to gel the fuel; thus it is the lowest temperature at which the fuel can flow. PP is the lowest temperature at which the oil specimen can still be moved. It is determined according to ASTM D97-96. These two properties are used to specify the cold temperature usability of a fuel. Two cooling baths with different cooling temperatures were used. Triglycerides have higher CP and PP compared to conventional diesel fuel (Prakash, 1998).

4.5.5.6   Distillation Range

The distillation range of a fuel affects its performance and safety. It is an important criterion for an engine's start and warm up. It is also needed in the estimation of the cetane index. The distillation range of the liquid product is determined by a test method (ASTM D2887-97) that covers the determination of the boiling range distribution of liquid fuels.

When the ASTM D86 procedure was used to distil the vegetable oils, they were cleaved into a two-phase distillate. Preliminary data indicate a complex mixture of products including alkanes, alkenes, and carboxylic compounds (Goering *et al.*, 1982). Typically, it is not possible to distil all of the vegetable oil and some brownish residue remained in the distillation flask. However, the soaps obtained from the vegetable oils can be distilled into hydrocarbon-rich products with higher yields. The findings from distillation ranges of vegetable oils are given in the literature (Barsic and Humke, 1981).

4.5.5.7   Heat of Combustion

The heat of combustion measures the energy content in a fuel. This property is also referred to as the calorific value or heating value. Although the cetane number determines the combustion performance, it is the heating value along with thermodynamic criteria that sets the maximum possible output of power (Song, 2000). The higher heating values (HHVs) of oil samples are measured in a bomb calorimeter according to the ASTM D2015 standard method.

The ultimate analysis of a vegetable oil provides the weight percentages of carbon, hydrogen, and oxygen. The carbon, hydrogen, and oxygen contents of various common vegetable oils are 74.5–78.4, 10.6–12.4, and 10.8–12.0 wt%, respectively. The HHV of vegetable oils (Goering *et al.*, 1982) ranges from 37.27–40.48 MJ/kg. The HHVs of different vegetable oils vary by <9%.

The saponification value (SV) of an oil decreases with increase of its molecular weight. On the other hand, the percentages of carbon and hydrogen in oil increase with decrease in molecular weight. The increase in iodine value (IV) (*i.e.*, carbon–carbon double bond, $-C=C-$, content) results in a decrease in the heat content of an oil. Therefore, for calculation of the HHVs (MJ/kg) of vegetable oils, Eq. 4.18 was suggested by Demirbas (1998):

$$HHV = 49.43 - [0.041(SV) + 0.015(IV)] \qquad (4.18)$$

4.5.5.8 Water Content

The water content of the fuel is required to accurately measure the net volume of actual fuel in sales, taxation, exchanges, and custody transfer (Srivastava and Prasad, 2000). Various methods are used for determination of water content in the oil samples such as evaporation methods, distillation methods, the xylene method, the Karl–Fischer titration method, *etc*. Evaporation methods rely on measuring the mass of water in a known mass of sample. The moisture content is determined by measuring the mass of an oil sample before and after the water is removed by evaporation. Distillation methods are based on direct measurement of the amount of water removed from an oil sample by evaporation. The Karl–Fischer titration is often used for determining the moisture content of oils that have low water contents.

4.5.6 Triglyceride Economy

High petroleum prices demand the study of biofuel production. Lower-cost feedstocks are needed since biodiesel from food-grade oils is not economically competitive with petroleum-based diesel fuel. Inedible plant oils have been found to be promising crude oils for the production of biodiesel.

The cost biofuel and demand of vegetable oils can be reduced by inedible oils and used oils, instead of edible vegetable oil. In the world a large amount of inedible oil plants are available in nature.

Vegetable oil is traditionally used as a natural raw material for linoleum, paint, lacquers, cosmetics, and washing powder additives. In the technical range there is a growing market in the field of lubricants, hydraulic oils, and special applications. The energetic use of pure plant oil in motors is an option to replace fossil fuels. Nowadays the technique is tested and well established. Pure plant oil-fuel has the advantages of low sulfur and aromatics contents, and safer handling. Using cold pressed plant oil instead of fossil diesel, there is a reduction in production of the green house gas CO_2.

Everybody is able to produce his own fuel. The cold-pressing process does not require complicated machinery. The characteristics of this process are low energy requirement without any use of chemical extractive agents.

4.6 Biodiesel

Biodiesel (Greek, bio, life + diesel from Rudolf Diesel) refers to a diesel-equivalent, processed fuel derived from biological sources. Biodiesel fuels are attracting increasing attention worldwide as a blending component or a direct replacement for diesel fuel in vehicle engines. Biodiesel is known as monoalkyl, such as methyl and ethyl, esters of fatty acids (FAME) derived from a renewable lipid feedstock, such as vegetable oil or animal fat. Biodiesel typically comprises alkyl fatty acid (chain length C_{14}–C_{22}) esters of short-chain alcohols, primarily, methanol, or ethanol.

The scarcity of known petroleum reserves will make renewable energy resources more attractive (Sheehan *et al.*, 1998). The most feasible way to meet this growing demand is by utilizing alternative fuels. Biodiesel is defined as the monoalkyl esters of vegetable oils or animal fats. Biodiesel is the best candidate for diesel fuels in diesel engines. The biggest advantage that biodiesel has over petroleum diesel is its environmental friendliness. Biodiesel burns similarly to petroleum diesel as it concerns regulated pollutants. On the other hand biodiesel probably has better efficiency than gasoline. One such fuel for compression-ignition engines that exhibits great potential is biodiesel. Diesel fuel can also be replaced by biodiesel made from vegetable oils. Biodiesel is now mainly being produced from soybean, rapeseed, and palm oils. The higher heating values of biodiesels are relatively high. The higher heating values of biodiesels (39–41 MJ/kg) are slightly lower than that of gasoline (46 MJ/kg), petrodiesel (43 MJ/kg) or petroleum (42 MJ/kg), but higher than coal (32–37 MJ/kg). Biodiesel is over double the price of petrodiesel. The major economic factor to consider for input costs of biodiesel production is the feedstock, which is about 80% of the total operating cost. The high price of biodiesel is in large part due to the high price of the feedstock. Economic benefits of a biodiesel industry would include value added to the feedstock, an increased number of rural manufacturing jobs, increased income taxes, and investments in plant and equipment. The production and utilization of biodiesel is facilitated firstly through the agricultural policy of subsidizing the cultivation of non-food crops. Secondly, biodiesel is exempt from the oil tax. The European Union accounted for nearly 89% of all biodiesel production worldwide in 2005. By 2010, the United States is expected to become the world's largest single biodiesel market, accounting for roughly 18% of world biodiesel consumption, followed by Germany.

Experts suggest that current oil and gas reserves would suffice to last only a few more decades. To exceed the rising energy demand and decreasing petroleum reserves, fuels such as biodiesel and bioethanol are in the forefront of the alternative technologies. Accordingly, the viable alternative for compression-ignition engines is biodiesel.

It is well known that transport is almost totally dependent on fossil particularly petroleum-based fuels such as gasoline, diesel fuel, liquefied petroleum gas (LPG), and natural gas (NG).

An alternative fuel to petrodiesel must be technically feasible, economically competitive, environmentally acceptable, and easily available. This current alternative diesel fuel can be termed as biodiesel. Biodiesel use may improve emissions levels of some pollutants and deteriorate others. However, for quantifying the effect of biodiesel it is important to take into account several other factors such as raw material, driving cycle, vehicle technology, *etc.* Usage of biodiesel will allow a balance to be sought between agriculture, economic development, and the environment (Meher *et al.*, 2006c).

Environmental and political concerns are generating a growing interest in alternative engine fuels such as biodiesel. Biodiesel is a renewable energy source produced from natural oils and fats, which can be used as a substitute for petroleum diesel without the need for diesel engine modification. In addition to being biodegradable and non-toxic, biodiesel is also essentially free of sulfur and aromatics, producing lower exhaust emissions than conventional gasoline whilst providing similar properties in terms of fuel efficiency. The greatest drawback of using pure vegetable oils as fuels are their high viscosity, although this can be reduced by techniques such as dilution, microemulsification, pyrolysis, or transesterification. Of these processes, the transesterification of vegetable oil triglycerides in supercritical methanol has been shown to be particularly promising, producing high yields of low-viscosity methyl esters without the need of a catalyst. Furthermore, these methyl esters have considerably lower flash points than those of pure vegetable oils.

Biodiesel is a clear amber-yellow liquid with a viscosity similar to petrodiesel, the industry term for diesel produced from petroleum. It can be used as an additive in formulations of diesel to increase the lubricity of pure ultra-low sulfur petrodiesel (ULSD) fuel. Much of the world uses a system known as the "B" factor to state the amount of biodiesel in any fuel mix, in contrast to the "BA" system used for bioalcohol mixes. For example, fuel containing 20% biodiesel is labeled B20. Pure biodiesel is referred to as B100. The common international standard for biodiesel is EN 14214. Biodiesel refers to any diesel-equivalent biofuel usually made from vegetable oils or animal fats. Several different kinds of fuels are called biodiesel: usually biodiesel refers to an ester, or an oxygenate, made from the oil and methanol, but alkane (non-oxygenate) biodiesel, that is, biomass-to-liquid (BTL) fuel is also available. Sometimes even unrefined vegetable oil is called "biodiesel". Unrefined vegetable oil requires a special engine, and the quality of petrochemical diesel is higher. In contrast, alkane biodiesel is of a higher quality than petrochemical diesel, and is actually added to petro-diesel to improve its quality.

Biodiesel has physical properties very similar to petroleum-derived diesel fuel, but its emission properties are superior. Using biodiesel in a conventional diesel engine substantially reduces emissions of unburned hydrocarbons, carbon monoxide, sulfates, polycyclic aromatic hydrocarbons, nitrated polycyclic aromatic hydrocarbons, and particulate matter. Diesel blends containing up to 20% biodiesel can be used in nearly all diesel-powered equipment, and higher-level blends and pure biodiesel can be used in many engines with little or no modification. Lower-level blends are compatible with most storage and distribution equipments, but special handling is required for higher-level blends.

4.6.1 The History of Biodiesel

The process for making fuel from biomass feedstock used in the 1800s is basically the same as that used today. The history of biodiesel is more political and economical than technological. The early 20th century saw the introduction of gasoline-powered automobiles. Oil companies were obliged to refine so much crude oil to supply gasoline that they were left with a surplus of distillate, which is an excellent fuel for diesel engines and much less expensive than vegetable oils. On the other hand, resource depletion has always been a concern with regard to petroleum, and farmers have always sought new markets for their products. Consequently, work has continued on the use of vegetable oils as fuel.

Biodiesel from vegetable oils is not a new process. Conversion of vegetable oils or animal fats to the monoalkyl esters or biodiesel is called as transesterification. Transesterification of triglycerides in oils is not a new process. Transesterification of a vegetable oil was conducted as early as 1853, by scientists E. Duffy and J. Patrick, many years before the first diesel engine became functional. Life for the diesel engine began in 1893 when the famous German inventor Dr. Rudolf Diesel published a paper entitled "The theory and construction of a rational heat engine". What the paper described was a revolutionary engine in which air would be compressed by a piston to a very high pressure thereby causing a high temperature. Dr. Diesel designed the original diesel engine to run on vegetable oil.

Dr. Diesel was educated at the predecessor school to the Technical University of Munich, Germany. In 1878, he was introduced to the work of Sadi Carnot, who theorized that an engine could achieve much higher efficiency than the steam engines of the day. Diesel sought to apply Carnot's theory to the internal combustion engine. The efficiency of the Carnot cycle increases with the compression ratio – the ratio of gas volume at full expansion to its volume at full compression. Nicklaus Otto invented an internal combustion engine in 1876 that was the predecessor to the modern gasoline engine. Otto's engine mixed fuel and air before their introduction to the cylinder, and a flame or spark was used to ignite the fuel-air mixture at the appropriate time. However, air gets hotter as it is compressed, and if the compression ratio is too high, the heat of compression will ignite the fuel prematurely. The low compression ratios needed to prevent premature ignition of the fuel-air mixture limited the efficiency of the Otto engine. Dr. Diesel wanted to build an engine with the highest possible compression ratio. He introduced fuel only when combustion was desired and allowed the fuel to ignite on its own in the hot compressed air. Diesel's engine achieved efficiency higher than that of the Otto engine and much higher efficiency than that of the steam engine. Diesel received a patent in 1893 and demonstrated a workable engine in 1897. Today, diesel engines are classified as "compression-ignition" engines, and Otto engines are classified as "spark-ignition" engines.

Dr. Diesel's prime model, a single 3 m iron cylinder with a flywheel at its base, ran on its own power for the first time in Augsburg, Germany on August 10, 1893. In remembrance of this event, August 10 has been declared International Biodiesel Day. Diesel later demonstrated his engine and received the "Grand Prix" (highest prize) at the World Fair in Paris, France in 1900. This engine stood as an example of Diesel's vision because it was powered by peanut oil, a biofuel, though not strictly biodiesel, since it was not transesterified. He believed that the utilization of a biomass fuel was the real future of his engine. In a 1912 speech, Dr. Diesel said "the use of vegetable oils for engine fuels may seem insignificant today, but such oils may become, in the course of time, as important as petroleum and the coal-tar products of the present time".

The use of vegetable oils as alternative renewable fuel competing with petroleum was proposed in the beginning of 1980s. The advantages of vegetable oils as diesel fuel are: liquid nature-portability, ready availability, renewability, higher heat content (about 88% of No. 2 petroleum diesel fuel), lower sulfur content, lower aromatic content, and biodegradability. The energy supply concerns of the 1970s renewed interest in biodiesel, but commercial production did not begin until the late 1990s.

Dr. Diesel believed that engines running on plant oils had potential, and that these oils could one day be as important as petroleum-based fuels. Since the 1980s, biodiesel plants have opened in many European countries, and some cities have run busses on biodiesel, or blends of petro and bio diesels. More recently, Renault and Peugeot have approved the use of biodiesel in some of their truck engines. Recent environmental and domestic economic concerns have prompted resurgence in the use of biodiesel throughout the world. In 1991, The European Community (EC) proposed a 90% tax deduction for the use of biofuels, including biodiesel. Biodiesel plants are now being built by several companies in Europe; each of these plants will produce up to 1.5 million gallons of fuel *per* year. The European Union accounted for nearly 89% of all biodiesel production worldwide in 2005.

4.6.2   Definitions of Biodiesel

Bio-diesel or biodiesel refers to a diesel-equivalent, processed fuel derived from biological sources. Biodiesel is the name for a variety of ester-based oxygenated fuel from renewable biological sources. It can be made from processed organic oils and fats.

Chemically, biodiesel is defined as the monoalkyl esters of long chain fatty acids derived from renewable biolipids. Biodiesel is typically produced through the reaction of a vegetable oil or animal fat with methanol or ethanol in the presence of a catalyst to yield (m)ethyl esters (biodiesel) and glycerin (Demirbas, 2002). Fatty acid (m)ethyl esters or biodiesels are produced from natural oils and fats. Generally, methanol is preferred for transesterification, because it is less expensive than ethanol.

The general definition of biodiesel is as follows: Biodiesel is a domestic, renewable fuel for diesel engines derived from natural oils like soybean oil, and which meets the specifications of ASTM D 6751.

Biodiesel is technically defined by using ASTM D 6751. Biodiesel is a diesel engine fuel comprised of monoalkyl esters of long chain fatty acids derived from vegetable oils or animal fats, designated B100, and meeting the requirements of ASTM D 6751. Biodiesel, defined as the monoalkyl esters of fatty acids derived from vegetable oil or animal fat, in application as an extender for combustion in

CIEs (diesel), has demonstrated a number of promising characteristics, including reduction of exhaust emissions. Chemically, biodiesel is referred to as the monoalkyl-esters especially (m)ethylester of long-chain-fatty acids derived from renewable lipid sources *via* transesterification process.

Biodiesel is a mixture of methyl esters of long chain fatty acids like lauric, palmitic, steric, oleic, *etc.* Typical examples are rapeseed oil, canola oil, soybean oil, sunflower oil, and palm oil and its derivatives from vegetable sources. Beef and sheep tallow and poultry oil from animal sources and cooking oil are also the sources of raw materials. The chemistry of conversion to biodiesel is essentially the same. Oil or fat react with methanol or ethanol in the presence of catalyst sodium hydroxide or potassium hydroxide to form biodiesel, (m)ethylesters, and glycerin. Biodiesel is technically competitive with or offers technical advantages compared to conventional petroleum diesel fuel. The biodiesel esters have been characterized for their physical and fuel properties including density, viscosity, iodine value, acid value, cloud point, pure point, gross heat of combustion, and volatility. The biodiesel fuels produce slightly lower power and torque, and higher fuel consumption than No. 2 diesel fuel. Biodiesel is better than diesel fuel in terms of sulfur content, flash point, aromatic content, and biodegradability. Some technical properties of biodiesels are shown in Table 4.25.

The cost of biodiesels varies depending on the base stock, geographic area, variability in crop production from season to season, the price of the crude petroleum, and other factors. Biodiesel has over double the price of petroleum diesel. The high price of biodiesel is in large part due to the high price of the feedstock. However, biodiesel can be made from other feedstocks, including beef tallow, pork lard, and yellow grease.

Biodiesels are biodegradable and non-toxic, and have significantly fewer emissions than petroleum-based diesel (petro-diesel) when burnt. Biodiesel functions in current diesel engines, and is a possible candidate to replace fossil fuels as the world's primary transport energy source. With a flash point of 433 K, biodiesel is

Table 4.25 Some technical properties of biodiesels

Common name	Bio-diesel or biodiesel
Common chemical name	Fatty acid (m)ethyl ester
Chemical formula range	$C_{14} - C_{24}$ Methyl esters or $C_{15-25} H_{28-48} O_2$
Kinematic viscosity range (mm²/s, at 313 K)	3.3–5.2
Density range (kg/m³, at 288 K)	860–894
Boiling point range (K)	>475
Flash point range (K)	428–453
Distillation range (K)	470–600
Vapor pressure (mm Hg, at 295 K)	<5
Solubility in water	Insoluble in water
Physical appearance	Light to dark yellow, clear liquid
Odor	Light musty/soapy odor
Biodegradability	Higher biodegradable than petroleum diesel
Reactivity	Stable, but avoid strong oxidizing agents

classified as a non-flammable liquid by the Occupational Safety and Health Administration. This property makes a vehicle fueled by pure biodiesel far safer in an accident than one powered by petroleum diesel or the explosively combustible gasoline. Precautions should be taken in very cold climates, where biodiesel may gel at higher temperatures than petroleum diesel.

4.6.3 Biodiesel from Triglycerides via Transesterification

The possibility of using vegetable oils as fuel has been recognized since the beginning of diesel engines. Vegetable oil has too high a viscosity for use in most existing diesel engines as a straight replacement fuel oil. There are a number of ways to reduce vegetable oil's viscosity. Dilution, microemulsification, pyrolysis, and transesterification are the four techniques applied to solve the problems encountered with the high fuel viscosity. One of the most common methods used to reduce oil viscosity in the biodiesel industry is called transesterification. Chemical conversion of the oil to its corresponding fatty ester is called transesterification (Bala, 2005).

Transesterification (also called alcoholysis) is the reaction of a fat or oil triglyceride with an alcohol to form esters and glycerol. Figure 4.11 shows the transesterification reaction of triglycerides. A catalyst is usually used to improve the reaction rate and yield. Because the reaction is reversible, excess alcohol is used to shift the equilibrium to the products side.

Figure 4.12 shows the enzymatic biodiesel production by interesterification with methyl acetate in the presence lipase enzyme as catalyst.

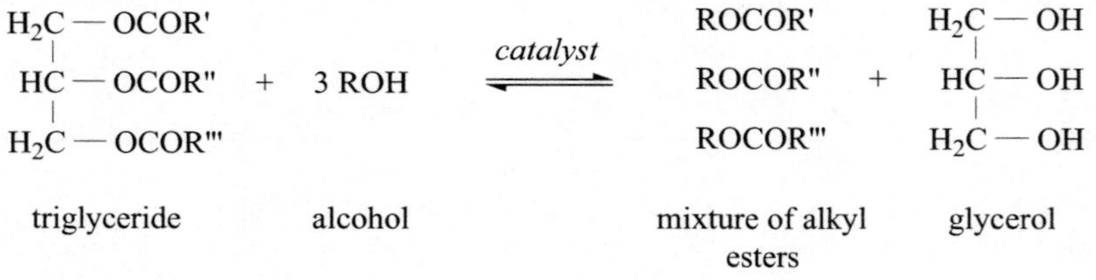

$$
\begin{array}{c}
H_2C-OCOR' \\
| \\
HC-OCOR'' \quad + \quad 3\,ROH \quad \underset{catalyst}{\rightleftharpoons} \quad
\begin{array}{c}
ROCOR' \\
ROCOR'' \\
ROCOR'''
\end{array}
\quad + \quad
\begin{array}{c}
H_2C-OH \\
| \\
HC-OH \\
| \\
H_2C-OH
\end{array} \\
| \\
H_2C-OCOR'''
\end{array}
$$

triglyceride   alcohol   mixture of alkyl   glycerol
                                                   esters

Fig. 4.11 Transesterification of triglycerides with alcohol

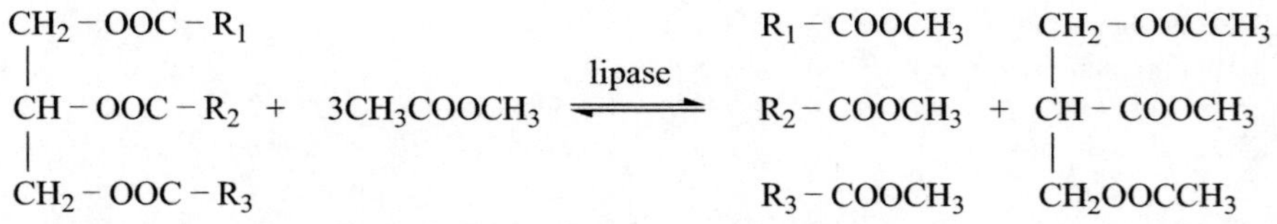

$$
\begin{array}{c}
CH_2-OOC-R_1 \\
| \\
CH-OOC-R_2 \quad + \quad 3CH_3COOCH_3 \quad \underset{lipase}{\rightleftharpoons} \quad
\begin{array}{c}
R_1-COOCH_3 \\
R_2-COOCH_3 \\
R_3-COOCH_3
\end{array}
\quad + \quad
\begin{array}{c}
CH_2-OOCCH_3 \\
| \\
CH-COOCH_3 \\
| \\
CH_2OOCCH_3
\end{array} \\
| \\
CH_2-OOC-R_3
\end{array}
$$

Fig. 4.12 Enzymatic biodiesel production by interesterification with methyl acetate

The biodiesel reaction requires a catalyst such as sodium hydroxide to split the oil molecules and an alcohol (methanol or ethanol) to combine with the separated esters. The main byproduct is glycerin. The process reduces the viscosity of the end product. Transesterification is widely used to reduce vegetable oil viscosity (Pinto *et al.*, 2005). Biodiesel is a renewable fuel source. It can be produced from oil from plants or from animal fats that are byproducts in meat processing.

One popular process for producing biodiesel from the fats/oils is *trans-esterification* of triglyceride by methanol (methanolysis) to make methyl esters of the straight chain fatty acid. The purpose of the transesterification process is to lower the viscosity of the oil. The transesterification reaction proceeds well in the presence of some homogeneous catalysts such as potassium hydroxide (KOH) and sodium hydroxide (NaOH) and sulfuric acid, or heterogeneous catalysts such as metal oxides or carbonates. Sodium hydroxide is very well accepted and widely used because of its low cost and high product yield (Demirbas, 2003a).

Transesterification is the general term used to describe the important class of organic reactions where an ester is transformed into another through interchange of the alkoxy moiety. When the original ester is reacted with an alcohol, the trans-esterification process is called alcoholysis. In this text, the term transesterification will be used as synonymous for alcoholysis of carboxylic esters, in agreement with most publications in this field. The transesterification is an equilibrium reaction and the transformation occurs essentially by mixing the reactants. However, the presence of a catalyst accelerates considerably the adjustment of the equilibrium. In order to achieve a high yield of the ester, the alcohol has to be used in excess. Transesterification is the process of exchanging the alkoxy group of an ester com-pound by another alcohol. These reactions are often catalyzed by the addition of a base and acid. Bases can catalyze the reaction by removing a proton from the alcohol, thus making it more reactive, while acids can catalyze the reaction by donating a proton to the carbonyl group, thus making it more reactive. The trans-esterification reaction proceeds with a catalyst or without any catalyst by using primary or secondary monohydric aliphatic alcohols having 1–8 carbon atoms.

The physical properties of the primary chemical products of transesterification are given in Table 4.26. The high viscosity of the vegetable oils has been the cause of severe operational problems, such as engine deposits. This is a major reason

Table 4.26 Physical properties of chemicals related to transesterification

Name	Specific gravity (g/mL)	Melting point (°C)	Boiling point (°C)	Solubility (<10%)
Methyl myristate	0.875	18.2	–	–
Methyl palmitate	0.825	30.6	196.0	Benzene, EtOH, Et_2O
Methyl stearate	0.850	38.0	215.0	Et_2O, chloroform
Methyl oleate	0.875	−20.1	190.0	EtOH, Et_2O
Methanol	0.792	−104.0	64.7	H_2O, ether, EtOH
Ethanol	0.789	−112.0	78.4	H_2O, ether
Glycerol	1.260	−18.1	290.0	H_2O, ether

Table 4.27 Comparison of various methanolic transesterification methods

Method (min)	Reaction temperature (^{o}C)	Reaction time
Acid or alkali catalytic process	30–35	60–360
Boron trifluoride–methanol	87–117	20–50
Sodium methoxide–catalyzed	20–25	4–6
Non-catalytic supercritical methanol	250–300	6–12
Catalytic supercritical methanol	250–300	0.5–1.5

why neat vegetable oils have largely been abandoned as alternative diesel fuels in favor of monoalkyl esters such as methyl esters. The comparisons various methanolic transesterification methods are tabulated in Table 4.27.

4.6.3.1 Catalytic Transesterification Methods

Vegetable oils can be transesterified by heating them with a large excess of anhydrous methanol and a catalyst. The transesterification reaction can be catalyzed by alkalis (Zhang et al., 2003), acids, or enzymes (Noureddini et al., 2005). Different studies have been carried out using different oils as raw material, different alcohol (methanol, ethanol, butanol), as well as different catalysts, homogeneous ones such as sodium hydroxide, potassium hydroxide, sulfuric acid and supercritical fluids, and heterogeneous ones such as lipases (Bala, 2005).

Sulfuric acid, hydrochloric acid, and sulfonic acid are usually preferred as acid catalysts. The catalyst is dissolved into methanol by vigorous stirring in a small reactor. The oil is transferred into the biodiesel reactor and then the catalyst/alcohol mixture is pumped into the oil.

The transesterification is carried out with the acidic reagent, which is 5% (w/v) anhydrous hydrogen chloride in methanol. It is most often prepared by bubbling hydrogen chloride gas into dry methanol. The hydrogen chloride gas is commercially available in cylinders or can be prepared by dropping concentrated sulfuric acid slowly onto fused ammonium chloride or into concentrated hydrochloric acid. This method is best suited to bulk preparation of the reagent. The hydrogen chloride gas can be obtained by adding acetyl chloride (5 mL) slowly to cooled dry methanol (50 mL).

Vegetable oils are transesterified very rapidly by heating in 10% sulfuric acid in methanol until the reflux. A solution of 1–2% concentrated sulfuric acid in methanol has almost identical properties to 5% methanolic hydrogen chloride, and is very easy to prepare.

Boron trifluoride-catalyzed transesterification of vegetable oils one of the most popular methods. For transesterification of vegetable oils boron trifluoride (BF_3) in methanol (15–20% w/v) is used.

In the alkali catalytic methanol transesterification method, the catalyst (KOH or NaOH) is dissolved into methanol by vigorous stirring in a small reactor. The oil is transferred into the biodiesel reactor and then the catalyst/alcohol mixture is pumped into the oil. The final mixture is stirred vigorously for 2 hours at 67°C in ambient pressure. A successful transesterification reaction produces two liquid phases: ester and crude glycerin.

The catalyst is dissolved into methanol by vigorous stirring in a small reactor. The oil is transferred into the biodiesel reactor and then the catalyst/alcohol mixture is pumped into the oil. The final mixture is stirred vigorously for 2 hours at 71°C in ambient pressure. A successful transesterification reaction produces two liquid phases: ester and crude glycerin. Crude glycerin, the heavier liquid, will collect at the bottom after several hours of settling. Phase separation can be observed within 10 minutes and can be complete within 2 hours of settling. Complete settling can take as long as 20 hours. After settling is complete, water is added at the rate of 5.5% by volume of the methyl ester of oil and then stirred for 5 minutes, and the glycerin is allowed to settle again. Washing the ester is a two-step process, which is carried out with extreme care. A water wash solution at the rate of 28% by volume of oil and 1 gram of tannic acid *per* liter of water is added to the ester and gently agitated. Air is carefully introduced into the aqueous layer while simultaneously stirring very gently. This process is continued until the ester layer becomes clear. After settling, the aqueous solution is drained and water alone is added at 28% by volume of oil for the final washing (Ma and Hanna, 1999; Demirbas, 2002).

Washing the methyl ester is a two-step process which is carried out with extreme care. A water wash solution at the rate of 28% by volume of oil and 1 g of tannic acid *per* liter of water is added to the methyl ester and gently agitated. Air is carefully introduced into the aqueous layer while simultaneously stirring very gently. This procedure is continued until the methyl ester layer becomes clear. After settling, the aqueous solution is drained and water alone is added at 28% by volume of oil for the final washing. The resulting biodiesel fuel when used directly in a diesel engine will burn up to 75% more cleanly than petroleum No. 2 diesel fuel.

For sodium methoxide-catalyzed transesterification, 100 g of vegetable oil is transesterified in toluene (80 mL) and methanol (200 mL) containing fresh sodium (0.8 g) in 10 minutes at reflux, and a related procedure is used to transesterify liter quantities of oils.

CH_2N_2 reacts rapidly with free fatty acids to give methyl esters. The CH_2N_2 is generally prepared in ethereal solution by the action of alkali (a 30% solution of KOH) on a nitrosamide, *e.g.*, *N*-methyl-*N*-nitroso-*p*-toluene-sulfonamide or nitroso-methyl-urea.

4.6.3.2   Non-catalytic Supercritical Methanol Transesterification

In general, methyl and ethyl alcohols are used in supercritical alcohol transesterification. In the conventional transesterification of animal fats and vegetable oils for biodiesel production, free fatty acids and water always produce negative effects, since the presence of free fatty acids and water causes soap formation, consumes the catalyst, and reduces catalyst effectiveness, all of which result in a low conversion. The transesterification reaction may be carried out using either basic or acidic catalysts, but these processes are relatively time consuming and complicate separation of the product and the catalyst, which results in high production costs and energy consumption. In order to overcome these problems, Saka and Kusdiana (2001) and Demirbas (2002, 2003a) first proposed that biodiesel fuels may be prepared from vegetable oil *via* non-catalytic transesterification with supercritical methanol (SCM). A novel process of biodiesel fuel production has been developed by a non-catalytic supercritical methanol method. Supercritical methanol is believed to solve the problems associated with the two-phase nature of normal methanol/oil mixtures by forming a single phase as a result of the lower value of the dielectric constant of methanol in the supercritical state. As a result, the reaction was found to be complete in a very short time. Compared with the catalytic processes under barometric pressure, the supercritical methanol process is non-catalytic, purification of products is much simpler, the reaction time is lower, it is more environmentally friendly and uses less energy. However, the reaction requires temperatures of 520–670 K and pressures of 35–60 MPa (Saka and Kusdiana, 2001; Demirbas, 2003a).

The non-catalytic supercritical methanol transesterification is performed in a stainless steel cylindrical reactor (autoclave) at 525 K (Demirbas, 2002). In a typical run, the autoclave is charged with a given amount of vegetable oil and liquid methanol with changed molar ratios. After each run, the gas is vented, and the autoclave is poured into a collecting vessel. The rest of the contents is removed from the autoclave by washing with methanol.

The most important variables affecting the methyl ester yield during transesterification reaction are the molar ratio of alcohol to vegetable oil and the reaction temperature. Viscosities of the methyl esters from vegetable oils were slightly higher than that of No. 2 diesel fuel.

In the transesterification process, the vegetable oil should have an acid value less than 1 and all materials should be substantially anhydrous. If the acid value was greater than 1, more NaOH or KOH was spent to neutralize the free fatty acids. Water also caused soap formation and frothing (Demirbas, 2002).

The stoichiometric ratio for transesterification reaction requires three moles of alcohol and one mole of triglyceride to yield three moles of fatty acid ester and one mole of glycerol. Higher molar ratios result in greater ester production in a shorter time. The vegetable oils were transesterified with 1:6–1:40 vegetable oil–alcohol molar ratios in catalytic and supercritical alcohol conditions (Demirbas, 2002).

Table 4.28 shows critical temperatures and critical pressures of various alcohols. Table 4.29 shows the comparisons between catalytic methanol method and supercritical methanol method for biodiesel from vegetable oils by transesterification. The supercritical methanol process is non-catalytic, simpler purification, has a lower reaction time and lower energy use. Therefore, the supercritical methanol

Table 4.28 Critical temperatures and critical pressures of various alcohols

Alcohol	Critical temperature (K)	Critical pressure (MPa)
Methanol	512.2	8.1
Ethanol	516.2	6.4
1-Propanol	537.2	5.1
1-Butanol	560.2	4.9

Table 4.29 Comparisons between the catalytic methanol (MeOH) method and the supercritical methanol (SCM) method for biodiesel from vegetable oils by transesterification

	Catalytic MeOH process	SCM method
Methylating agent	Methanol	Methanol
Catalyst	Alkali (NaOH or KOH)	None
Reaction temperature (K)	303–338	523–573
Reaction pressure (MPa)	0.1	10–25
Reaction time (min)	60–360	7–15
Methyl ester yield (wt%)	96	98
Removal for purification	Methanol, catalyst, glycerol, soaps	Methanol
Free fatty acids	Saponified products	Methyl esters, water
Exhaust smelling	Soap smelling	Sweet smelling

method should be more effective and efficient than the common commercial process (Kusdiana and Saka, 2004).

The parameters affecting the methyl esters formation are reaction temperature, pressure, molar ratio, water content, and free fatty acid content. It is evident that at the subcritical state of alcohol, the reaction rate is low and gradually increases as either pressure or temperature rises. It was observed that increasing the reaction temperature, especially to supercritical conditions, had a favorable influence on the yield of ester conversion. The yield of alkyl ester increased with increasing the molar ratio of oil to alcohol (Demirbas, 2002). In the supercritical alcohol transesterification method, the yield of conversion rises by 50–95% during the first 10 min. Figure 4.13 shows the plots for changes in fatty acids alkyl esters conversion from triglycerides as treated in supercritical alcohols at 575 K.

Water content is an important factor in the conventional catalytic transesterification of vegetable oil. In the conventional transesterification of fats and vegetable oils for biodiesel production, free fatty acids and water always produce negative effects since the presence of free fatty acids and water causes soap formation, consumes the catalyst, and reduces catalyst effectiveness. In catalyzed methods, the presence of water has negative effects on the yields of methyl esters. However, the presence of water affected positively the formation of methyl esters in our supercritical methanol method. Figure 4.14 shows the plots for yields of methyl esters as a function of water content in transesterification of triglycerides. Figure 4.15 shows the plots for yields of methyl esters as a function of free fatty acid content in biodiesel production.

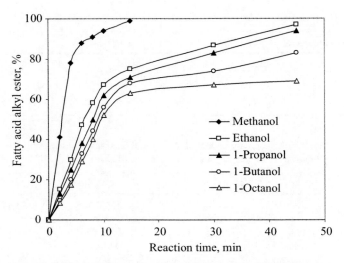

Fig. 4.13 Plots for changes in fatty acids alkyl esters conversion from triglycerides as treated in supercritical alcohols at 575 K
Source: Demirbas, 2008

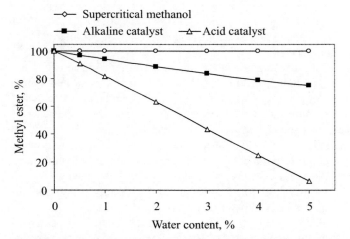

Fig. 4.14 Plots for yields of methyl esters as a function of water content in transesterification of triglycerides
Source: Demirbas, 2008

Figure 4.16 shows the effect of molar ratio of sunflower seed oil to ethanol on yield of ethyl ester at 518 K. The sunflower seed oil was transesterified with 1:1, 1:3, 1:9, 1:20, and 1:40 vegetable oil-ethanol molar ratios in supercritical ethanol conditions. It was observed that an increasing molar ratio had a favorable influence on ester conversion.

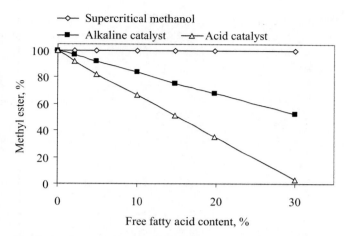

Fig. 4.15 Plots for yields of methyl esters as a function of free fatty acid content in biodiesel production
Source: Demirbas, 2008

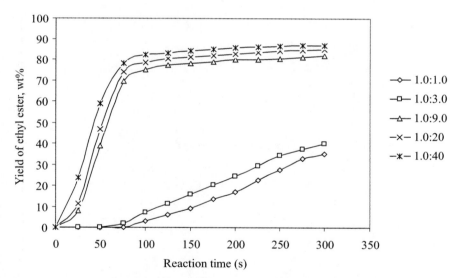

Fig. 4.16 Effects of the molar ratio of vegetable oil to ethanol on yield of ethyl ester. Temperature: 518 K
Source: Balat, 2005

Ethyl esters of vegetable oils have several outstanding advantages among other new-renewable and clean engine fuel alternatives. The variables affecting the ethyl ester yield during transesterification reaction, such as molar ratio of alcohol to vegetable oil and reaction temperature have been investigated. Viscosities of the ethyl esters from vegetable oils were twice as high as that of No. 2 diesel fuel (Balat, 2005).

Catalytic supercritical methanol transesterification is carried out in the autoclave in the presence of 1–5% NaOH, CaO and MgO as a 4catalyst at 423 K. In the catalytic supercritical methanol transesterification method, the yield of conversion rises 60–90% during the first minute.

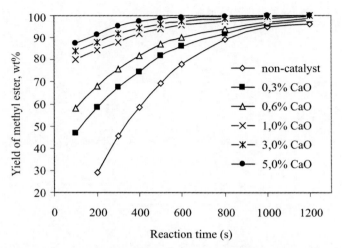

Fig. 4.17 Effect of the CaO content on the methyl ester yield. Temperature: 523 K, molar ratio of methanol to sunflower oil: 41:1
Source: Demirbas, 2008

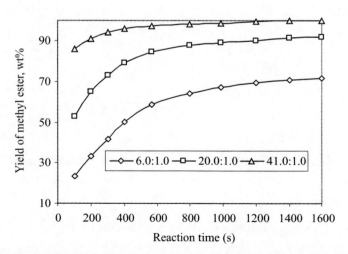

Fig. 4.18 Effect of the molar ratio of methanol to sunflower oil on the methyl ester yield for catalytic (3% CaO) transesterification in supercritical methanol at 523 K
Source: Demirbas, 2008

Figure 4.17 shows the relationship between the reaction time and the catalyst content. It can be affirmed that CaO can accelerate the methyl ester conversion from sunflower oil in 423 K and 24 MPa even if a little catalyst (0.3% of the oil) is added. The transesterification speed obviously improved as the content of CaO increased from 0.3% to 3%. However, when the catalyst content was further enhanced to 5% there was a small increase in the methyl ester yield.

Biodiesel can be obtained from biocatalytic transesterification methods (Noureddini *et al.*, 2005). Methyl acetate, a novel acyl acceptor for biodiesel production has been developed, and a comparative study on Novozym 435-catalyzed transesterification of soybean oil for biodiesel production with different acyl acceptors has been studied (Noureddini *et al.*, 2005). Figure 4.18 shows the effect of the molar ratio of methanol to sunflower oil on the methyl ester yield for catalytic (3% CaO) transesterification in supercritical methanol at 523 K.

4.6.4   Recovery of Glycerol

Glycerol is a byproduct from biodiesel and soap industries. The principal byproduct of biodiesel production is crude glycerol, which is about 10%wt of vegetable oil. Glycerol is the inevitable byproduct of the transesterification process. Direct addition of glycerol to the fuel is not possible. However, derivatives of glycerol such as ethers have potential for use as additives with biodiesel, diesel, or biodiesel-diesel blends. Glycerol can be converted into commercially valued oxygenate products using for the diesel fuel blends. Glycerol alkyl ethers (GAEs) are easily synthesized using glycerol, which is reacted with isobutylene in the presence of an acid catalyst. Etherification of glycerol with isobutylene can be effectively accomplished using Amberlyst-15 as catalyst under a variety of reaction conditions. The ethers of glycerol can be effectively added to biodiesel fuels, providing a 5 K reduction in the cloud point and an 8% reduction in viscosity.

The standards make sure that the following important factors in the biodiesel production process by transesterification are satisfied: (a) complete transesterification reaction, (b) removal of the catalyst, (c) removal of alcohol, (d) removal of glycerol, and (e) complete esterification of free fatty acids. The following transesterification procedure is for methyl ester production. The catalyst is dissolved into the alcohol by vigorous stirring in a small reactor. The oil is transferred into the biodiesel reactor and then the catalyst/methanol mixture is pumped into the oil and final mixture stirred vigorously for two hours. A successful reaction produces two liquid phases: ester and crude glycerol. The entire mixture then settles and glycerol is left on the bottom and methyl esters (biodiesel) are left on top. Crude glycerol, the heavier liquid, will collect at the bottom after several hours of settling. Phase separation can be observed within 10 minutes and can be complete within two hours after stirring has stopped. Complete settling can be taken as long as 18 hours. After settling is complete, water was added at the rate of 5.0% by volume of the oil and then stirred for 5 minutes, and the glycerol allowed to settle again. After settling is complete the glycerol is drained and the ester layer remains (Bala, 2005).

The recovery of high quality glycerol as a biodiesel byproduct is primary options to be considered to lower the cost of biodiesel. With neutralizing the free fatty acids, removing the glycerol, and creating an alcohol ester transesterification occurs. This is accomplished by mixing methanol with sodium hydroxide to make sodium methoxide. This dangerous liquid is then mixed into vegetable oil. Washing the methyl ester is a two-step process, which is carried out with extreme care. This procedure is continued until the methyl ester layer becomes clear. After settling, the aqueous solution is drained and water alone is added at 28% by volume of oil for the final washing. The resulting biodiesel fuel when used directly in a diesel engine will burn up to 75% cleaner than No. 2 diesel fuel (Bala, 2005).

The process of converting vegetable oil into biodiesel fuel is called transesterification and is luckily less complex than it sounds. Chemically, transesterification means taking a triglyceride molecule or a complex fatty acid, neutralizing the free fatty acids, removing the glycerin, and creating an alcohol ester. This is accomplished by

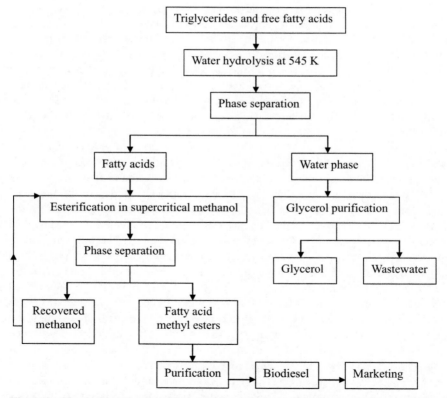

Fig. 4.19 Simple flow diagram of glycerol from continuous biodiesel production process with subcritical water and supercritical methanol stages

mixing methanol with sodium hydroxide to make sodium methoxide. This dangerous liquid is then mixed into vegetable oil. The entire mixture then settles. Glycerin is left on the bottom and methyl esters, or biodiesel, is left on top. The glycerin can be used to make soap (or any one of 1,600 other products) and the methyl esters are washed and filtered. The resulting biodiesel fuel when used directly in a diesel engine will burn up to 75% cleaner than petroleum No. 2 diesel fuel. The recovery of high quality glycerol as a biodiesel byproduct is primary options to be considered to lower the cost of biodiesel. Figure 4.19 shows a flow diagram of glycerol from continuous biodiesel production process with subcritical water and supercritical methanol stages.

4.6.5 Reaction Mechanism of Transesterification

Triacylglycerols (vegetable oils and fats) are esters of long-chain carboxylic acids combined with glycerol. Carboxylic acids $\{R–C(=O)–O–H\}$ can be converted to methyl esters $\{R–C(=O)–O–CH_3\}$ by the action of the transesterification agent. The parameters affecting the methyl esters formation are reaction temperature, pressure, molar ratio, water content, and free fatty acid content. It has been observed that increasing the reaction temperature, has a favorable influence on the yield of ester conversion. The yield of alkyl ester increases with increasing the molar ratio of oil to alcohol (Demirbas, 2002).

$$\text{Fatty acid } (R_1COOH) + \text{Alcohol } (ROH) \leftrightarrows \text{Ester } (R_1COOR)$$
$$+ \text{ Water } (H_2O) \tag{4.19}$$

$$\text{Triglyceride} + ROH \leftrightarrows \text{Diglyceride} + RCOOR_1 \tag{4.20}$$

$$\text{Diglyceride} + ROH \leftrightarrows \text{Monoglyceride} + RCOOR_2 \tag{4.21}$$

$$\text{Monoglyceride} + ROH \leftrightarrows \text{Glycerol} + RCOOR_3 \tag{4.22}$$

Transesterification consists of a number of consecutive, reversible reactions. The triglyceride is converted stepwise to diglyceride, monoglyceride and finally glycerol (Eqs. 4.19–4.22), in which 1 mol of alkyl esters is removed in each step. The reaction mechanism for alkali-catalyzed transesterification was formulated as three steps. The formation of alkyl esters from monoglycerides is believed to be a step that determines the reaction rate, since monoglycerides are the most stable intermediate compound (Ma and Hanna, 1999).

Several aspects, including the type of catalyst (alkaline, acid or enzyme), alcohol/vegetable oil molar ratio, temperature, purity of the reactants (mainly water content) and free fatty acid content have an influence on the course of the transesterification. In the conventional transesterification of fats and vegetable oils for biodiesel production, free fatty acid and water always produce negative effects, since the presence of free fatty acids and water causes soap formation, consumes the catalyst, and reduces catalyst effectiveness, all of which result in a low conversion (Kusdiana and Saka, 2004).

Transesterification is the general term used to describe the important class of organic reactions where an ester is transformed into another through interchange of the alkoxy moiety. When the original ester is reacted with an alcohol, the transesterification process is called alcoholysis. The transesterification is an equilibrium reaction and the transformation occurs essentially by mixing the reactants. In the transesterification of vegetable oils, a triglyceride reacts with an alcohol in the presence of a strong acid or base, producing a mixture of fatty acids alkyl esters and glycerol. The stoichiometric reaction requires 1 mol of a triglyceride and 3 mol of the alcohol. However, an excess of the alcohol is used to increase the yields of the alkyl esters and to allow their phase separation from the glycerol formed.

The base-catalyzed transesterification of vegetable oils proceeds faster than the acid-catalyzed reaction. The first step is the reaction of the base with the alcohol, producing an alkoxide and the protonated catalyst. The nucleophilic attack of the alkoxide at the carbonyl group of the triglyceride generates a tetrahedral intermediate, from which the alkyl ester and the corresponding anion of the diglyceride are formed. The latter deprotonates the catalyst, thus regenerating the active species, which is now able to react with a second molecule of the alcohol, starts another catalytic cycle. Diglycerides and monoglycerides are converted by the same mechanism to a mixture of alkyl esters and glycerol. Alkaline metal alkoxides (as CH_3ONa for the methanolysis) are the most active catalysts, since they give very high yields (>98%) in short reaction times (30 min) even if they are applied at low molar concentrations (0.5 mol%). However, they require the absence of water, which makes them inappropriate for typical industrial processes.

Alkaline metal hydroxides, such as KOH and NaOH, are cheaper than metal alkoxides, but less active. The presence of water gives rise to hydrolysis of some of the produced ester, with consequent soap formation. The undesirable saponification reaction reduces the ester yields and results in making the recovery of the glycerol very difficult due to the formation of emulsions. Potassium carbonate, used in a concentration of 2% or 3 mol% gives high yields of fatty acid alkyl esters and reduces the soap formation.

There are a number of detailed recipes for sodium methoxide-catalyzed transesterification. The methodology can be used on quite a large scale if need be. The reaction between sodium methoxide in methanol and a vegetable oil is very rapid. It has been shown that triglycerides can be completely transesterified in 2–5 minutes at room temperature. The methoxide anion prepared by dissolving the clean metals in anhydrous methanol. Sodium methoxide (0.5–2M) in methanol effects transesterification of triglycerides much more rapidly than other transesterification agents. At equivalent molar concentrations with the same triglyceride samples, potassium methoxide effects complete esterification more quickly than does sodium methoxide. Because of the dangers inherent in handling metallic potassium, which has a very high heat of reaction with methanol, it is preferred to use sodium methoxide in methanol. The reaction is generally slower with alcohols of higher molecular weight. As with acidic catalysis, inert solvents must be added to dissolve simple lipids before methanolysis will proceed.

4.6.5.1 Parameters Affecting the Formation of Methyl Esters

The parameters affecting the formation of methyl esters are reaction temperature, pressure, molar ratio, water content, and free fatty acid content. It is evident that at subcritical state of alcohol, the reaction rate is so low and gradually increases as either pressure or temperature rises. It has been observed that increasing the reaction temperature, especially to supercritical conditions, has a favorable influence on the yield of ester conversion. The yield of alkyl ester increases with increasing the molar ratio of oil to alcohol (Demirbas, 2002). In the supercritical alcohol transesterification method, the yield of conversion raises 50–95% for the first 10 minutes. Figure 4.20 shows the changes in yield percentage of ethyl esters as treated with subcritical and supercritical ethanol at different temperatures as a function of the reaction time.

Water content is an important factor in the conventional catalytic transesterification of vegetable oil. In the conventional transesterification of fats and vegetable oils for biodiesel production, free fatty acids and water always produce negative effects since the presence of free fatty acids and water causes soap formation, consumes the catalyst, and reduces catalyst effectiveness. In catalyzed methods, the presence of water has negative effects on the yields of methyl esters. However, the presence of water affects positively the formation of methyl esters in our supercritical methanol method.

In the supercritical alcohol transesterification method, the yield of conversion increases 50–95% during first 8 minutes. In the catalytic supercritical methanol

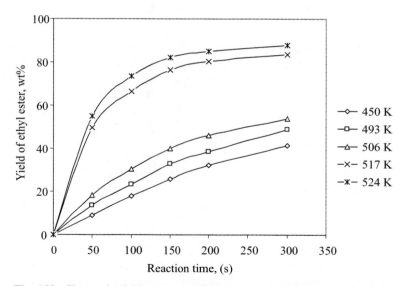

Fig. 4.20 Changes in yield percentage of ethyl esters as treated with subcritical and supercritical ethanol at different temperatures as a function of reaction time. Molar ratio of vegetable oil to ethyl alcohol: 1:40

transesterification method, the yield of conversion increases 60–90% during the first minute.

The transesterification can be carried out chemically or enzymatically. In recent work, three different lipases (*Chromobacterium viscosum, Candida rugosa,* and Porcine pancreas) were screened for a transesterification reaction of *Jatropha* oil in a solvent-free system to produce biodiesel; only lipase from *Chromobacterium viscosum* was found to give an appreciable yield (Shah *et al.*, 2004). Immobilization of lipase (*Chromobacterium viscosum*) on Celite-545 enhanced the biodiesel yield to 71% from 62% yield obtained by using free tuned enzyme preparation with a process time of 8 h at 113 K. Immobilized *Chromobacterium viscosum* lipase can be used for ethanolysis of oil. It was seen that immobilization of lipases and optimization of transesterification conditions resulted in adequate yield of biodiesel in the case of the enzyme-based process (Shah *et al.*, 2004).

Although the enzyme-catalyzed transesterification processes are not yet commercially developed, new results have been reported in recent articles and patents. The common aspects of these studies consist in optimizing the reaction conditions (solvent, temperature, pH, type of microorganism that generates the enzyme, *etc.*) in order to establish suitable characteristics for an industrial application. However, the reaction yields as well as the reaction times are still unfavorable compared to the base-catalyzed reaction systems. Due to their ready availability and the ease with which they can be handled, hydrolytic enzymes have been widely applied in organic synthesis.

4.6.6   Current Biodiesel Production Technologies

Transesterification is basically a sequential reaction. Triglycerides are first reduced to diglycerides during the transesterification. The diglycerides are subsequently reduced to monoglycerides. The monoglycerides are finally reduced to fatty acid esters. The order of the reaction changes with the reaction conditions. Transesterification is extremely important for biodiesel. Biodiesel as it is defined today is obtained by transesterifying the triglycerides with methanol. Methanol is the preferred alcohol for obtaining biodiesel because it is the cheapest alcohol. Base catalysts are more effective than acid catalysts and enzymes (Ma and Hanna, 1999). Methanol is made to react with the triglycerides to produce methyl esters (biodiesel) and glycerol (Eq. 4.23).

$$C_3H_5(OOCR)_3 \ + \ 3CH_3OH \ \rightarrow \ 3RCOOCH_3 \ + \ C_3H_5(OH)_3 \qquad (4.23)$$

Triglyceride   Methanol   Methyl ester   Glycerin

The production processes for biodiesel are well known. There are four basic routes to biodiesel production from oils and fats: (a) base catalyzed transesterification, (b) direct acid catalyzed transesterification, (c) conversion of the oil to its fatty acids and then to biodiesel, and (d) non-catalytic transesterification of oils and fats.

The basic catalyst is typically sodium hydroxide (caustic soda) or potassium hydroxide (caustic potash). It is dissolved in the alcohol using a standard agitator or mixer. The methyl alcohol and catalyst mix is then charged into a closed reactor and the oil or fat is added. The reaction mix is kept just above the boiling point of the alcohol (around 344 K) to speed up the reaction and the transesterification reaction takes place. Recommended reaction time varies from 1 to 8 hours, and the optimal reaction time is about 2 hours (Van Gerpen *et al.*, 2004). Excess alcohol is normally used to ensure total conversion of the fat or oil to its esters. After the reaction is complete, two major products form: glycerin and biodiesel. Each has a substantial amount of the excess methanol that was used in the reaction. The reacted mixture is sometimes neutralized at this step if needed. The glycerin phase is much denser than biodiesel phase, and the two can be gravity separated with glycerin simply drawn off the bottom of the settling vessel. In some cases, a centrifuge is used to separate the two materials faster. The biodiesel product is sometimes purified by washing gently with warm water to remove residual catalyst or soaps, dried, and sent to storage (Ma and Hanna, 1999; Demirbas, 2002). For an alkali-catalyzed transesterification, the triglycerides and alcohol must be substantially anhydrous because water makes the reaction partially change to saponification, which produces soap. The soap lowers the yield of esters and renders the separation of ester and glycerol and the water washing difficult. Low free fatty acid content in triglycerides is required for alkali-catalyzed transesterification. If more water and free fatty acids are in the triglycerides, acid catalyzed transesterification can be used.

When an alkali catalyst is present, the free fatty acid will react with alkali catalyst to form soap. It is common for oils and fats to contain small amounts of water. When water is present in the reaction it generally manifests itself through excessive soap production. The soaps of saturated fatty acids tend to solidify at ambient temperatures so a reaction mixture with excessive soap may gel and form a semi-solid mass that is very difficult to recover. When water is present, particularly at high temperatures, it can hydrolyze the triglycerides to diglycerides and form a free fatty acid.

If an oil or fat containing a free fatty acid such as oleic acid is used to produce biodiesel, the alkali catalyst typically used to encourage the reaction will react with this acid to form soap. This reaction is undesirable because it binds the catalyst into a form that does not contribute to accelerating the reaction. Excessive soap in the products can inhibit later processing of the biodiesel, including glycerol separation and water washing. Water in the oil or fat can also be problem (Van Gerpen *et al.*, 2004).

In some systems the biodiesel is distilled in an additional step to remove small amounts of color bodies to produce a colorless biodiesel. Once the glycerin and biodiesel phases have been separated, the excess alcohol in each phase is removed with a flash evaporation process or by distillation. The glycerin byproduct contains unused catalyst and soaps that are neutralized with sulfuric acid and sent to storage as crude glycerin. In most cases the salt is left in the glycerin. Water and alcohol are removed to produce 90% pure glycerin that is ready to be sold as crude glycerin. Before to use as a commercial fuel, the finished biodiesel must be analyzed using sophisticated analytical equipment to ensure it meets ASTM specifications.

The primary raw materials used in the production of biodiesel are vegetable oils, animal fats, and recycled greases. These materials contain triglycerides, free fatty acids, and other impurities. The primary alcohol used to form the ester is the other major feedstock. Most processes for making biodiesel use a catalyst to initiate the esterification reaction. The catalyst is required because the alcohol is sparingly soluble in the oil phase. The catalyst promotes an increase in solubility to allow the reaction to proceed at a reasonable rate. The most common catalysts used are strong mineral bases such as sodium hydroxide and potassium hydroxide. After the reaction, the base catalyst must be neutralized with a strong mineral acid. Figure 4.21 shows the simplified flow diagram of base catalyzed biodiesel processing.

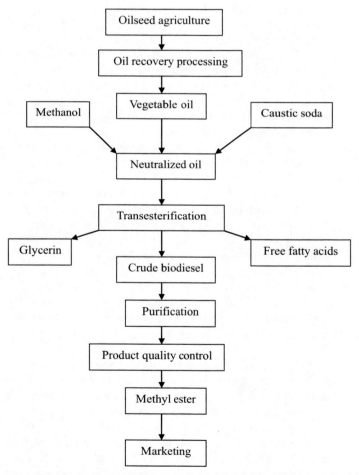

Fig. 4.21   Simplified flow diagram of base catalyzed biodiesel processing

Table 4.30 shows the typical proportions for the chemicals used to make biodiesel. The quantitized transesterification reaction of triolein obtained for methyl oleate (biodiesel) is given in Eq. 4.24.

Triolein + 6 Methanol → Methyl oleate + Glycerol + 3 Methanol
885.46 g   192.24 g   (Catalyst) 889.50 g   92.10 g   96.12 g

$$(4.24)$$

The most important aspects of biodiesel production to ensure trouble free operation in diesel engines are:

- Complete transesterification reaction
- Removal of glycerol
- Removal of catalyst
- Removal of alcohol
- Removal of free fatty acids.

These parameters are all specified through the biodiesel standard, ASTM D 6751. This standard identifies the parameters the pure biodiesel (B100) must meet before being used as a pure fuel or being blended with petroleum-based diesel fuel. Biodiesel, B100, specifications (ASTM D 6751–02 requirements) are tabulated in Table 4.31.

Table 4.30 Typical proportions of the chemicals used to make biodiesel

Reactants	Amount (kg)
Fat or oil	100
Primary alcohol (methanol)	10
Catalyst (sodium hydroxide)	0.30
Neutralizer (sulfuric acid)	0.36

Table 4.31 Biodiesel, B100, specifications (ASTM D 6751-02 requirements)

Property	Method	Limits	Units
Flash point	D 93	130 min	K
Water and sediment	D 2709	0.050 max	% volume
Kinematic viscosity at 313 K	D 445	1.9–6.0	mm2/s
Sulfated ash	D 874	0.020 max	wt%
Total sulfur	D 5453	0.05 max	wt%
Copper strip corrosion	D 130	No. 3 max	–
Cetane number	D 613	47 min	–
Cloud point	D 2500	Report	K
Carbon residue	D 4530	0.050 max	wt%
Acid number	D 664	0.80 max	mg KOH/g
Free glycerin	D 6584	0.020	wt%
Total glycerin	D 6584	0.240	wt%
Phosphorus	D 4951	0.0010	wt%
Vacuum distillation end point	D 1160	633 K max, at 90% distilled	K

Table 4.32 International standard (EN 14214) requirements for biodiesel

Property	Units	Lower limit	Upper limit	Test-Method
Ester content	% (m/m)	96.5	–	Pr EN 14103d
Density at 288 K	kg/m³	860	900	EN ISO 3675/ EN ISO 12185
Viscosity at 313 K	mm²/s	3.5	5.0	EN ISO 3104
Flash point	K	>374	–	ISO CD 3679e
Sulfur content	mg/kg	–	10	–
Tar remnant (at 10% distillation remnant)	% (m/m)	–	0.3	EN ISO 10370
Cetane number	–	51.0	–	EN ISO 5165
Sulfated ash content	% (m/m)	–	0.02	ISO 3987
Water content	mg/kg	–	500	EN ISO 12937
Total contamination	mg/kg	–	24	EN 12662
Copper band corrosion (3 hours at 323 K)	rating	Class 1	Class 1	EN ISO 2160
Oxidation stability, 383 K	hours	6	–	pr EN 14112k
Acid value	mg KOH/g	–	0.5	pr EN 14104
Iodine value	–	–	120	pr EN 14111
Linoleic acid methyl ester	% (m/m)	–	12	pr EN 14103d
Polyunsaturated (>= 4 double bonds) methylester	% (m/m)	–	1	–
Methanol content	% (m/m)	–	0.2	pr EN 141101
Monoglyceride content	% (m/m)	–	0.8	pr EN 14105m
Diglyceride content	% (m/m)	–	0.2	pr EN 14105m
Triglyceride content	% (m/m)	–	0.2	pr EN 14105m
Free glycerine	% (m/m)	–	0.02	pr EN 14105m/ pr EN 14106
Total glycerine	% (m/m)	–	0.25	pr EN 14105m
Alkali metals (Na+K)	mg/kg	–	5	pr EN 14108/ pr EN 14109
Phosphorus content	mg/kg	–	10	pr EN14107p

EN 14214 is an international standard that describes the minimum requirements for biodiesel that has been produced from rapeseed fuel stock (also known as rapeseed methyl esters). Table 4.32 shows international standard (EN 14214) requirements for biodiesel.

4.6.7   Biodiesel Production Processes

The main factors affecting transesterification are the molar ratio of glycerides to alcohol, catalyst, reaction temperature and pressure, reaction time, and the contents of free fatty acids and water in oils. The choice of oils or fats to be used in producing biodiesel is an important aspect of economic decision. The cost of oils or fats directly affects on the cost of biodiesel cost by 70–80%. Crude vegetable oils contain some free fatty acids and phospholipids. The phospholipids are removed in a *degumming* step, and the free fatty acids are removed in a *refining* step. Excess free fatty acids can be removed as soaps in a later transesterification or caustic stripping step.

The most desirable vegetable oils sources are soybean, canola, palm, and rape. Main animal fat sources are beef tallow, lard, poultry fat, and fish oils. Yellow greases can be mixtures of vegetable and animal sources. The free fatty acid content affects the type of biodiesel process used, and the yield of fuel from that process. Other contaminants present can affect the extent of feedstock preparation necessary to use a given reaction chemistry.

From the viewpoint of chemical reaction, refined vegetable oil is the best starting material to produce biodiesel because the conversion of pure triglyceride to fatty acid methyl ester is high, and the reaction time is relatively short. Nevertheless, waste cooking oil, if no suitable treatment is available, would be discharged and cause environmental pollution, but waste cooking oil can be collected for further purification and then biodiesel processing. This collected material is a good commercial choice to produce biodiesel due to its low cost (Zhang *et al.*, 2003).

The most commonly used primary alcohol used in biodiesel production is methanol, although other alcohols, such as ethanol, isopropanol, and butyl, can be used. A key quality factor for the primary alcohol is the water content. Water interferes with transesterification reactions and can result in poor yields and high levels of soap, free fatty acids, and triglycerides in the final fuel.

The stoichiometric ratio for transesterification reaction requires three moles of alcohol and one mole of triglyceride to yield three moles of fatty acid ester and one mole of glycerol. Higher molar ratios result in greater ester production in a shorter time. The commonly accepted molar ratios of alcohol to vegetable oils are 3:1–6:1. Excess methanol (such as the 21:1 ratio) is generally necessary in batch reactors where water accumulates. Another method is to approach the reaction in two stages: fresh methanol and sulfuric acid is reacted, removed, and replaced with fresher reactant. Much of the water is removed in the first round and the fresh reactant in the second round drives the reaction closer to completion. The reason for using extra alcohol is that it drives the reaction closer to the 99.7% yield needed to meet the total glycerol standard for fuel grade biodiesel. The unused alcohol must be recovered and recycled back into the process to minimize operating costs and environmental impacts (Van Gerpen *et al.*, 2004).

The most commonly used catalyst materials for converting triglycerides to biodiesel are sodium hydroxide, potassium hydroxide, and sodium methoxide. Most base catalyst systems use vegetable oils as a feedstock. The base catalysts are highly hygroscopic and they form chemical water when dissolved in the alcohol reactant. They also absorb water from the air during storage.

Acid catalysts can be used for transesterification, however, they are generally considered to be too slow for industrial processing. Acid catalysts are more commonly used for the directly esterification of free fatty acids. Acid catalysts include sulfuric acid and phosphoric acid.

There is continuing interest in using lipases as enzymatic catalysts for the production of alkyl fatty acid esters. Some enzymes work on the triglycerides, converting them to methyl esters; and some work on the fatty acids.

Neutralizers are used to remove the base or acid catalyst from the product biodiesel and glycerol. If a base catalyst is used, the neutralizer is typically an acid, and *visa versa*. If the biodiesel is being washed, the neutralizer can be added to the wash water. While hydrochloric acid is a common choice to neutralize base catalysts, as mentioned earlier, if phosphoric acid is used, the resulting salt has value as a chemical fertilizer.

4.6.7.1 Biodiesel Production with Batch Processing

The simplest method for producing alcohol esters is to use a batch, stirred tank reactor. Alcohol to triglyceride ratios from 4:1–20:1 (mole:mole) have been reported, with a 6:1 ratio being most common. The reactor may be sealed or equipped with a reflux condenser. The operating temperature is usually about 340 K, although temperatures from 298–358 K have been reported (Ma and Hanna, 1999; Demirbas, 2002; Bala, 2005). The most commonly used catalyst is sodium hydroxide, with potassium hydroxide also being used. Typical catalyst loadings range from 0.3% to about 1.5%. Completions of transesterification of 85–95% have been reported. Higher temperatures and higher alcohol:oil ratios also can enhance the percent completion. Typical reaction times range from 20 minutes to more than 1 hour.

The oil is first charged to the system, followed by the catalyst and methanol. The system is agitated during the reaction time. Then agitation is stopped. In some processes, the reaction mixture is allowed to settle in the reactor to give an initial separation of the esters and glycerol. In other processes the reaction mixture is pumped into a settling vessel, or is separated using a centrifuge (Van Gerpen *et al.*, 2004).

The alcohol is removed from both the glycerol and ester stream using an evaporator or a flash unit. The esters are neutralized, washed gently using warm, slightly acid water to remove residual methanol and salts, and then dried. The finished biodiesel is then transferred to storage. The glycerol stream is neutralized and washed with soft water. The glycerol is than sent to the glycerol refining unit.

High free fatty acid feedstocks will react with the catalyst and form soaps if they are fed to a base catalyzed system. The maximum amount of free fatty acids acceptable in a base catalyzed system is less than 2%, and preferably less than 1%.

The Lurgi process is shown as a two-step reactor. Most of the glycerin is recovered after the first stage where a rectifying column leads to separation of the excess methanol and crude glycerin. The methyl ester output of the second stage is purified to some extent of residual glycerin and methanol by a wash column. Table 4.33 shows the inputs and mass requirements for the Lurgi process.

Table 4.33 Inputs and mass requirements for the Lurgi process

Input	Requirement/ton biodiesel
Feedstock	1,000 kg vegetable oil
Steam requirement	415 kg
Electricity	12 kWh
Methanol	96 kg
Catalyst	5 kg
Hydrochloric acid (37%)	10 kg
Caustic soda (50%)	1.5 kg
Nitrogen	1 Nm^3
Process water	20 kg

4.6.7.2 Biodiesel Production with Continuous Processes

Transesterification can be conventionally performed using alkaline, acid, or en-zyme catalysts. As alkali-catalyzed systems are very sensitive to both water and free fatty acids contents, the glycerides and alcohol must be substantially anhy-drous because water makes the reaction partially change to saponification, which produces soaps, thus consuming the catalyst and reducing the catalytic efficiency, as well as causing an increase in viscosity, formation of gels, and difficulty in separations (Ma and Hana, 1999; Zhang *et al.*, 2003).

There are several processes that use intense mixing, either from pumps or mo-tionless mixers, to initiate the esterification reaction. A popular variation of the batch process is the use of continuous stirred tank reactors in series. Instead of allowing time for the reaction in an agitated tank, the reactor is tubular. The reac-tion mixture moves through this type of reactor in a continuous plug, with little mixing in the axial direction. The result is a continuous system that requires rather short residence times, as low as 6–10 min, for near completion of the reaction.

4.6.7.3 Biodiesel Production with Non-catalyzed Transesterification

There are two non-catalyzed transesterification processes. These are BIOX co-solvent process and supercritical methanol process.

The BIOX (co-)process is a new Canadian process developed originally by Pro-fessor David Boocock of the University of Toronto, which has attracted consider-able attention. Dr. Boocock has transformed the production process through the selection of inert co-s that generate an oil-rich one-phase system. This reaction is over 99% complete in seconds at ambient temperatures, compared to previous processes that required several hours. BIOX is a technology development company that is a joint venture of the University of Toronto Innovations Foundation and Madison Ventures Ltd. BIOX's patented production process converts first the free fatty acids (by way of acid esterification) up to 10% FFA content and then the triglycerides (by way of transesterification), through the addition of a co-solvent, in a two-step, single phase, continuous process at atmospheric pressures and near-ambient temperatures. The co-solvent is then recycled and reused continuously in the process. The unique feature of the BIOX process is that it uses inert reclaimable co-s in a single-pass reaction taking only seconds at ambient temperature and pressure. The developers are aiming to produce biodiesel that is cost competitive with petrodiesel. The BIOX process handles not only grain-based feedstocks but also waste cooking greases and animal fats (Van Gerpen *et al.*, 2004). The BIOX process uses a co-solvent, tetrahydrofuran, to solubilize the methanol. Co-options are designed to overcome slow reaction times caused by the extremely low solubility of the alcohol in the triglyceride phase. The result is a fast reaction, on the order of 5–10 min, and no catalyst residues in either the ester or the glycerol phase.

In the conventional transesterification of animal fats and vegetable oils for biodiesel production, free fatty acids and water always produce negative effects since the presence of free fatty acids and water causes soap formation, consumes the catalyst, and reduces catalyst effectiveness, all of which results in a low conversion. The transesterification reaction may be carried out using either basic or acidic catalysts, but these processes require relatively time-consuming and complicated separation of the product and the catalyst, which results in high production costs and energy consumption. To overcome these problems, Kusdiana and Saka (2001) and Demirbas (2002) have proposed that biodiesel fuels may be prepared from vegetable oil *via* non-catalytic transesterification with supercritical methanol (SCM). A novel process of biodiesel fuel production has been developed by a non-catalytic supercritical methanol method. Supercritical methanol is believed to solve the problems associated with the two-phase nature of normal methanol/oil mixtures by forming a single phase as a result of the lower value of the dielectric constant of methanol in the supercritical state. As a result, the reaction was found to be complete in a very short time. Compared with the catalytic processes under barometric pressure, the supercritical methanol process is non-catalytic, involves a much simpler purification of products, has a lower reaction time, is more environmentally friendly, and requires lower energy use. However, the reaction requires temperatures of 525–675 K and pressures of 35–60 MPa.

The dynamic transesterification reaction of peanut oil in supercritical methanol media has been investigated. The reaction temperature and pressure were in the range of 523–583 K and 10.0–16.0 MPa, respectively. The molar ratio of peanut oil to methanol was 1:30. It was found that the yield of methyl esters was higher than 90% under the supercritical methanol. The apparent reaction order and activation energy of transesterification was 1.5 and 7.472 kJ/mol, respectively. In this method, the reaction time was shorter and the processing was simpler than that of the common acid catalysis transesterification.

Figure 4.22 shows the two-stage continuous biodiesel production process with subcritical water and supercritical methanol. The oil and the alcohol stream are fed into the supercritical reactor at 1 atm and 298 K. In the first stage triglycerides rapidly hydrolyze to free fatty acids under 10 MPa pressure at 445 K temperature in the supercritical reactor 1. Methanol become supercritical at a pressure of 10 pa and

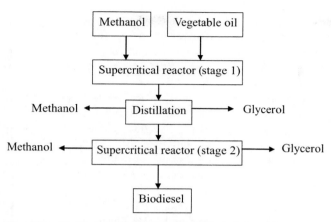

Fig. 4.22 Biodiesel production by supercritical methanol in two stages

a temperature of 445 K, and the supercritical conditions favor rapid formation of methyl esters from the free fatty acids. After the supercritical reactor 1, a distillation column is required to separate the methanol, which will be recycled from the rest. This is done to have a better global separation in the following decanter. When the pressure is reduced the mixture promptly separates into two phases and the water phase can be separated to recover glycerol. In the last equipment, the glycerin is separated from the oil phase; this last one is fed to supercritical reactor 2.

Catalytic supercritical methanol transesterification is carried out in an autoclave in the presence of 1 to 5% NaOH, CaO, and MgO as catalyst at 520 K. In the catalytic supercritical methanol transesterification method, the yield of conversion rises to 60–90% for the first minute. The transesterification reaction of the crude oil of rapeseed with supercritical and subcritical methanol in the presence of a relatively low amount (1%) of NaOH was successfully carried out, where soap formation did not occur.

4.6.8 Basic Plant Equipment Used in Biodiesel Production

The basic plant equipment used in a biodiesel production are reactors, pumps, settling tanks, centrifuges, distillation columns, and storage tanks. The reactor is the only place in the process where chemical conversion occurs. Reactors can be placed into two broad categories, batch reactors and continuous reactors. In the batch reactor, the reactants are charged into the reactor in the determined amount. The reactor is then closed and taken to the desired reaction conditions. The chemical composition within the reactor changes with time. The materials of construction are an important consideration for the reactor and storage tanks. For the base-catalyzed transesterification reaction, stainless steel is the preferred material for the reactor (Van Gerpen *et al.*, 2004).

Key reactor variables that dictate conversion and selectivity are temperature, pressure, reaction time (residence time), and degree of mixing. In general, increasing the reaction temperature increases the reaction rate and, hence, the conversion for a given reaction time. Increasing temperature in the transesterification reaction does impact the operating pressure.

Two reactors within the continuous reactor category are continuous stirred tank reactors (CSTRs) and plug flow reactors (PFRs). For CSTRs, the reactants are fed into a well-mixed reactor. The composition of the product stream is identical to the composition within the reactor. The hold-up time in a CSTR is given by a residence time distribution. For PFRs, the reactants are fed into one side of the reactor. The chemical composition changes as the material moves in plug flow through the reactor (Van Gerpen *et al.*, 2004).

The pumps play the key role in moving chemicals through the manufacturing plant. The most common type of pump in the chemical industry is a centrifugal pump. The primary components of a centrifugal pump are (1) a shaft, (2) a coupling attaching the shaft to a motor, (3) bearings to support the shaft, (4) a seal around the shaft to prevent leakage, (5) an impeller, and (6) a volute, which converts the kinetic energy imparted by the impeller into feet of head. Gear pumps are generally used in biodiesel plants. There are a number of different types of positive displacement pumps, including gear pumps (external and internal) and lobe pumps. External gear pumps generally have two gears with an equal number of teeth located on the outside of the gears, whereas, internal gear pumps have one larger gear with internal teeth and a smaller gear with external teeth (Van Gerpen *et al.*, 2004).

The separation of biodiesel and glycerin can be achieved using a settling tank. While a settling tank may be cheaper, a centrifuge can be used to increase the rate of separation relative to a settling tank. Centrifuges are most typically used to separate solids and liquids, but they can also be used to separate immiscible liquids of different densities. In a centrifuge the separation is accomplished by exposing the mixture to a centrifugal force. The denser phase will be preferentially separated to the outer surface of the centrifuge. The choice of appropriate centrifuge type and size are predicated on the degree of separation needed in a specific system.

An important separation device for miscible fluids with similar boiling points (*e.g.*, methanol and water) is the distillation column. Separation in a distillation column is predicated on the difference in volatilities (boiling points) between chemicals in a liquid mixture. In a distillation column the concentrations of the more volatile species are enriched above the feed point, and the less volatile species are enriched below the feed point.

4.6.9   Fuel Properties of Biodiesels

The fuel properties of biodiesels are characterized by determining their viscosity, density, cetane number, cloud and pour points, distillation range, flash point, ash content, sulfur content, carbon residue, acid value, copper corrosion, and higher heating value. The most important variables affecting the ester yield during the transesterification reaction are the molar ratio of alcohol to vegetable oil and the reaction temperature. The viscosity values of vegetable oil methyl esters greatly decreases after transesterification process. Compared to No. 2 diesel fuel, all vegetable oil methyl esters are slightly viscous. The flash point values of vegetable oil methyl esters are much lower than those of vegetable oils. The flash point values of vegetable oil methyl esters are much lower than those of vegetable oils. There is a large regression between density and viscosity values vegetable oil methyl esters. The relationships between viscosity and flash point for vegetable oil methyl esters are very regular.

The parameters affecting the formation of methyl esters are reaction temperature, pressure, molar ratio, water content, and free fatty acid content. It is evident that at the subcritical state of alcohol, the reaction rate is very low and gradually increased as either pressure or temperature rises. It has been observed that increasing the reaction temperature, especially to supercritical conditions, has a favorable influence on the yield of ester conversion. The yield of alkyl ester increased with increasing the molar ratio of oil to alcohol.

4.6.9.1 Physical Properties of Biodiesel Fuels

The physical properties and some fuel properties of biodiesel are close to those of diesel fuels. Biodiesel has been characterized by determining its viscosity, density, cetane number, cloud and pour points, characteristics of distillation, flash and combustion points, and higher heating value (HHV) according to ISO norms.

Viscosity is the most important property of biodiesel since it affects the operation of the fuel injection equipment, particularly at low temperatures when the increase in viscosity affects the fluidity of the fuel. High viscosity leads to poorer atomization of the fuel spray and less accurate operation of the fuel injectors. The lower the viscosity of the biodiesel, the easier it is to pump and atomize and achieve finer droplets. The conversion of triglycerides into methyl or ethyl esters through the transesterification process reduces the molecular weight to one-third that of the triglyceride reduces the viscosity by a factor of about 8. Viscosities show the same trends as temperatures, with lard and tallow biodiesels having higher viscosities than soybean and rapeseed biodiesels. Biodiesel has a viscosity close to that of diesel fuels. As the temperature of oil is increase, its viscosity decreases. Table 4.34 shows some fuel properties of six methyl ester biodiesels given by different researchers.

Table 4.35 shows viscosity, density and flash point measurements of nine oil methyl esters. The viscosity, density and flash point values of methyl esters considerably decrease *via* the transesterification process.

Vegetable oils can be used as fuel for combustion engines, but their viscosity is much higher than usual diesel fuel and requires modifications of the engines. The major problem associated with the use of pure vegetable oils as fuels for diesel

Table 4.34 Some fuel properties of six methyl ester biodiesels

Source	Viscosity cSt at 40°C	Density g/mL at 15.5°C	Cetane Number
Sunflower	4.6	0.880	49
Soybean	4.1	0.884	46
Palm	5.7	0.880	62
Peanut	4.9	0.876	54
Babassu	3.6	–	63
Tallow	4.1	0.877	58

Source: Demirbas, 2008

Table 4.35 Viscosity, density and flash point measurements of nine oil methyl esters

Methyl ester	Viscosity mm^2/s (at 313 K)	Density kg/m^3 (at 288 K)	Flash point K
Cottonseed oil	3.75	870	433
Hazelnut kernel oil	3.59	860	422
Linseed oil	3.40	887	447
Mustard oil	4.10	885	441
Palm oil	3.94	880	431
Rapeseed oil	4.60	894	453
Safflower oil	4.03	880	440
Soybean oil	4.08	885	441
Sunflower oil	4.16	880	439

Source: Demirbas, 2008

engines is caused by high fuel viscosity in compression ignition. Therefore, vegetable oils are converted into their methyl esters (biodiesel) by transesterification. The viscosity values of vegetable oils are between 27.2 mm^2/s and 53.6 mm^2/s, whereas those of vegetable oil methyl esters are between 3.6 mm^2/s and 4.6 mm^2/s. The viscosity values of vegetable oil methyl esters highly decreases after transesterification process. The viscosity of No. 2 diesel fuel is 2.7 mm^2/s at 313 K (Demirbas, 2003a). Compared to No. 2 diesel fuel all of the vegetable oil methyl esters are slightly viscous.

The flash point values of vegetable oil methyl esters are much lower than those of vegetable oils. The flash point values of vegetable oil methyl esters are much lower than those of vegetable oils. An increase in density from 860 kg/m^3 to 885 kg/m^3 for vegetable oil methyl esters or biodiesels increases the viscosity from 3.59 mm^2/s to 4.63 mm^2/s and the increases are highly regular. There is high regression between density and viscosity values vegetable oil methyl esters. The relationships between viscosity and flash point for vegetable oil methyl esters are irregular.

The cetane number (CN) is based on two compounds, namely hexadecane with a cetane of 100 and heptamethylnonane with a cetane of 15. CN is a measure of ignition quality of diesel fuels and high CN implies short ignition delay. The CN of CSO samples were the range of 41–44.0. The CN of biodiesel is generally higher than conventional diesel. The longer the fatty acid carbon chains and the more saturated the molecules, the higher the CN. The CN of biodiesel from animal fats is higher than those of vegetable oils (Bala, 2005).

Two important parameters for low temperature applications of a fuel are the cloud point (CP) and the pour point (PP). CP is the temperature at which wax first becomes visible when the fuel is cooled. PP is the temperature at which the amount of wax out of solution is sufficient to gel the fuel, thus it is the lowest temperature at which the fuel can flow. Biodiesel has higher CP and PP compared to conventional diesel.

4.6.9.2 Higher Combustion Efficiency of Biodiesel

Biodiesel is an oxygenated fuel. Oxygen in the biodiesel structure improves its combustion process and decreases its oxidation potential. The structural oxygen content of a fuel improves combustion efficiency due to an increase in the homogeneity of oxygen with the fuel during combustion. Because of this the combustion efficiency of biodiesel is higher than petrodiesel, and the combustion efficiency methanol/ethanol is higher than that of gasoline. A visual inspection of the injector types would indicate no difference between the biodiesel fuels when tested on petrodiesel. The overall injector coking is considerably low. Biodiesel contains 11% oxygen by weight and contains no sulfur. The use of biodiesel can extend the life of diesel engines because it is more lubricating than petroleum diesel fuel. Biodiesel has better lubricant properties than petrodiesel.

The higher heating values of biodiesels are relatively high. The higher heating values (HHVs) of biodiesels (39–41 MJ/kg) are slightly lower than those of gasoline (46 MJ/kg), petrodiesel (43 MJ/kg), or petroleum (42 MJ/kg), but higher than coal (32–37 MJ/kg). Table 4.36 shows a comparison of chemical properties and the HHVs between biodiesel and petrodiesel fuels.

Table 4.36 Comparison of chemical properties and higher heating values (HHVs) between biodiesel and No. 2 diesel fuels

Chemical property	Biodiesel (methyl ester)	No. 2 diesel fuel
Ash (wt%)	0.002–0.036	0.006–0.010
Sulfur (wt%)	0.006–0.020	0.020–0.050
Nitrogen (wt%)	0.002–0.007	0.0001–0.003
Aromatics (vol%)	0	28–38
Iodine number	65–156	0
HHV (MJ/kg)	39.2–40.6	45.1–45.6

Source: Demirbas, 2008

4.6.9.3 Water Content

Water content is an important factor in the conventional catalytic transesterification of vegetable oil. In the conventional transesterification of fats and vegetable oils for biodiesel production, free fatty acids and water always produce negative effects since the presence of free fatty acids and water causes soap formation, consumes the catalyst, and reduces catalyst effectiveness. Soap can prevent the separation of biodiesel from glycerol fraction (Madras *et al.*, 2004). In catalyzed methods, the presence of water has negative effects on the yields of methyl esters. However, the presence of water positively affects the formation of methyl esters in the supercritical methanol method (Kusdiana and Saka, 2004).

4.6.9.4 Comparison of the Fuel Properties of Methyl and Ethyl Alcohols and Their Esters

The main goals of the alcohol studies were to better understand the fuel properties of alcohol and basic principles of conversion in order to provide the representative cross-section of converting diesel engines and gasoline engines to accepting blended fuel.

The alcohols are oxygenated fuels where the alcohol molecule has one or more oxygen atoms, which decreases to the combustion heat. Practically, any of the organic molecules of the alcohol family can be used as a fuel. The alcohols that can be used for motor fuels are methanol, ethanol, propanol, and butanol. However, only two of the alcohols are technically and economically suitable as fuels for internal combustion engines (ICEs). The main production facilities of methanol and ethanol are tabulated in Table 4.37. Methanol is produced by a variety of processes, the most common being the following: distillation of liquid products from wood and coal, natural gas, and petroleum gas. Ethanol is produced mainly from biomass bioconversion (Bala, 2005).

In general, the physical and chemical properties and the performance of ethyl esters are comparable to those of methyl esters. Methyl and ethyl esters have almost the same heat content. The viscosities of ethyl esters are slightly higher and the cloud and pour points are slightly lower than those of methyl esters. Engine tests demonstrated that methyl esters produce slightly higher power and torque than ethyl esters (Encinar *et al.*, 2002). Some desirable attributes of the ethyl esters over methyl esters are: significantly lower smoke opacity, lower exhaust temperatures, and lower pour point. Ethyl esters tend to have more injector coking than methyl esters.

Many alcohols have been used to make biodiesel. Issues such as the cost of the alcohol, the amount of alcohol needed for the reaction, the ease of recovering and recycling the alcohol, fuel tax credits, and global warming issues influence the choice of alcohol. Some alcohols also require slight technical modifications to the production process such as higher operating temperatures, longer or slower mixing times, or lower mixing speeds. Since the reaction to form the esters is on a molar

Table 4.37 Main production facilities of methanol and ethanol for biodiesel production

Product	Production process
Methanol	
	Distillation of liquid from wood pyrolysis
	Gaseous products from biomass gasification
	Distillation of liquid from coal pyrolysis
	Synthetic gas from biomass and coal
	Natural gas
	Petroleum gas
Ethanol	
	Fermentation of sugars and starches
	Bioconversion of cellulosic biomass
	Hydration of alkanes
	Synthesis from petroleum
	Synthesis from coal
	Enzymatic conversion of synthetic gas

Source: Demirbas, 2008

basis, and alcohol is purchased on a volume basis, their properties make a significant difference in raw material price. It takes three moles of alcohol to react completely with one mole of triglyceride.

Ethanol is produced a more environmentally benign fuel. The systematic effect of ethyl alcohol differs from that of methyl alcohol. Ethyl alcohol is rapidly oxidized in the body to carbon dioxide and water, and in contrast to methyl alcohol no cumulative effect occurs. Ethanol is also a preferred alcohol in the transesterification process compared to methanol because it is derived from agricultural products and is renewable and biologically less objectionable in the environment. Table 4.37 shows the main production facilities of methanol and ethanol for biodiesel production.

Methanol use in current-technology vehicles has some distinct advantages and disadvantages. On the plus side, methanol has a higher octane rating than gasoline. Methanol vaporizes at such a temperature that it results in lower peak flame temperatures than gasoline and lower nitrogen oxide emissions. Its greater tolerance to lean combustion, higher air-to-fuel equivalence ratio results in generally lower overall emissions and higher energy efficiency. However, several disadvantages must be studied and overcome before neat methanol is considered a viable alternative to gasoline. The energy density of methanol is about half that of gasoline, reducing the range a vehicle can travel on an equivalent tank of fuel.

There are some important differences in the combustion characteristics of alcohols and hydrocarbons. Alcohols have higher flame speeds and extended flammability limits. Pure methanol is very flammable, and its flame is colorless when ignited. The alcohols mix in all proportions with water due to the polar nature of the OH group. Low volatility is indicated by a high boiling point and a high flash point. Combustion of alcohol in the presence of air can be initiated by an intensive source of localized energy, such as a flame or a spark and, also, the mixture can be ignited by application of energy by means of heat and pressure, such as happens in the compression stroke of a piston engine. The high latent heat of vaporization of alcohols cools the air entering the combustion chamber of the engine, thereby increasing the air density and mass flow. This leads to increased volumetric efficiency and reduced compression temperatures. The oxygen contents of alcohols depress the heating value of the fuel in comparison with hydrocarbon fuels. The heat of combustion *per* unit volume of alcohol is approximately half that of isooctane.

Methanol is not miscible with hydrocarbons and separation ensues readily in the presence of small quantities of water, particularly with reduction in temperature. On the other hand, anhydrous ethanol is completely miscible in all proportions with gasoline, although separation may be effected by water addition or by cooling. If water is already present, the water tolerance is higher for ethanol than for methanol, and can be improved by the addition of higher alcohols, such as butanol. Also benzene or acetone can be used. The wear problem is believed to be caused by formic acid attack, when methanol is used or acetic acid attack when ethanol is used.

Methanol is considerably easier to recover than ethanol. Ethanol forms an azeotrope with water, so it is expensive to purify the ethanol during recovery. If the water is not removed it will interfere with the reactions. Methanol recycles more easily because it does not form an azeotrope. These two factors are the reason that even though methanol is more toxic, it is the preferred alcohol for producing biodiesel. Methanol has a flash point of 283 K, while the flash point of ethanol is 281 K, so both are considered highly flammable. One should never let methanol come into contact with skin or eyes, as it can be readily absorbed. Excessive exposure to methanol can cause blindness and other detrimental health effects.

Dry methanol is very corrosive to some aluminum alloys, but additional water at 1% almost completely inhibits corrosion. It must be noted that methanol with additional water at more than 2% becomes corrosive again. Ethanol always contains some acetic acid and is particularly corrosive to aluminum alloys.

Since alcohols, especially methanol, can be readily ignited by hot surfaces, preignition can occur. It must be emphasized here that preignition and knocking in alcohol using engines is a much more dangerous condition than gasoline engines. Other properties, however, are favorable to the increase of power and reduction of fuel consumption. Such properties are as follows: (1) the number of molecules or products is more than that of reactants; (2) extended limits of flammability; (3) high octane number; (4) high latent heat of vaporization; (5) constant boiling temperature; and (6) high density.

Figure 4.23 shows the distillation curves of the No 2. diesel fuel and linseed oil methyl and ethyl esters. As seen in Fig. 4.23, the most volatile fuel was No 2. diesel fuel. The volatility of methyl esters was higher than that of ethyl ester at all temperatures.

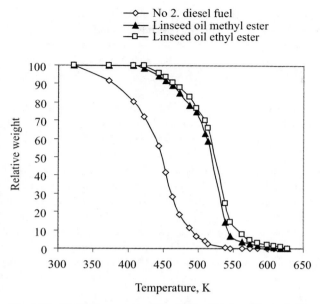

Fig. 4.23 Distillation curves of No 2. diesel fuel and linseed oil methyl and ethyl esters

4.6.10 Advantages of Biodiesels

The advantages of biodiesel as diesel fuel are liquid nature-portability, ready availability, renewability, higher combustion efficiency, lower sulfur and aromatic content (Ma and Hanna, 1999; Knothe *et al.*, 2006), higher cetane number, and higher biodegradability (Zhang *et al.*, 2003). The main advantages of biodiesel given in the literature include domestic origin, reducing the dependency on imported petroleum, biodegradability, high flash point, and inherent lubricity in the neat form (Knothe *et al.*, 2005).

Biodiesel is the only alternative fuel for unmodified conventional diesel engines using low concentration biodiesel-diesel blends. It can be stored anywhere that petroleum diesel fuel is stored. Biodiesel can be made from domestically produced, renewable oilseed crops such as soybean, rapeseed, and sunflower. The risks of handling, transporting, and storing biodiesel are much lower than those associated with petrodiesel. Biodiesel is safe to handle and transport because it is as biodegradable as sugar and has a high flashpoint compared to petroleum diesel fuel. Biodiesel can be used alone or mixed in any ratio with petroleum diesel fuel. The most common blend is a mix of 20% biodiesel with 80% petroleum diesel, or B20 according to recent scientific investigations; however, in Europe the current regulation foresees a maximum 5.75% biodiesel.

Combustion of biodiesel alone provides over a 90% reduction in total unburned hydrocarbons, and a 75–90% reduction in polycyclic aromatic hydrocarbons (PAHs). Biodiesel further provides significant reductions in particulates and carbon monoxide compared to petroleum diesel fuel. Biodiesel provides a slight increase or decrease in nitrogen oxides depending on the engine family and testing procedures.

Fuel characterization data show some similarities and differences between biodiesel and petrodiesel fuels. The sulfur content of petrodiesel is 20–50 times that of biodiesels. Several municipalities are considering mandating the use of low levels of biodiesel in diesel fuel on the basis of several studies that have found hydrocarbon (HC) and particulate matter (PM) benefits from the use of biodiesel.

The use of biodiesel to reduce N_2O is attractive for several reasons. Biodiesel contains little nitrogen, as compared with petrodiesel, which is also used as a reburning fuel. The N_2O reduction was strongly dependent on initial N_2O concentration and only slightly dependent upon temperature, where increased temperature increased N_2O reduction. This results in lower N_2O production from fuel nitrogen species for biodiesel. In addition, biodiesel contains virtually trace amounts of sulfur, so SO_2 emissions are reduced in direct proportion to the petrodiesel replacement.

Biodiesel has demonstrated a number of promising characteristics, including the reduction of exhaust emissions. Vegetable oil fuels have not been acceptable because they are more expensive than petroleum fuels. With recent increases in petroleum prices and uncertainties concerning petroleum availability, there is renewed interest in vegetable oil fuels for compression ignition (CIE or diesel) engines. Alternative fuels for CIE have become increasingly important due to increased environmental concerns, and several socio-economic aspects. In this sense, vegetable oils and animal fats represent a promising alternative to conventional diesel fuel (Demirbas, 2008).

One of the most common blends of biodiesel contains 20 volume percent biodiesel and 80 volume percent conventional diesel. For soybean-based biodiesel at this concentration, the estimated emission impacts for percent change in emissions of NO_x, PM, HC, and CO were +20%, −10.1%, −21.1%, and −11.0%, respectively (EPA, 2002). The use of blends of biodiesel and diesel oil are preferred in engines, in order to avoid some problems related to the decrease of power and torque and to the increase of NO_x emissions (a contributing factor in the localized formation of smog and ozone) with increasing content of pure biodiesel in the blend. Emissions of all pollutants except NO_x appear to decrease when biodiesel is used. The fact that NO_x emissions increase with increasing biodiesel concentration may be a detriment in areas that are out of attainment for ozone. Figure 4.24 shows the average emission impacts of vegetable oil-based biodiesel for CIE. Figure 4.25 shows average emission impacts of animal-based biodiesel for CIE.

Using biodiesel in a conventional diesel engine substantially reduces emissions of unburned hydrocarbons, carbon dioxide, carbon monoxide, sulfates, polycyclic aromatic hydrocarbons, nitrated polycyclic aromatic hydrocarbons, ozone-forming hydrocarbons, and particulate matter. The net contribution of carbon dioxide from

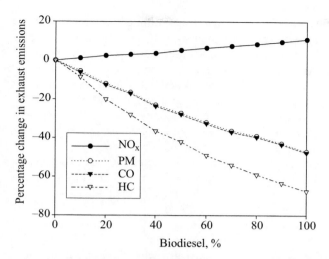

Fig. 4.24 Curves for average emission impacts of vegetable oil-based biodiesel for CIE

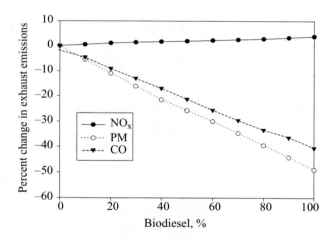

Fig. 4.25 Curves for average emission impacts of animal-based biodiesel for CIE

biomass combustion is small. Reductions in net carbon dioxide emissions are estimated at 77–104 g/MJ of diesel displaced by biodiesel. These reductions increase as the amount of biodiesel blended into diesel fuel increases. The best emissions reductions are seen with biodiesel.

The exhaust emissions of commercial biodiesel and petrodiesel have been studied in a 2003 model year heavy-duty 14 L six-cylinder diesel engine with EGR (Knothe *et al.*, 2006). The commercial biodiesel fuel significantly reduced PM exhaust emissions (75–83%) compared to the petrodiesel base fuel. However, NO_x exhaust emissions were slightly increased with commercial biodiesel compared to the base fuel. The chain length of the compounds had little effect on NO_x and PM exhaust emissions, while the influence was greater on HC and CO, the latter being reduced with decreasing chain length. Unsaturation in the fatty compounds causes an increase in NO_x exhaust emissions (Knothe et al., 2006).

The exhaust emissions of commercial biodiesel and petrodiesel were studied in a 2003 model year heavy-duty 14 L six-cylinder diesel engine with EGR (Knothe et al., 2006). The commercial biodiesel fuel significantly reduced PM exhaust emissions (75–83%) compared to the petrodiesel base fuel. However, NO_x exhaust emissions were slightly increased with commercial biodiesel compared to the base fuel. The chain length of the compounds had little effect on NO_x and PM exhaust emissions, while the influence was greater on HC and CO, the latter being reduced with decreasing chain length. Unsaturation in the fatty compounds causes an increase in NO_x exhaust emissions (Knothe et al., 2006).

Biodiesel fuels can be used as a renewable energy source to substitute conventional petroleum diesel in compression ignition engines. When degradation is caused by biological activity, especially by enzymatic action, it is called biodegradation. Biodegradability of biodiesel has been proposed as a solution for the waste problem. Biodegradable fuels such as biodiesels have an expanding range of potential applications and they are environmentally friendly. Therefore, there is growing interest in degradable diesel fuels that degrade more rapidly than conventional diesel fuels.

The advantages of biodiesel include domestic origin, reducing the dependency on imported petroleum, biodegradability, high flash point, and inherent lubricity in the neat form (Knothe et al., 2005; Mittelbach and Gangl, 2001). Recently, biodiesel has become more attractive because of its environmental benefits and the fact that it is made from renewable resources (Ma and Hanna, 1999).

Biodiesel is non-toxic and degrades about four times faster than petrodiesel. Its oxygen content improves the biodegradation process, leading to a decreased level of quick biodegradation. In comparison with petrodiesel, biodiesel shows better emission parameters. It improves the environmental performance of road transport, including decreased greenhouse emissions (mainly of carbon dioxide).

As biodiesel fuels are becoming commercialized their existence in the environment is an area of concern since petroleum oil spills constitute a major source of contamination of the ecosystem. Among these concerns, water quality is one of the most important issues for living systems. It is important to examine the biodegradability of biodiesel fuels and their biodegradation rates in natural waterways in case they enter the aquatic environment in the course of their use or disposal. Chemicals from biodegradation of biodiesel can be release into the environment. With the increasing interest in biodiesel, the health and safety aspects are of utmost importance, including determination of their environmental impacts in transport, storage, or processing (Ma and Hanna, 1999).

Biodegradation is degradation caused by biological activity, particularly by enzyme action leading to significant changes in the material chemical structure. There are many methods for biodegradation. Among them, the carbon dioxide (CO_2) evolution method is relatively simple, economical, and environmentally safe. Another method is to measure the biochemical oxygen demand (BOD) with a respirometer (Demirbas, 2008).

The biodegradabilities of several biodiesels in the aquatic environment show that all biodiesel fuels are readily biodegradable. After 28 days all biodiesel fuels were 77–89% biodegraded; diesel fuel was only 18% biodegraded (Zhang, 1996). The enzymes responsible for the dehydrogenation/oxidation reactions that occur in the process of degradation recognize oxygen atoms and attack them immediately (Zhang *et al.*, 1998).

The biodegradability data of petroleum and biofuels available in the literature are presented in Table 4.38. Heavy fuel oil has low biodegradation of 11%, in 28-day laboratory studies, due to its higher proportion of high molecular weight aromatics (Mulkins-Phillips and Stewart, 1974; Walker *et al.*, 1976). Gasoline is less biodegradable (28%) after 28 days. Vegetables oils and their derived methyl esters (biodiesels) are rapidly degraded to reach biodegradation of between 76–90% (Zhang *et al.*, 1998). In their studies Zhang *et al.* (1998) have shown that vegetables oils are slightly less degraded than their modified methyl ester.

Many of the vegetable oils contain polyunsaturated fatty acid chains that are methylene-interrupted rather than conjugated. The double bond of unsaturated fatty acids restricts the rotation of the hydrogen atoms attached to them. Therefore, an unsaturated fatty acid with a double bond can exist in two forms. The *cis* form in which the two hydrogens are on the same "side" and the *trans* form in which the hydrogen atoms are on opposite sides.

Trans unsaturated fatty acids, or *trans* fats, are solid fats produced artificially by heating liquid vegetable oils in the presence of metal catalysts and hydrogen. This process, called partial hydrogenation, causes carbon atoms to bond in a straight configuration and remain in a solid state at room temperature.

Physical properties that are sensitive to the effects of fatty oil oxidation include viscosity, the refractive index, and the di-electric constant. In oxidative instability, the methylene group ($-CH_2-$) carbons between the olefinic carbons are the sites of first attack.

Oxidation to CO_2 of biodiesel results in the formation of hydroperoxides. The formation of the hydroperoxide follows a well known peroxidation chain mechanism. Oxidative lipid modifications occur through lipid peroxidation mechanisms

Table 4.38 Biodegradability data of petroleum and biofuels

Fuel sample	Degradation in 28 days (%)	References
Gasoline (91 octane)	28	Speidel *et al.*, 2000
Heavy fuel (Bunker C oil)	11	Mulkins-Phillips and Stewart, 1974; Walker *et al.*, 1976
Refined rapeseed oil	78	Zhang *et al.*, 1998
Refined soybean oil	76	Zhang *et al.*, 1998
Rapeseed oil methyl ester	88	Zhang *et al.*, 1998
Sunflower seed oil methyl ester	90	Zhang *et al.*, 1998

in which free radicals and reactive oxygen species abstract a methylene hydrogen atom from polyunsaturated fatty acids, producing a carbon-centered lipid radical. Spontaneous rearrangement of the 1,4-pentadiene yields a conjugated diene, which reacts with molecular oxygen to form a lipid peroxyl radical. Abstraction of a proton from neighboring polyunsaturated fatty acids produces a lipid hydroperoxide (LOOH) and regeneration of a carbon-centered lipid radical, thereby propagating the radical reaction. After hydrogen is removed from such carbons oxygen rapidly attacks, and a LOOH is ultimately formed where the polyunsaturation has been isomerized to include a conjugated diene. This reaction is a chain mechanism that can proceed rapidly once an initial induction period has occurred. The greater the level of unsaturation in a fatty oil or ester, the more susceptible it will be to oxidation. Once the LOOHs have formed, they decompose and inter-react to form numerous secondary oxidation products including higher molecular weight oligomers, often called polymers.

4.6.11 Disadvantages of Biodiesel as Motor Fuel

The major disadvantages of biodiesel are higher viscosity, lower energy content, higher cloud point and pour point, higher nitrogen oxides (NO_x) emissions, lower engine speed and power, injector coking, engine compatibility, high price, and higher engine wear.

Table 4.39 shows the fuel ASTM Standards of biodiesel and petroleum diesel fuels. Important operating disadvantages of biodiesel in comparison with petro-diesel are cold start problems, the lower energy content, higher copper strip corro-

Table 4.39 ASTM Standards of biodiesel and petrodiesel fuels

Property	Test Method	ASTM D975 (Petroleum Diesel)	ASTM D6751 (Biodiesel, B100)
Flash Point	D 93	325 K min	403 K
Water and Sediment	D 2709	0.05 max %vol	0.05 max %vol
Kinematic Viscosity (at 313 K)	D 445	1.3–4.1 mm²/s	1.9–6.0 mm²/s
Sulfated Ash	D 874	–	0.02 max %wt
Ash	D 482	0.01 max %wt	–
Sulfur	D 5453	0.05 max %wt	–
Sulfur	D 2622/129	–	0.05 max %wt
Copper Strip Corrosion	D 130	No. 3 max	No 3 max
Cetane Number	D 613	40 min	47 min
Aromaticity	D 1319	35 max %vol	–
Carbon Residue	D 4530	–	0.05 max %mass
Carbon Residue	D 524	0.35 max %mass	–
Distillation Temp (90% Volume Recycle)	D 1160	555 K min–611 K max	–

sion, and fuel pumping difficulty due to higher viscosity. This increases fuel consumption when biodiesel is used in comparison with application of pure petrodiesel, in proportion to the share of the biodiesel content. Taking into account the higher production value of biodiesel as compared to petrodiesel, this increase in fuel consumption in addition raises the overall cost of application of biodiesel as an alternative to petrodiesel.

Biodiesel has a higher cloud point and a higher pour point compared to conventional diesel. Neat biodiesel and biodiesel blends increase nitrogen oxides (NO_x) emissions compared with petroleum-based diesel fuel used in an unmodified diesel engine (EPA, 2002). The peak torque is less for biodiesel than petroleum diesel but occurs at lower engine speed, and generally the torque curves are flatter. The biodiesels on the average decrease power by 5% compared to diesel at the rated load.

4.6.12 Engine Performance Tests

Biodiesel, commercial diesel fuel, and their blends were used as the fuel of a direct injection (DI) compression ignition engine. Criteria pollutant emissions from biodiesel blends are now becoming a relevant subject due to the increase in consumption of this renewable fuel worldwide. Biodiesel is the first and only alternative fuel to commercial diesel to have a complete evaluation of emission results. The exhaust gas emissions are reduced with an increase in biodiesel concentration of the fuel blend. The emission forming gases such as carbon dioxide and carbon monoxide from combustion of biodiesel hydrocarbons are generally less than those of diesel fuel. The lower percentage of biodiesel blends emits very low amount of carbon dioxide (CO_2) in comparison with diesel. The engine emits more carbon monoxide (CO) using diesel as compared to that of biodiesel blends under all loading conditions. With increasing biodiesel percentage, CO emission level decreases. Sulfur emissions are essentially eliminated with pure biodiesel. The torque values of commercial diesel fuel are greater than those of biodiesel. The brake power of biodiesel is nearly the same as with diesel fuel. The specific fuel consumption values of biodiesel are greater than those of commercial diesel fuel. The effective efficiency and effective pressure values of commercial diesel fuel are greater than those of biodiesel. Biodiesel methyl esters improve the lubrication properties of the diesel fuel blend. Biodiesel reduced long-term engine wear in diesel engines. Even biodiesel levels below 1% can provide up to a 30% increase in lubricity.

Biodiesel and petroleum-based diesel have similar fuel properties, *e.g.*, viscosity, heating value, boiling temperature, cetane number, *etc*. For this reason, biodiesel may be used in standard diesel engines. The only modifications required are a two-to-three degree retardation of injection timing and replacement of all natural rubber seals with synthetic ones, due to the solvent characteristics of biodiesel. The kinematic viscosity values of biodiesels are between 3.6 mm^2/s and 4.6 mm^2/s.

From an operational point of view, biodiesel has about 90% of the energy content of petroleum diesel, measured on a volumetric basis. Due to this fact, on average basis the use of biodiesel reduces the fuel economy and power by about 10% in comparison with petroleum diesel. The reason for this improvement stems mainly from the oxygen content of biodiesel, the ensuing better combustion process, and the improved lubricity, which partly compensate the impact of the lower energy content. Biodiesel is an oxygenated compound. Oxygenates are just preused hydrocarbons having a structure that provides a reasonable antiknock value. Also, as they contain oxygen, fuel combustion is more efficient, reducing hydrocarbons in exhaust gases. The only disadvantage is that oxygenated fuel has less energy content. For the same efficiency and power output, more fuel has to be burned. On the other hand, biodiesel blends are safer than pure petroleum diesel, because biodiesel has a higher flash point. Most operational disadvantages of pure biodiesel, *e.g.*, replacement of natural rubber seals, cold start problems, *etc.*, do not arise when using blended kinds of biodiesel.

Fuel characterization data show some similarities and differences between biodiesel and diesel fuels (Shay, 1993):

- The specific weight is higher for biodiesel, heat of combustion is lower, and viscosities are 1.3–1.6 times that of No. 2 diesel fuel.
- Pour points for biodiesel fuels vary from 1–25°C higher for biodiesel fuels, depending on the feedstock.
- The sulfur content for biodiesel fuel is 20–50% that of No. 2 diesel fuel.
- The methyl esters all have higher levels of injector coking than No. 2 diesel fuel.

In cities across the globe, the personal automobile is the single greatest polluter, as emissions from millions of vehicles on the road add up to a planet-wide problem. The biodiesel impacts on exhaust emissions vary depending on the type of biodiesel and on the type of conventional diesel. Blends of up to 20% biodiesel mixed with petroleum diesel fuels can be used in nearly all diesel equipment and are compatible with most storage and distribution equipment. Using biodiesel in a conventional diesel engine substantially reduces emissions of unburned hydrocarbons, carbon monoxide, sulfates, polycyclic aromatic hydrocarbons, nitrated polycyclic aromatic hydrocarbons, and particulate matter. These reductions increase as the amount of biodiesel blended into diesel fuel increases. In general, biodiesel increases NO_x emissions when used as fuel in a diesel engine. The fact that NO_x emissions increase with increasing biodiesel concentration may be a detriment in areas that are out of attainment for ozone.

Biodiesel provides significant lubricity improvement over petroleum diesel fuel. Lubricity results of biodiesel and petroleum diesel using industry test methods indicate that there is a marked improvement in lubricity when biodiesel is added to conventional diesel fuel. Even biodiesel levels below 1% can provide up to a 30% increase in lubricity.

In general, biodiesel will soften and degrade certain types of elastomers and natural rubber compounds over time. Using high percent blends can impact fuel system components (primarily fuel hoses and fuel pump seals), that contain elastomer compounds incompatible with biodiesel. Manufacturers recommend that natural or butyl rubbers not be allowed to come in contact with pure biodiesel. Biodiesel will lead to degradation of these materials over time, although the effect is lessened with biodiesel blends.

Biodiesel has demonstrated a number of promising characteristics, including reduction of exhaust emissions. The combustion of biodiesel fuel in CIEs in general results in lower smoke, particulate matter, carbon monoxide, and hydrocarbon emissions compared to standard diesel fuel combustion, while the engine efficiency is either unaffected or improved. The increased in-cylinder pressure and temperature lead to increased NO_x emissions, while the more advanced combustion assists in the reduction of smoke compared to pure diesel combustion. The advanced rapeseed methyl ester combustion resulted in the reduction of smoke, HC, and CO, while both NOx emissions and fuel consumption were increased.

The commercial biodiesel fuel significantly reduces PM exhaust emissions (75–83%) compared to the petrodiesel base fuel. However, NO_x exhaust emissions increases slightly with commercial biodiesel compared to the base fuel. The chain length of the compounds has little effect on NO_x and PM exhaust emissions, while the influence is greater on HC and CO, the latter being reduced with decreasing chain length. Non-saturation in the fatty compounds causes an increase in NO_x exhaust emissions.

In a previous study (Canakci, 2007), the combustion characteristics and emissions of two different petroleum diesel fuels (No. 1 and No. 2) and biodiesel from soybean oil were compared. Then experimental results compared with No. 2 diesel fuel showed that biodiesel provided significant reductions in PM, CO, and unburned HC; the NO_x increased by 11.2%. Biodiesel had a 13.8% increase in brake-specific fuel consumption due to its lower heating value. However, using No. 1 diesel fuel gave better emission results, NO_x, and brake-specific fuel consumption reduced by 16.1% and 1.2%, respectively.

PM-10 emissions and power of a diesel engine fueled with crude and refined biodiesel from salmon oil has been investigated. The results indicate a maximum power loss of about 3.5% and also near 50% of PM-10 reduction with respect to diesel when a 100% of refined biodiesel is used. Previous research has shown that biodiesel-fueled engines produce less carbon monoxide (CO), unburned hydrocarbon (HC), and particulate emissions compared to mineral diesel fuel, but higher NO_x emissions (Demirbas, 2008). The addition of tobacco seed oil methyl ester to the diesel fuel reduced CO and SO_2 emissions while causing slightly higher NO_x emissions. Meanwhile, it was found that the power and the efficiency increased slightly with the addition of tobacco seed oil methyl ester. Biodiesel fueled engines produce less carbon monoxide, unburned hydrocarbons, and particulate emissions compared to diesel fuel.

The combustion of biodiesel alone provides over a 90% reduction in total unburned hydrocarbons and a 75–90% reduction in polycyclic aromatic hydrocarbons. Biodiesel further provides significant reductions in particulates than petroleum diesel fuel. This reduction is mainly caused by reduced soot formation and enhanced soot oxidation. The oxygen content and the absence of aromatic content in biodiesel have been pointed out as being the main reasons for this.

The use of biodiesel to reduce N_2O is attractive for several reasons. Biodiesel contains little nitrogen, as compared with petrodiesel, which is also used as a re-burning fuel. The N_2O reduction is strongly dependent upon initial N_2O concentrations and only slightly dependent upon temperature, where increased temperature increases N_2O reduction. This results in lower N_2O production from fuel nitrogen species for biodiesel. In addition, biodiesel contains trace amounts of sulfur, so SO_2 emissions are reduced in direct proportion to the petrodiesel replacement.

Different scenarios for the use of agricultural residues as fuel for heat or power generation have been analyzed. Reductions in net CO_2 emissions are estimated at 77–104 g/MJ of diesel displaced by biodiesel. The predicted reductions in CO_2 emissions are much greater than values reported in recent studies on biodiesel derived from other vegetable oils, due both to the large amount of potential fuel in the residual biomass and to the low-energy inputs in traditional coconut farming techniques. Unburned hydrocarbon emissions from biodiesel fuel combustion decrease compared to regular petroleum diesel. Figure 4.26 shows a plot for decreasing of unburned hydrocarbon *versus* biodiesel in the mixture of diesel and biodiesel (Bala, 2005).

Biodiesel, produced from different vegetable oils, seems very interesting for several reasons: It can replace diesel oil in boilers and internal combustion engines without major adjustments, only a small decrease in performances is reported, almost zero emissions of sulfates, a small net contribution of CO_2 when the whole life-cycle is considered, the emission of pollutants is comparable with that of diesel oil. For these reasons, several campaigns have been planned in many countries to introduce and promote the use of biodiesel.

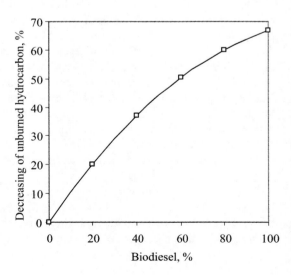

Fig. 4.26 Plot for decreasing of unburned hydrocarbon (%) *versus* biodiesel (%) in the mixture of diesel and biodiesel

The use of blends of biodiesel and diesel oil are preferred in engines, in order to avoid some problems related to the decrease of power and torque, and to the increase of NO_x emissions with increasing content of pure biodiesel in the blend.

One of the most common blends of biodiesel contains 20 volume percent biodiesel and 80 volume percent conventional diesel (B20). For soybean-based biodiesel at this concentration, the estimated emission impacts for percent change in emissions of NO_x, PM, HC, and CO were +20%, −10.1%, −21.1%, and −11.0%, respectively (EPA, 2002). The use of blends of biodiesel and diesel oil are preferred in engines in order to avoid some problems related to the decrease of power and torque, and to the increase of NO_x emissions (a contributing factor in the localized formation of smog and ozone) that occurs with an increase in the content of pure biodiesel in a blend. Emissions of all pollutants except NO_x appear to decrease when biodiesel is used.

The lower percentage of biodiesel blends emits very low amount of CO_2 in comparison with diesel. B10 emits very low level of CO_2 emissions. Using higher concentration biodiesel blends as the fuel, CO_2 emission is found to increase. But, its emission level is lower than that of the diesel mode. B100 emits more amount of CO_2, as compared to that of diesel operation. A larger amount of CO_2 in exhaust emission is an indication of the complete combustion of fuel. This supports the higher value of the exhaust gas temperature.

Emissions of regulated air pollutants, including CO, HC, NOx, particulate matter (PM), and polycyclic aromatic hydrocarbons (PAHs) have been measured and results show that B20 use can reduce both PAH emission and its corresponding carcinogenic potency.

The typical engine combustion reaction is:

$$\text{Fuel} + \text{Air} \rightarrow CO_2 + CO + H_2O + N_2 + O_2 + (HC) + O_3 + NO_2 + SO_2 \quad (4.25)$$

Carbon dioxide (CO_2) is a colorless, odorless, non-poisonous gas that results from fossil fuel combustion and is a normal constituent of ambient air. CO_2 does not directly impair human health, but it is a "greenhouse gas" that traps the Earth's heat and contributes to the potential for global warming.

Carbon monoxide (CO) is a colorless, odorless, toxic gas produced by the incomplete combustion of carbon-containing substances. One of the major air pollutants, it is emitted in large quantities by exhaust from petroleum fuel-powered vehicles. CO is emitted directly from vehicle tailpipes. Incomplete combustion is most likely to occur at low air-to-fuel ratios in the engine. These conditions are common during vehicle starting when air supply is restricted, when cars are not tuned properly, and at altitude, where "thin" air effectively reduces the amount of oxygen available for combustion. Two-thirds of the carbon monoxide emissions come from transportation sources, with the largest contribution coming from highway motor vehicles. In urban areas, the motor vehicle contribution to carbon monoxide pollution can exceed 90%.

Under the high pressure and temperature conditions in an engine, nitrogen and oxygen atoms in the air react to form various nitrogen oxides, collectively known as NO_x. Nitrogen oxides, like hydrocarbons, are precursors to the formation of ozone. They also contribute to the formation of acid rain. Nitric oxide (NO) is precursor of ozone, NO_2, and nitrate; usually emitted from combustion processes. Converted to nitrogen dioxide (NO_2) in the atmosphere, it then becomes involved in the photochemical process and/or particulate formation. Nitrogen oxides (NO_x) are gases formed in great part from atmospheric nitrogen and oxygen when combustion takes place under conditions of high temperature and high pressure; they are considered a major air pollutant and precursor of ozone.

Hydrocarbon (HC) emissions result when fuel molecules in an engine burn only partially. Some of the hydrocarbon compounds are major air pollutants; they may be active participants in the photochemical process or affect health. HCs react in the presence of nitrogen oxides and sunlight to form ground-level ozone, a major component of smog. Ozone irritates the eyes, damages the lungs, and aggravates respiratory problems. It is our most widespread and intractable urban air pollution problem. A number of exhaust hydrocarbons are also toxic, with the potential to cause cancer. Hydrocarbon pollutants also escape into the air through fuel evaporation; evaporative losses can account for a majority of the total hydrocarbon pollution from current model cars on hot days when ozone levels are highest.

Sulfur oxides (SO_x) are pungent, colorless gases formed primarily by the combustion of sulfur-containing fossil fuels, especially coal and oil. Considered major air pollutants, SO_x may impact human health and damage vegetation.

Ozone (O_3) is a pungent, colorless, toxic gas. Close to the Earth's surface it is produced photochemically from hydrocarbons, oxides of nitrogen, and sunlight, and is a major component of smog. At very high altitudes, it protects the Earth from harmful ultraviolet radiation.

Particulate matter (PM) is tiny solid or liquid particles of soot, dust, smoke, fumes, and aerosols. The size of the particles (10 microns or smaller) allows them to easily enter the air sacs in the lungs where they may be deposited, resulting in adverse health effects. PM also causes visibility reduction and is a criteria air pollutant.

Smog is a term used to describe many air pollution problems. Smog is a contraction of smoke and fog. Soot is very fine carbon particles that appear black when visible.

Combustion is a basic chemical process that releases energy from a fuel and air mixture. For combustion to occur, fuel, oxygen, and heat must be present together. The combustion is the chemical reaction of a particular substance with oxygen. Combustion represents a chemical reaction, during which from certain matter other simple matter are produced, this is a combination of inflammable matter with oxygen of the air accompanied by heat release. The quantity of heat evolved when one mole of a hydrocarbon is burned to carbon dioxide, and water is called the heat of combustion. Combustion to carbon dioxide and water is characteristic of organic compounds; under special conditions it is used to determine their carbon and hydrogen content. During combustion the combustible part of fuel is subdivided into volatile part and solid residue. During heating it evaporates together with a part of carbon in the form of hydrocarbon combustible gases and carbon monoxide release by thermal degradation of the fuel. Carbon monoxide is mainly formed the following reactions: (a) from reduction of CO_2 with unreacted C,

$$CO_2 + C \rightarrow 2CO \tag{4.26}$$

and (b) from degradation of carbonyl fragments (–CO) in the fuel molecules at 600–750 K temperature.

The combustion process is started by heating the fuel above its ignition temperature in the presence of oxygen or air. Under the influence of heat, the chemical bonds of the fuel are cleaved. If complete combustion occurs, the combustible elements (C, H, and S) react with the oxygen content of the air to form CO_2, H_2O, and mainly SO_2.

If not enough oxygen is present or the fuel and air mixture is insufficient, then the burning gases are partially cooled below the ignition temperature, and the combustion process stays incomplete. The flue gases then still contain combustible components, mainly carbon monoxide (CO), unburned carbon (C), and various hydrocarbons (C_xH_y).

The standard measure of the energy content of a fuel is its heating value (HV), sometimes called the calorific value or heat of combustion. In fact, there are multiple values for HV, depending on whether it measures the enthalpy of combustion (ΔH) or the internal energy of combustion (ΔU), and whether for a fuel containing hydrogen product water is accounted for in the vapor phase or the condensed (liquid) phase. With water in the vapor phase, the lower heating value (LHV) at constant pressure measures the enthalpy change due to combustion. The heating value is obtained by the complete combustion of a unit quantity of solid fuel in an oxygen-bomb calorimeter under carefully defined conditions. The gross heat of combustion or higher heating value (GHC or HHV) is obtained by oxygen-bomb calorimeter method as the latent heat of moisture in the combustion products is recovered.

The methyl ester of vegetable oil has been evaluated as a fuel in CIEs by researchers (Kusdiana and Saka, 2001), who concluded that the performance of the esters of vegetable oil does not differ greatly from diesel fuel. The brake power of biodiesel was nearly the same as with diesel fuel, while the specific fuel consumption was higher than that of No. 2 diesel. Based on crankcase oil analysis, engine wear rates were low but some oil dilution did occur. Carbon deposits inside the engine were normal with the exception of intake valve deposits. The results showed that the transesterification treatment decreased the injector coking to a level significantly lower than that observed with No. 2 diesel fuel (Demirbas, 2007). Although most researchers agree that vegetable oil ester fuels are suitable for use in CIEs, a few contrary results have also been obtained. The results of these studies point out that most vegetable oil esters are suitable as diesel substitutes but that more long-term studies are necessary for commercial utilization to become practical.

Blends of up to 20% biodiesel mixed with petroleum diesel fuels can be used in nearly all diesel equipment and are compatible with most storage and distribution equipment. Higher blends, even B100, can be used in many engines built with little or no modification. Transportation and storage, however, require special management. Material compatibility and warrantee issues have not been resolved with higher blends.

Because alcohols have limited solubility in diesel, stable emulsion must be formed that will allow it to be injected before separation occurs. A hydro-shear emulsification unit can be used to produce emulsions of diesel-alcohol. However, the emulsion can only remain stable for 45 seconds. In addition, 12% alcohol (energy basis) is the maximum percentage. This kind of method has several problems, which are as follows: (a) specific fuel consumption at low speed increases, (b) high cost, and (c) instability. Therefore, other methods have been developed.

To reduce the high viscosity of vegetable oils, microemulsions with immiscible liquids such as methanol and ethanol and ionic or non-ionic amphiphiles have been studied (Ma and Hanna, 1999). The short engine performances of both ionic and non-ionic microemulsions of ethanol in soybean oil were nearly as good as that of No. 2 diesel fuel (Goering *et al.*, 1982).

All microemulsions with butanol, hexanol, and octanol met the maximum viscosity requirement for No. 2 diesel fuel. 2-octanol was found an effective amphiphile in the micellar solubilization of methanol in triolein and soybean oil (Ma and Hanna, 1999).

A fumigation system injects a gaseous or liquid fuel into the intake air stream of a compression ignited engine. This fuel burns and becomes a part-contributor to the power producing fuel. While alcohol and gasoline may be used, gaseous fumigation seems to exhibit the best overall power yields, performance, and emissions benefits. Rudolf Diesel's 1901 patent mentions the diesel fumigation process. It was not until the 1940s that there were any commercial fumigation applications. Fumigation with propane has been studied as a means to reduce injector coking.

Fumigation is a process of introducing alcohol into the diesel engine (up to 50%) by means of a carburetor in the inlet manifold. At the same time, the diesel pump operates at a reduced flow. In this process, No. 2 diesel fuel is used for generating a pilot flame and alcohol is used as a fumigated fuel.

Alcohol is used as a fumigated fuel. At low loads, the quantity of alcohol must be reduced to prevent misfire. On the other hand, at high loads, the quantity of alcohol must also be reduced to prevent preignition.

In dual injection systems, a small amount of diesel is injected as a pilot fuel for the ignition source. A large amount of alcohol is injected as the main fuel. It must be noted that the pilot fuel must be injected prior to injection of the alcohol. Some ideal results can be achieved when this method is used. Thermal efficiency is better and, at the same time, NO_x emission is lower. Moreover, CO emissions and HC emissions are the same. However, the system requires two fuel pumps, thus, leading to a high cost. Moreover, alcohol needs additives for lubricity.

A visual inspection of the injector types would indicate no difference between the biodiesels when tested on No. 2 diesel fuel. The overall injector coking is considerably low. Linear regression is used to compare injector coking, viscosity, percent of biodiesel, total glycerol, and heat of combustion data with the others.

Alcohol can ignite with hot surfaces. For this reason, glow-plugs can be utilized as a source of ignition for alcohol. In this system, specific fuel consumption depends on glow-plug positions and temperatures. It must be noted that the temperature of glow-plugs must vary with load. However, the glow-plug becomes inefficient at a high load. In addition, the specific fuel consumption is higher than that of diesel.

The peak torque is less for biodiesel fuels than for No. 2 diesel fuel but occurs at lower engine speed, and generally the torque curves are flatter. Testing includes the power and torque of the methyl esters and diesel fuel, and ethyl esters *versus* diesel fuel. Biodiesels on the average decrease power by 5% compared to that of diesel at rated load.

When a spark plug is used, a diesel engine can be converted to an Otto cycle engine. In this case, the compression ratio should be reduced from 16:1 to 10.5:1. There are two types of this kind of conversion. They are as follows: (a) The original fuel injection system is maintained. Alcohol needs additive for lubricity. Besides, both distributor and sparkplug need to be installed, thus leading to a high cost of conversion. It is critical to adjust an ideal injection and spark-time for this kind of conversion. (b) Original fuel injection is eliminated. But, a carburetor, a spark-plug, and a distributor need to be installed, which increases the cost of conversion. In this conversion, spark timing is critical.

The effects of oxidative degradation caused by contact with ambient air (autoxidation) during long-term storage present a legitimate concern in terms of maintaining the fuel quality of biodiesel. Oxidative degradation reactions of biodiesel fuels have been conducted in the laboratory under varying time and temperature conditions (Dunn, 2002). Results showed that the reaction time significantly affects kinematic viscosity (v). With respect to increasing reaction temperature, v, acid value (AV), peroxide value (PV), and specific gravity (SG) increased significantly, whereas cold flow properties were minimally affected for temperatures up to 423 K. Antioxidants tert-butylhydroquinone (TBHQ) and α-tocopherol showed beneficial effects on retarding the oxidative degradation of methyl soyate (biodiesel) under the conditions of this study. Results indicated that v and AV have the best potential as parameters for timely and easy monitoring of biodiesel fuel quality during storage. TBHQ is the most effective antioxidant for highly unsaturated vegetable oils and many animal fats.

4.6.12.1   Evaluation of Engine Performance Using Biodiesel

Through its combustion of fossil fuels the average vehicle emits greenhouse gases like carbon dioxide, methane, nitrous oxide, and chlorofluorocarbons that surround the Earth's atmosphere like a clear thermal blanket allowing the sun's warning rays in and trapping the hear close to Earth's surface. By reducing vehicular emission alternate fuels help combat both air pollution and global climate change. There is a growing concern that world may run out of petroleum bases fuel resources. All these issues make it imperative that search for alternative fuels is taken earnestly.

Biodiesels are monoalkyl esters containing approximately 10% oxygen by weight. The oxygen improves the efficiency of combustion, but it takes up space in the blend and therefore slightly increases the apparent fuel consumption rate observed while operating an engine with biodiesel. Current investigations reveal that the engine performance with diesel–biodiesel blends is comparable with diesel fuel operation. Technical aspects of biodiesel are approached, such as the physical and chemical characteristics of methyl esters related to its performance in CIEs. When testing an engine in a test bench, equivalent performance requires attaining the same engine speed and torque, regardless of the fuel used. A meaningful comparison of emissions and fuel consumption is only possible if tests are carried out under the same operation mode.

The high combustion temperature at high engine speed becomes the dominant factor, making both heated and unheated fuel to acquire the same temperature before fuel injection. Various methods of using vegetable oil (jatropha oil) and methanol such as blending, transesterification and dual fuel operation have been studied experimentally (He and Bao, 2003). Brake thermal efficiency was better in the dual fuel operation and with the methyl ester of jatropha oil as compared to the blend. It increased form 27.4% with neat jatropha oil to a maximum of 29% with the methyl ester and 28.7% in the dual fuel operation (He and Bao, 2003).

The performance of the esters of vegetable oil does not differ greatly from that of diesel fuel. The brake power is nearly the same as with diesel fuel, while the specific fuel consumption is higher than that of diesel fuel. Carbon deposits inside the engine were normal, with the exception of intake valve deposits. It was shown that biodiesel decreased the injector coking to a level significantly lower than that observed with No 2 diesel fuel.

Various engine performance parameters such as thermal efficiency, brake specific fuel consumption (BSFC), and brake specific energy consumption (BSEC), *etc.*, can be calculated from the acquired data. The torque, brake power and fuel consumption values associated with CIE fuels were determined under certain operating conditions. In an earlier study (Shay, 1993), the physical and chemical properties of ethyl ester produced from sunflower oil were similar to diesel fuel. After the test was conducted, it was observed that sunflower oil ethyl ester (ethylic biodiesel) gives similar results at short period performance tests. However, there was a huge difference between the specific fuel consumption values of fuels at the same speed values. In the studies, the measured difference in torque values for diesel fuel and the ethyl ester was calculated as 8% at $2400\,min^{-1}$, where the maximum engine moment was measured. The difference in torque values of these fuels was calculated as 18% at $3900\,min^{-1}$, where the maximum power was measured. The difference of effective powers was calculated as 10% and 17%, respectively, at same speed values. The differences of smoke values were calculated as 14% at speed level where the maximum moment was measured, and 22% at speed level where the maximum power was obtained. The ethylic biodiesel has lower noise level than that of the diesel fuel at all engine speed (Shay, 1993). Calculations of the fuel conversion efficiency, based on the lower heating value of the fuel, showed that all of the biodiesel fuels had the same efficiency. This indicates that regardless of the degree of saturation of the ester, there are no fundamental changes in the way the different esters burn in the engine.

Kaplan *et al.* (2006) compared sunflower oil biodiesel and diesel fuels at full and partial loads and at different engine speeds in a 2.5153 kW engine. The loss of torque and power ranged between 5% and 10%, and particularly at full load, the loss of power was closer to 5% at low speed and to 10% at high speed.

Petroleum diesel fuel and ethylic biodiesel fuel were tested in a naturally aspirated direct injected (DI) diesel engine in four steady operation modes, defined by their engine speed and brake mean effective pressure, the latter being proportional to the effective torque (Kaplan *et al.*, 2006). In earlier studies, the operation modes selected tried to simulate representative engine conditions, often taking as reference certification cycles, which in the case of heavy-duty engines cover the whole load range (concentrating mainly on 25%, 50%, 75%, and 100% of maximum torque) at various speeds and, in the case of vehicle engines, are concentrated around the low–medium load (Puhan *et al.*, 2005).

In an earlier study (Canakci *et al.*, 2006), the tests were performed with commercial diesel fuel and biodiesel. The experimental results of the commercial diesel fuel are listed in Table 4.40. Table 4.41 shows the experimental results of the biodiesel fuel (Ozkan *et al.*, 2005). Using Tables 4.40 and 4.41, the performance curves for the fuels are shown together for comparison purposes. The maximum brake power values of biodiesel and diesel were 4.390 kW and 5.208 kW obtained at 2750 and 2500 rpm, respectively. According to these values, the commercial diesel fuel has the greatest brake power. Compared to diesel fuel, a 25% power loss occurred with biodiesel. The maximum torque values are about 21.0 Nm at 1500 rpm for the diesel fuel and 19.7 Nm at 1500 rpm for the biodiesel. The torque values of commercial diesel fuel are greater than those of biodiesel. The specific fuel consumption values of biodiesel are greater than those of commercial diesel fuel. The effective efficiency and effective pressure values of commercial diesel fuel are greater than those of biodiesel (Canakci *et al.*, 2006).

Fuel consumption at full load condition and low speeds generally is high. Fuel consumption first decreases and then increases with increasing speed. The reason is that the produced power in low speeds is low and the main part of fuel is consumed to overcome the engine friction (Ozkan *et al.*, 2005).

The trend of the performance curves concerning power and torque are very much like those mentioned in valid concerned literature. The range of speed was selected between 1200 and 3600 rpm. Considering power and torque performance with fuel blends, one can say that the trend of these parameters *versus* speed is perfectly similar to diesel fuel. Engine test results with diesel fuel showed that the maximum torque was 64.2 Nm, which occurred at 2400 rpm. The maximum power was 18.12 kW at 3200 rpm (Najafi *et al.*, 2007).

Biodiesel methyl esters improve the lubrication properties of the diesel fuel blend. Fuel injectors and some types of fuel pumps rely on fuel for lubrication. Biodiesel reduced long-term engine wear in test diesel engines to less than half of

Table 4.40 Experimental results of commercial diesel fuel

Engine speed (Rpm)	Torque (Nm)	Power (kW)	SFC[a] (g/kWh)	Effective efficiency (%)	Effective pressure (bar)
1000	20.2	2.119	380.498	22.6	5.600
1250	20.8	2.716	395.718	22.7	5.744
1500	21.0	3.300	346.706	24.8	5.816
1750	19.9	3.641	324.722	26.5	5.500
2000	19.9	4.156	304.825	28.2	5.493
2250	19.8	4.651	282.468	30.4	5.464
2500	19.9	5.208	272.411	31.6	5.507
2750	16.9	4.877	290.880	29.6	4.688
3000	13.5	4.229	349.462	24.6	3.726

[a]SFC: Specific fuel consumption.

Table 4.41 Experimental results of biodiesel fuel

Engine speed (Rpm)	Torque (Nm)	Power (kW)	SFC[a] (g/kWh)	Effective efficiency (%)	Effective pressure (bar)
1000	17.8	1.861	471.859	18.23	4.919
1250	17.2	2.246	433.236	19.85	4.749
1500	17.2	2.695	417.441	20.60	4.749
1750	17.7	3.234	359.089	23.95	4.885
2000	18.0	3.768	341.232	25.20	4.980
2250	18.4	4.331	307.841	27.94	5.089
2500	15.9	4.171	331.977	25.91	4.410
2750	15.3	4.390	341.664	25.17	4.220
3000	9.6	3.018	372.715	23.07	2.660

[a]SFC: Specific fuel consumption

what was observed in engines running on current low sulfur diesel fuel. Lubricity properties of fuel are important for reducing friction wear in engine components, normally lubricated by the fuel rather than crankcase oil (Ma and Hanna, 1999; Demirbas, 2003a). Lubricity results of biodiesel and petroleum diesel using industry test methods indicate that there is a marked improvement in lubricity when biodiesel is added to conventional diesel fuel. Even biodiesel levels below 1% can provide up to a 30% increase in lubricity (Demirbas, 2008).

Using biodiesel in a conventional diesel engine substantially reduces emissions of unburned hydrocarbons, carbon monoxide, sulfates, polycyclic aromatic hydrocarbons, nitrated polycyclic aromatic hydrocarbons, and particulate matter. These reductions increase as the amount of biodiesel blended into diesel fuel increases. The best emission reductions are seen with B100 (100% biodiesel). The biodiesel sulfur content is another interesting advantage of the produced fuel, which is 18 ppm only. Comparing the 18 ppm sulfur content of the produced biodiesel with the 500 ppm sulfur content of the diesel fuel, the advantage of the biodiesel over the diesel fuel in terms of the environmental benefits can be justified (Ozkan *et al.*, 2005).

Biodiesel fuels can be performance improving additives in compression ignition engines. Performance testing showed that while the power decreased and the brake specific fuel consumption increased for all of the biodiesel samples, compared with No. 2 diesel fuel, the amounts of the changes were in direct proportion to the lower energy content of the biodiesel.

Biodiesel is a biodegradable and renewable fuel. It contributes no net carbon dioxide or sulfur to the atmosphere and emits less gaseous pollutants than normal diesel. Carbon monoxide, aromatics, polycyclic aromatic hydrocarbons (PAHs), and partially burned or unburned hydrocarbon emissions are all reduced in vehicles operating on biodiesel.

4.7 Bio-oils from Biorenewables

The term bio-oil is used mainly to refer to liquid fuels from biorenewable feedstocks. Biomass is heated in the absence of oxygen, or partially combusted in a limited oxygen supply, to produce an oil-like liquid, a hydrocarbon rich gas mixture and a carbon rich solid residue. Pyrolysis dates back to at least ancient Egyptian times, when tar for caulking boats and certain embalming agents were made by pyrolysis. In the 1980s, researchers found that the pyrolysis liquid yield could be increased using fast pyrolysis where a biomass feedstock is heated at a rapid rate and the vapors produced are also condensed rapidly (Mohan *et al.*, 2006).

In wood derived pyrolysis oil, specific oxygenated compounds are present in relatively large amounts. A current comprehensive review focuses on the recent developments in the wood/biomass pyrolysis and reports the characteristics of the resulting bio-oils, which are the main products of fast wood pyrolysis (Mohan *et al.*, 2006). Sufficient hydrogen added to the synthesis gas to convert all of the biomass carbon into methanol carbon would more than double the methanol produced from the same biomass base.

Pyrolysis is the simplest and almost certainly the oldest method of processing one fuel in order to produce a better one. Pyrolysis can also be carried out in the presence of a small quantity of oxygen ("gasification"), water ("steam gasification"), or hydrogen ("hydrogenation"). One of the most useful products is methane, which is a suitable fuel for electricity generation using high-efficiency gas turbines. The main pyrolysis methods and their variants are listed in Table 4.42.

Biomass pyrolysis is attractive because solid biomass and wastes can be readily converted into liquid products. These liquids, such as crude bio-oil or slurry of charcoal of water or oil, have advantages in transport, storage, combustion, retrofitting, and flexibility in production and marketing.

The pyrolysis of biomass is a thermal treatment that results in the production of charcoal, liquid, and gaseous products. Among the liquid products, methanol is one of the most valuable products. The liquid fraction of the pyrolysis products

Table 4.42 Pyrolysis methods and their variants

Method	Residence time	Temperature, K	Heating rate	Products
Carbonation	days	675	very low	charcoal
Conventional	5–30 min	875	low	oil, gas, char
Fast	0.5–5 s	925	very high	bio-oil
Flash-liquid[a]	<1 s	<925	high	bio-oil
Flash-gas[b]	<1 s	<925	high	chemicals, gas
Hydropyrolysis[c]	<10 s	<775	high	bio-oil
Methano-pyrolysis[d]	<10 s	>975	high	chemicals
Ultra pyrolysis[e]	<0.5 s	1275	very high	chemicals, gas
Vacuum pyrolysis	2–30 s	675	medium	bio-oil

[a]Flash-liquid: liquid obtained from flash pyrolysis accomplished in a time of <1 s.
[b]Flash-gas: gaseous material obtained from flash pyrolysis within a time of <1 s.
[c]Hydropyrolysis: pyrolysis with water.
[d]Methanopyrolysis: pyrolysis with methanol.
[e]Ultra pyrolysis: pyrolyses with very high degradation rate.

consists of two phases: an aqueous phase containing a wide variety of organo-oxygen compounds of low molecular weight and a non-aqueous phase containing insoluble organics of high molecular weight. This phase is called tar and is the product of greatest interest. The ratios of acetic acid, methanol, and acetone of the aqueous phase were higher than those of the non-aqueous phase. The point where the cost of producing energy from fossil fuels exceeds the cost of biomass fuels has been reached. With a few exceptions, energy from fossil fuels will cost more money than the same amount of energy supplied through biomass conversion.

Table 4.43 shows the fuel properties of diesel, biodiesel and biomass pyrolysis oil. The kinematic viscosity of pyrolysis oil varies from as low as 11 cSt to as high as 115 mm^2/s (measured at 313 K), depending on nature of the feedstock, temperature of pyrolysis process, thermal degradation degree and catalytic cracking, the water content of the pyrolysis oil, the amount of light ends that have collected, and the pyrolysis process used. The pyrolysis oils have a water content of typically 15–30 wt% of the oil mass, which cannot be removed by conventional methods like distillation. Phase separation may partially occur above certain water contents. The water content of pyrolysis oils contributes to their low energy density, lowers the flame temperature of the oils, leads to ignition difficulties, and, when preheating the oil, can lead to premature evaporation of the oil and resultant injection difficulties. The higher heating value (HHV) of pyrolysis oils is below 26 MJ/kg (compared to 42–45 MJ/kg for compared to values of 42–45 MJ/kg for conventional petroleum fuel oils). In contrast to petroleum oils, which are non-polar and in which water is insoluble, biomass oils are highly polar and can readily absorb over 35% water (Demirbas, 2007).

The pyrolysis oil (bio-oil) from wood is typically a liquid, almost black through dark red to brown. The density of the liquid is about 1200 kg/m^3, which is higher than that of fuel oil and significantly higher than that of the original biomass. Bio-oils have a water content of typically 14–33 wt.%, which cannot be removed by

Table 4.43 Fuel properties of diesel, biodiesel and biomass pyrolysis oil

Property	Test method	ASTM D975 (Diesel)	ASTM D6751 (Biodiesel, B100)	Pyrolysis Oil (Bio-oil)
Flash Point	D 93	325 K min	403 K	–
Water and sediment	D 2709	0.05 max %vol	0.05 max %vol	0.01–0.04
Kinematic viscosity (at 313 K)	D 445	1.3–4.1 mm^2/s	1.9–6.0 mm^2/s	25–1000
Sulfated ash	D 874	–	0.02 max %wt	–
Ash	D 482	0.01 max %wt	–	0.05–0.01 %wt
Sulfur	D 5453	0.05 max %wt	–	–
Sulfur	D 2622/129	–	0.05 max %wt	0.001–0.02%wt
Copper strip Corrosion	D 130	No 3 max	No 3 max	–
Cetane number	D 613	40 min	47 min	–
Aromaticity	D 1319	–	35 max %vol	–
Carbon residue	D 4530	–	0.05 max %mass	0.001–0.02%wt
Carbon residue	D 524	0.35 max %mass	–	–
Distillation temperature	D 1160	555 K min –611 K max	–	–

conventional methods like distillation. Phase separation may occur above certain water contents. The higher heating value (HHV) is below 27 MJ/kg (compared to 43–46 MJ/kg for conventional fuel oils).

The bio-oil formed at 725 K contained high concentrations of compounds such as acetic acid, 1-hydroxy-2-butanone, 1-hydroxy-2-propanone, methanol, 2,6-dimethoxyphenol, 4-methyl-2,6-dimethoxyphenol, 2-cyclopenten-1-one, *etc.* A significant characteristic of bio-oils was the high percentage of alkylated compounds especially methyl derivatives. As the temperature increased, some of these compounds were transformed *via* hydrolysis (Kuhlmann *et al.*, 1994). The formation of unsaturated compounds from biomass materials generally involves a variety of reaction pathways such as dehydration, cyclization, Diels–Alder cycloaddition reactions, and ring rearrangement. For example, 2,5-hexanedione can undergo cyclization under hydrothermal conditions to produce 3-methyl-2-cyclopenten-1-one with very high selectivity of up to 81% (An *et al.*, 1997).

The mechanism of pyrolysis reactions of biomass was extensively discussed in an earlier study (Demirbas, 2000). Water is formed by dehydration. In the pyrolysis reactions, methanol arises from the breakdown of methyl esters and/or ethers from decomposition of pectin-like plant materials. Methanol also arises from methoxyl groups of uronic acid (Demirbas and Güllü, 1998). Acetic acid is formed in the thermal decomposition of all three main components of wood. When the yield of acetic acid originating from the cellulose, hemicelluloses, and lignin is taken into account, the total is considerably less than the yield from the wood itself. Acetic acid comes from the elimination of acetyl groups originally linked to the xylose unit.

Furfural is formed by dehydration of the xylose unit. Quantitatively, 1-hydroxy-2-propanone and 1-hydroxy-2-butanone present high concentrations in the liquid products. These two alcohols are partly esterified by acetic acid. In conventional slow pyrolysis, these two products are not found in so great a quantity because of their low stability (Beaumont, 1985).

If wood is completely pyrolyzed, the resulting products are about what would be expected when pyrolyzing the three major components separately. The hemicelluloses would break down first at temperatures of 470 to 530 K. Cellulose follows in the temperature range 510–620 K, with lignin being the last component to pyrolyze at temperatures of 550–770 K. A wide spectrum of organic substances was contained in the pyrolytic liquid fractions given in the literature (Beaumont, 1985). Degradation of xylan yields eight main products: water, methanol, formic, acetic and propionic acids, 1-hydroxy-2-propanone, 1-hydroxy-2-butanone, and 2-furfuraldeyde. The methoxy phenol concentration decreased with increasing temperature, while phenols and alkylated phenols increased. The formation of both methoxy phenol and acetic acid was possibly as a result of the Diels–Alder cycloaddition of a conjugated diene and unsaturated furanone or butyrolactone.

Timell (1967) described the chemical structure of the xylan as the 4-methyl-3-acetylglucuronoxylan. It has been reported that the first runs in the pyrolysis of the pyroligneous acid consist of about 50% methanol, 18% acetone, 7% esters, 6% aldehydes, 0.5% ethyl alcohol, 18.5% water, and small amounts of furfural (Demirbas, 2000). Pyroligneous acids disappear in high-temperature pyrolysis.

The composition of the water soluble products was not ascertained but it has been reported to be composed of hydrolysis and oxidation products of glucose such as acetic acid, acetone, simple alcohols, aldehydes, sugars, etc. (Sasaki et al., 1998). Pyroligneous acids disappear in high-temperature pyrolysis. Levoglucosan is also sensitive to heat and decomposes to acetic acid, acetone, phenols, and water. Metha-

Table 4.44 Chemical and physical properties of biomass bio-oils and fuel oils

	Pine	Oak	Poplar	Hardwood	No. 2 oil	No. 6 oil
Elemental (wt% dry)						
C	56.3	55.6	59.3	58.8	87.3	87.7
H	6.5	5.0	6.6	9.7	12.0	10.3
O	36.9	39.2	33.8	31.2	0.0	1.2
N	0.3	0.1	0.2	0.2	<0.01	0.5
S	–	0.0	0.00	0.04	0.1	0.8
Proximate (wt%)						
ash	0.05	0.05	<0.001	0.07	<0.001	0.07
water	18.5	16.1	27.2	18.1	0.0	2.3
volatiles	66.0	69.8	67.4	–	99.8	94.1
fixed C	15.5	14.1	5.4	–	0.1	3.6
Viscosity (mm^2/s) at 313 K	44	115	11	58	2.6	–
Density (kg/m^3) at 313 K	1210	1230	–	1170	860	950
Pyrolysis yield (wt%)						
dry oil	58.9	55.3	41.0	–	–	–
water	13.4	10.4	14.6	–	–	–
char	19.6	12.2	16.4	–	–	–
gases	13.9	12.4	21.7	–	–	–

Source: Demirbas, 2007

nol arises from the methoxyl groups of aronic acid (Demirbas, 2000). Table 4.44
shows the chemical and physical properties of biomass bio-oils and fuel oils.

Table 4.45 shows the gas chromatographic analysis of bio-oil from beech wood
pyrolysis (wt% dry basis). The bio-oil formed at 725 K contained high concentra-

Table 4.45 Gas chromatographic analysis of bio-oil from beech wood pyrolysis (wt% dry basis)

Compound	Reaction temperature (K)					
	625	675	725	775	825	875
Acetic acid	16.8	16.5	15.9	12.6	8.42	5.30
Methyl acetate	0.47	0.35	0.21	0.16	0.14	0.11
1-hydroxy-2-propanone	6.32	6.84	7.26	7.66	8.21	8.46
Methanol	4.16	4.63	5.08	5.34	5.63	5.82
1-hydroxy-2-butanone	3.40	3.62	3.82	3.88	3.96	4.11
1-hydroxy-2-propane acetate	1.06	0.97	0.88	0.83	0.78	0.75
Levoglucosan	2.59	2.10	1.62	1.30	1.09	0.38
1-hydroxy-2-butanone acetate	0.97	0.78	0.62	0.54	0.48	0.45
Formic acid	1.18	1.04	0.84	0.72	0.60	0.48
Guaiacol	0.74	0.78	0.82	0.86	0.89	0.93
Crotonic acid	0.96	0.74	0.62	0.41	0.30	0.18
Butyrolactone	0.74	0.68	0.66	0.67	0.62	0.63
Propionic acid	0.96	0.81	0.60	0.49	0.41	0.34
Acetone	0.62	0.78	0.93	1.08	1.22	1.28
2,3-butanedione	0.46	0.50	0.56	0.56	0.58	0.61
2,3-pentanedione	0.34	0.42	0.50	0.53	0.59	0.64
Valeric acid	0.72	0.62	0.55	0.46	0.38	0.30
Isovaleric acid	0.68	0.59	0.51	0.42	0.35	0.26
Furfural	2.52	2.26	2.09	1.84	1.72	1.58
5-methyl-furfural	0.65	0.51	0.42	0.44	0.40	0.36
Butyric acid	0.56	0.50	0.46	0.39	0.31	0.23
Isobutyric acid	0.49	0.44	0.38	0.30	0.25	0.18
Valerolactone	0.51	0.45	0.38	0.32	0.34	0.35
Propanone	0.41	0.35	0.28	0.25	0.26	0.21
2-butanone	0.18	0.17	0.32	0.38	0.45	0.43
Crotonolactone	0.12	0.19	0.29	0.36	0.40	0.44
Acrylic acid	0.44	0.39	0.33	0.25	0.19	0.15
2-Cyclopenten-1-one	1.48	1.65	1.86	1.96	2.05	2.13
2-Methyl-2-cyclopenten-1-one	0.40	0.31	0.24	0.17	0.13	0.14
2-methyl-cyclopentenone	0.20	0.18	0.17	0.22	0.25	0.29
Cyclopentenone	0.10	0.14	0.16	0.23	0.27	0.31
Methyl-2-furancarboxaldehyde	0.73	0.65	0.58	0.50	0.44	0.38
Phenol	0.24	0.30	0.36	0.43	0.54	0.66
2,6-dimethoxyphenol	2.28	2.09	1.98	1.88	1.81	1.76
Dimethyl phenol	0.08	0.13	0.18	0.42	0.64	0.90
Methyl phenol	0.32	0.38	0.44	0.50	0.66	0.87
4-methyl-2,6-dimethoxyphenol	2.24	2.05	1.84	1.74	1.69	1.58

Source: Demirbas, 2007

tions of compounds such as acetic acid, 1-hydroxy-2-butanone, 1-hydroxy-2-pro-
panone, methanol, 2,6-dimethoxyphenol, 4-methyl-2,6-dimethoxyphenol, 2-cyclo-
penten-1-one, *etc*. A significant characteristic of the bio-oils was the high percentage of alkylated compounds, especially methyl derivatives.

The influence of temperature on the compounds existing in liquid products obtained from biomass samples *via* pyrolysis were examined in relation to the yield
and composition of the product bio-oils. The product liquids were analyzed by
a gas chromatography mass spectrometry combined system. The bio-oils were
composed of a range of cyclopentanone, methoxyphenol, acetic acid, methanol,
acetone, furfural, phenol, formic acid, levoglucosan, guaiocol, and their alkylated
phenol derivatives. Biomass structural components, such as cellulose, hemicelluloses, and were converted into liquids and gas products, as well as a solid residue
of charcoal by thermal depolymerization and decomposition. The structural components of the biomass samples mainly affect pyrolytic degradation products.
A reaction mechanism is proposed that describes a possible reaction route for the
formation of the characteristic compounds found in the oils. The supercritical
water extraction and liquefaction partial reactions also occur during the pyrolysis.
Acetic acid is formed in the thermal decomposition of all three main components
of biomass. In the pyrolysis reactions of biomass, water is formed by dehydration,
acetic acid comes from the elimination of acetyl groups originally linked to the
xylose unit, furfural is formed by dehydration of the xylose unit, formic acid proceeds from carboxylic groups of uronic acid, and methanol arises from methoxyl
groups of uronic acid (Demirbas, 2007).

Figure 4.27 shows the plot for the yield of bio-oils from liquefaction of corn
stover at subcritical and supercritical temperatures. The yields of bio-oils increase
with increasing temperature. The yield of bio-oils sharply increases between 650 K
and 680 K and then it reaches a plateau. This temperature range (650–680 K) is in

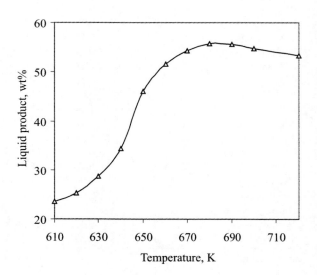

Fig. 4.27 Plot for the yield of liquid products from liquefaction of corn stover at subcritical and supercritical tempera-tures. Liquefaction time: 70 min. Water-to-solid ratio: 5:1

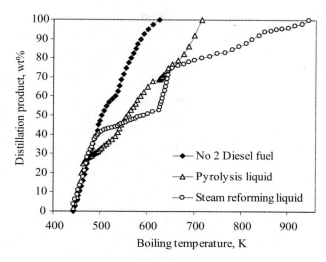

Fig. 4.28 Comparison of distillation curves of average distillation products obtained from pyrolysis and steam reforming of tallow to that of No. 2 diesel fuel

the supercritical zone. At temperatures higher than 690 K the yield of bio-oils are slightly lower. The maximum yield of bio-oil is obtained at reaction temperature 680 K. The highest bio-oil yield of 55.7% is obtained at reaction temperature 680 K, and the yield is 53.3% at 720 K.

Figure 4.28 shows a comparison of distillation curves of average distillation products obtained from pyrolysis and steam reforming of tallow to that of No 2 diesel fuel.

4.8 Other Alternate Liquid Fuels

4.8.1 Glycerol-Based Fuel Oxygenates for Biodiesel and Diesel Fuel Blends

Glycerol (1,2,3-propanetriol or glycerine) is a trihydric alcohol. It is a colorless, odorless, sweet-tasting, syrupy liquid. It melts at 291 K, boils with decomposition at 563 K, and is miscible with water and ethanol (Perry and Green, 1997). The chemical formula for glycerol is $OH–CH_2 –CH (OH)–CH_2 –OH$. Glycerol is present in the form of its esters (triglycerides) in vegetable oils and animal fats.

As early as 2800 BC, glycerol was isolated by heating fats in the presence of ash (to produce soap) and as an industrial chemical. Since the late 1940s, glycerol has been produced from epichlorohydrin obtained from propylene as large chemical companies forecasted a glycerol shortage and initiated its synthetic production (Pagliaro *et al.*, 2007). The global glycerine market is estimated to be 750,000–800,000 metric tons *per* year (Spooner-Wyman and Appleby, 2003).

Glycerol cannot be added directly to fuel because at high temperatures it polymerizes and it is partly oxidized to toxic acrolein (Noureddini *et al.*, 1998). Glycerol must be modified to fuel oxygenate derivatives using as additive in diesel and biodiesel fuel blends. The most obvious derivative of glycerol has an analogy in gasoline reformulation. Oxygenated gasolines are well recognized and the demand for methyl tertiary butyl ethers (MTBEs) has grown to an estimated 473,000 barrels *per* day worldwide (Saunders, 1997). Glycerol tertiary butyl ether (GTBE) may similarly be used for diesel and biodiesel reformulation.

Glycerol is a byproduct obtained during the production of biodiesel. As the biodiesel production is increasing exponentially, the crude glycerol generated from the transesterification of vegetables oils has also been generated in a large quantity (Pachauri and He, 2006). With the increasing production of biodiesel a glut of glycerol has been created, causing market prices to plummet. This situation warrants finding alternative uses for glycerol. Glycerol is directly produced with high purity levels (at least 98%) by biodiesel plants (Ma and Hanna, 1999; Bournay *et al.*, 2005). Research efforts to find new applications of glycerol as a low-cost feedstock for functional derivatives have led to the introduction of a number of selective processes for converting glycerol into commercially valued products (Pagliaro *et al.*, 2007). The principal by-product of biodiesel production is the crude glycerol, which is about 10%wt of vegetable oil. For every 9 kg of biodiesel produced, about 1 kg of a crude glycerol byproduct is formed (Dasari *et al.*, 2005).

Oxygenated compounds such as methyl tertiary butyl ether (MTBE) are used as valuable additives as a result of their antidetonant and octane-improving properties. In this respect, glycerol tertiary butyl ether (GTBE) is an excellent additive with a large potential for diesel and biodiesel reformulation.

Glycerol can be converted into higher value products. The products are 1,3-propanediol, 1,2-propanediol, dihydroxyacetones, hydrogen, polyglycerols, succinic acid, and polyesters. Main glycerol-based oxygenates are 1,3-propanediol, 1,2-propanediol, propanol, glycerol tert-butyl ethers, ethylene glycol, and propylene glycol.

In an earlier study, glycerol was pyrolyzed for the production of clean fuels such as H_2 or a feedstock such as syngas for additional transportation fuel *via* Fischer–Tropsch synthesis. The pyrolysis of glycerol was carried out at various flow rates of N_2 (30–70 mL/min), temperatures (650–800°C), and types and sizes of packing material in a tubular reactor at atmospheric pressure. The products were mostly gas, essentially consisting of CO, H_2, CO_2, CH_4 and C_2H_4 (Valliyappan *et al.*, 2007). Given the highly reduced nature of carbon in glycerol and the cost advantage of anaerobic processes, fermentative metabolism of glycerol is of special interest (Yazdani and Gonzalez, 2007). Glyceride esters have been enzymatically synthesized in solvent-free system (Freitas *et al.*, 2007). The influence of important factors that affect the synthesis of glyceride esters were found to be glycerol/fatty acid molar ratio, lipase source and activating agent of the support obtained by the sol-gel technique.

4.8.1.1   Glycerol Alkyl Ethers (GAE)

Glycerol alkyl ethers (GAEs) are easily synthesized using glycerol that is reacted with isobutylene in the presence of an acid catalyst, and the yield is maximized by carrying out the reaction in a two-phase reaction system, with one phase being a glycerol-rich polar phase (containing the acidic catalyst), and the other phase being an olefin-rich hydrocarbon phase from which the product ethers can be readily separated (Gupta, 1995). On the other hand, if the reaction is carried out over an amberlyst resin, methanol in the crude glycerol must be removed to avoid catalyst poisoning. The best results were obtained for catalyst loading >5 wt%, 1–2 h of reaction time and a glycerol to isobutylene ratio of approximately 3. Lower temperatures (353 K) under these conditions resulted in a higher concentration of di- and tri-ethers and lower byproduct concentrations (Noureddini et al., 1998). The optimization of glycerol ether formulations based on the results of engine tests is currently being carried out in Europe and in the US, where recently also dibutoxy glycerol was shown to act as an excellent fuel oxygenate (Spooner-Wyman and Appleby, 2003).

The etherification of glycerol with tert-butanol has been studied by Klepacova et al. (2003) at the presence of catex Amberlyst 15 as a catalyst. The maximum conversion of glycerol near 96% was reached at the temperature 363 K and at the molar ratio tert-butanol/glycerol = 4:1 after 180 min (Klepacova et al., 2003). The etherification of glycerol with isobutylene or tert-butyl alcohol without solvent in the liquid phase catalyzed by strong acid ion-exchange resins of Amberlyst type and by two large-pore zeolites H-Y (Si/Al = 15) and H-BEA (Si/Al = 12.5) has been studied. The reaction of glycerol with tert-butyl alcohol in the liquid phase on acid Amberlyst-type ion exchange resins has been studied (Klepacova et al., 2006). The best results of glycerol tert-butylation by isobutylene at 100% conversion of glycerol with selectivity to di- and tri-ethers larger than 92% were obtained over strong acid macroreticular ion-exchange resins. Glycerol tert-butyl ethers (GTBEs) or mono-, di- and tri-tert-butyl ethers of glycerol are potential oxygenates to diesel fuel (Klepacova et al., 2005).

The preparation of the glycerol alkyl ethers (GAE) by etherification (o-alkylation) by alkenes, and preferentially by isobutylene, is one of the possibilities of glycerol usage. The GAEs can be also obtained from alkenes with C_4-fraction from pyrolysis and fluid catalytic cracking at the presence of the acid catalyst. A mixture of mono-, di- and tri-alkyl glycerol ethers can be produced. These ethers (mainly di- and tri-alkyl glycerols) are the main products of glycerol etherification reaction and are excellent oxygen additives for diesel fuel (Jamroz et al., 2007). The addition of these ethers has a positive effect on the combustion of diesel fuel (high cetane number, CN), and preferentially ethers are active by reduction of fumes and particulate matters, carbon oxides, and carbonyl compounds in exhausts (Klepacova et al., 2003).

In particular, a mixture of 1,3-di-, 1,2-di-, and 1,2,3-tri-tertbutyl glycerol, when incorporated in standard 30–40% aromatic-containing diesel fuel, leads to significantly reduced emissions of particulate matter, hydrocarbons, carbon monoxide, and unregulated aldehydes (Kesling *et al.*, 1994).

4.8.1.2 Propylene Glycol

Glycerol can be converted to propylene glycol (1,2-propanediol) by dehydration followed by hydrogenation (hydrogenolysis) reaction. Propylene glycol is used in an antifreeze mixture (70% propylene glycol and 30% glycerol) that can be produced, refined, and marketed directly by existing biodiesel facilities. The method is based on hydrogenolysis of glycerol over a copper chromite catalyst at 475 K (Pagliaro *et al.*, 2007).

The reaction pathway proceeds *via* an acetol (hydroxyacetone) intermediate in a two-step process. The first step of forming acetol occurs at atmospheric pressure, while subsequent hydrogenation at 475 K and 10 bar hydrogen gas eventually affords propylene glycol in 73% yield at significantly lower cost than propylene glycol made from petroleum. A main advantage of the process is that the copper-chromite catalyst can be used to convert crude glycerol without further purification (Pagliaro *et al.*, 2007).

4.8.2 P-series Fuels

P-series fuel is a unique blend of liquefied petroleum gas liquids, ethanol, hydrocarbons, and methyltetrahydrofuran (MeTHF). P-series fuels are blends of ethanol, MeTHF, and pentanes, with butane added for blends that would be used in severe cold-weather conditions to meet cold start requirements. P-series fuel is made primarily from biorenewable resources and provides significant emissions benefits over reformulated gasoline. P-series fuel can be mixed with gasoline in any proportion and used in flexible fuel vehicles. P-series fuels are clear, colorless, 89–93 octane, liquid blends that are formulated to be used in flexible fuel vehicles (FFVs). Like gasoline, low vapor pressure formulations are produced to prevent excessive evaporation during summer and high vapor pressure formulations are used for easy starting in winter. P-series fuel is at least 60% non-petroleum. It also has many environmental benefits. Because a majority of the components that make up P-series fuels comes from domestically produced renewable resources, this alternative fuel promotes both energy security and environmental quality. P-series fuels could be 96% derived from domestic resources. P-series fuels could reduce fossil energy use by 49% to 57% and petroleum use by 80% relative to gasoline. Greenhouse gas emissions from the production and use of P-series are substantially better than those from gasoline. Each unit of P-series fuel emits approx 50% less carbon dioxide, 35% less hydrocarbons, and 15% less carbon monoxide than gasoline. It also has 40% less ozone forming potential.

4.8.3 Dimethyl Ether (DME)

Dimethyl ether (DME or CH_3-O-CH_3), is a new fuel that has attracted much attention recently. Today, DME is made from natural gas, it can also be produced by gasifying biomass. DME can be stored in liquid form at 5–10 bar pressure at normal temperature. A major advantage of DME is its naturally high cetane number, which means that self-ignition is easier. The high cetane rating makes DME the most suitable for use in diesel engines, which implies that the high level of efficiency of the diesel engine is retained when using DME. The energy content of DME is lower than in diesel.

DME can be produced effectively from biosyngas in a single-stage, liquid phase (LPDME) process. The origin of syngas includes a wide spectrum of feedstocks such as coal, natural gas, biomass, and others. Non-toxic, high density, liquid DME fuel can be easily stored at modest pressures. The production of DME is very similar to that of methanol. DME conversion to hydrocarbons, lower olefins in particular, is being studied using ZSM-5 catalysts with varying SiO_2/Al_2O_3 ratios, whereas the DME carbonization reaction to produce methyl acetate is studied over a variety of group VIII metal substituted heteropolyacid catalysts.

4.8.4 Fischer–Tropsch (FT) Liquid Fuel from Biomass

Fischer–Tropsch (FT) liquid fuel can be used as alternate diesel fuel. The FT catalytic conversion process can be used to synthesize diesel fuels from a variety of feedstocks, including coal, natural gas and biomass. The alternate fuel source is coal, indirectly converted to diesel fuel. The FT process uses various catalysts to produce linear hydrocarbons and oxygenates, including unrefined gasoline, diesel, and wax ranges.

Synthetic FT diesel fuels can have excellent autoignition characteristics. FT diesel is composed of only straight chain hydrocarbons and has no aromatics or sulfur. Reaction parameters are temperature, pressure, and H_2/CO ratio. FT product composition is strongly influenced by the catalyst composition: the yield of paraffins is higher with cobalt catalytic runs, and the yield of olefins and oxygenates is higher with iron catalytic runs.

Basic FT reactions are:

$$nCO + 2nH_2 \quad \rightarrow n(-CH_2-) + nH_2O \tag{4.27}$$

$$nCO + (2n+1)H_2 \rightarrow C_nH_{2n+2} + nH_2O \tag{4.28}$$

Catalysts and reactors have been extensively investigated for liquid phase Fischer–Tropsch synthesis (Davis, 2002). Synthetic Fischer–Tropsch diesel fuel can provide benefits in terms of both PM and NO_x emissions (May, 2003). Properties of FT and No. 2 diesel fuels are given in Table 4.46.

Table 4.46 Properties of Fischer–Tropsch (FT) diesel and No. 2 diesel fuels

Property diesel	Fischer–Tropsch diesel	No. 2 petroleum
Density, g/cm^3	0.7836	0.8320
Higher heating value, MJ/kg	47.1	46.2
Aromatics, %	0–0.1	8–16
Cetane number	76–80	40–44
Sulfur content, ppm	0–0.1	25–125

FT is most compatible with existing distribution for conventional diesel and only minimal adjustments are required to obtain optimal performance from existing diesel engines. Physical properties of FT are very similar to No. 2 diesel fuel, and its chemical properties are superior in that the FT process yields middle distillates that, if correctly processed (as through a cobalt-based catalyst), contain no aromatics or sulfur compounds.

4.8.5 Other Bio-oxygenated Liquid Fuels

Methanol and ethanol are not the only transportation fuels that might be made from wood. A number of possibilities exist for producing alternatives. The most promising bio-oxygenated fuels, and closest to being competitive in current markets without subsidy, are ethanol, methanol, ethyl-tert-butyl ether, and methyl-tert-butyl ether. Other candidates include isopropyl alcohol, sec-butyl alcohol, tert-butyl alcohol, mixed alcohols, and tert-amylmethyl ether.

Ethanol or grain alcohol is not restricted to grain as a feedstock. It can be produced from other agricultural crops and lignocellulosics such as wood. It has often been advocated as a motor fuel, and has been used frequently in times of gasoline scarcity.

Besides comparisons in production costs, there is a question whether ethanol at the same price *per* gallon as gasoline is of equal value. The heating value of ethanol is less, only 76,500 BTUs *per* gallon as compared to 124,800 for gasoline. However, ethanol is a higher octane number than gasoline, and for that reason it might attain about the same mileage *per* gallon as gasoline. The octane number is a value used to indicate the resistance of a motor fuel to knock. Octane numbers are based on a scale on which isooctane is 100 (minimal knock) and heptane is 0 (bad knock). We are therefore assuming a gallon of gasoline and a gallon of ethanol to be of equal value.

Another possibility for bio-oxygenated fuels is methanol. Methanol could conceivably be made from grain, but its most common source is natural gas. Use of natural gas is better for reducing carbon dioxide production in comparison to other fossil fuels, but use of renewable fuels instead of natural gas would be better still. It can be made from coal or wood with more difficulty and lower efficiency than from natural gas. The cost of making methanol from natural gas is around $ 0.40 *per* gallon. It could probably be sold as a motor fuel for about $ 0.60–$ 0.70 *per* gallon. This would be equivalent to gasoline selling at about $ 0.92–$ 1.03 *per* gallon. Methanol was once produced from wood as a byproduct of charcoal manufacture, but overall yields were low. To produce methanol from wood with a significantly higher yield would require production of synthesis gas in a process similar to that used for production of methanol from coal. Such processes for gasifying wood are less fully developed than the two-stage hydrolysis process for the production of ethanol.

The high-octane rating is characteristic of all oxygenated fuels, including ethanol, methanol, ethyl-tert-butyl ether, and methyl-tert-butyl ether (MTBE). MTBE is made by reacting isobutylene with methanol. Ethyl-tert-butyl ether (ETBE) is made by using ethanol instead of methanol. Thus either ethanol or methanol from either grain or wood could be a factor in making tert-butyl ether octane enhancers. The characteristics of ethers are generally closer to those of gasolines than those of alcohols. Ethers are benign in their effect on fuel system materials and are miscible in gasoline; therefore, they are not subject to phase separation in the presence of water, as are methanol and ethanol.

Alternative fuels from wood, as well as grain, have a potential for being competitive with gasoline and diesel motor fuels from petroleum, even without subsidization. For ethanol to compete directly, without subsidy, oil would probably have to sell for $40 or more *per* barrel. However, the environmental and octane-enhancing benefits of ethanol and other oxygenated fuels that may be produced from grain and wood may make them worth more than comparisons on fuel value alone would indicate.

Diesel fuels or gasoline from wood are possibilities through a number of approaches. The Fischer–Tropsch pyrolysis process, used successfully for converting coal to synthesis gas in South Africa could also be used to make synthesis gas from wood. Synthesis gas could then be used to make gasoline or diesel fuel. Or methanol could be produced from wood and then, by a catalyzed reaction known as the Mobil process, be transformed to gasoline.

In the two-stage hydrolysis process, for every 100 kg of oven dry wood feedstock about 20 kg carbohydrates suitable for processing to ethanol are obtained from the second stage. There are more carbohydrates derived from the first stage, about 24.9 kg, but many of these first stage carbohydrates are not necessarily fermentable to ethanol. Ethanol is a possibility if xylose can be fermented to ethanol economically. Fermentation of the xylose and glucose from the first stage could result in almost doubling the ethanol production as compared to only fermenting the glucose from the second stage. Other possible products from the first stage carbohydrates are single-cell protein, furfural, and feed molasses.

References

AICHE. 1997. Alternative transportation fuels: a comparative analysis. Report by AICHE Government Relations Committee, September.

ASAE, 1982. Vegetable oil fuels. Proceedings of the international conference on plant and vegetable oils as fuels. Leslie Backers, editor. ASAE, St Joseph, MI.

Ali, Y., Hanna, M.A., Cuppett, S.L. 1995. Fuel properties of tallow and soybean oil esters. JAOCS 72:1557–1564.

An, J., Bagnell, L., Cablewski, T., Strauss, C.R., Trainor, R.W. 1997. Applications of high-temperature aqueous media for synthetic organic reactions. J Org Chem 62:2505–2511.

Azam, M.M., Waris, A., Nahar, N.M. 2005. Prospects and potential of fatty acid methyl esters of some non-traditional seed oils for use as biodiesel in India. Biomass Bioenergy 29:293–302.

Azar, C., Lindgren, K., Andersson, B.A. 2003. Global energy scenarios meeting stringent CO2 constraints – Cost-effective fuel choices in the transportation sector. Energy Policy 31:961–976.

Bala, B.K. 2005. Studies on biodiesels from transformation of vegetable oils for diesel engines. Energy Edu Sci Technol 15:1–45.

Balat, M. 2005. Bio-diesel from vegetable oils via transesterification in supercritical ethanol. Energy Edu Sci Technol 16:45–52.

Balat, M., Balat, H., Oz, C. 2008. Progress in bioethanol processing. Progress Energy Combus Sci (in press).

Bajpai, S., Prajapati, S., Luthra, R., Sharma, S., Naqvi, A., Kumar, S. 1999. Variation in the seed and oil yields and oil quality in the Indian germplasm of opium poppy Papaver somniferum. Genetic Res Crop Evol 46:435–439.

Barsic, N.J., Humke, A.L. 1981. Performance and emissions characteristics of a naturally aspi-rated diesel engine with vegetable oil fuels. SAE paper no. 810262, Society of Automotive Engineers, Warrendale, PA.

Beaumont, O. 1985. Flash pyrolysis products from beech wood. Wood Fiber Sci. 17:228–239.

Bergeron, P. 1996. Environmental impacts of bioethanol, In Wyman CH (ed.), Handbook on bioethanol: Production and utilization. Taylor and Francis, Washington DC, pp. 163–178.

Bhatia, S., Heng, J.K., Lim, M.L., Mohamed, A.R. 1998. Production of biofuel by catalytic cracking of palm oil: Performance of different catalyst. Proc. Biofuel, PORIM Intl. Biofuel and Lubricant Conf., Malaysia, pp. 107–112.

Bhatia, S., Twaiq, F.A., Zabidi, N.A.M. 1999. Catalytic conversion of palm oil to hydrocarbons: Performance of various zeolite catalysts. Ind Eng Chem Res 38:3230–3237.

Bilgin, A., Durgun, O., Sahin Z. 2002. The effects of diesel-ethanol blends on diesel engine performance. Energy Sources 24:431–440.

Billaud, F., Dominguez, V., Broutin, P., Busson C. 1995. Production of hydrocarbons by pyroly-sis of methyl esters from rapeseed oil. J Am Oil Chem Soc 72:1149.

Bowen, D.A., Lau, F., Zabransky, R., Remick, R., Slimane, R., Doong, S. 2003.Techno-economic analysis of hydrogen production by gasification of biomass. NREL 2003 progress report, National Renewable Energy Laboratory, USA.

Brown, H.P., Panshin, A.J., Forsaith, C.C. 1952. Textbook of wood technology, Vol. II. McGraw-Hill, New York.

Canakci, M., Erdil, A., Arcaklioglu, E. 2006. Performance and exhaust emissions of a biodiesel engine. Applied Energy 83:594–605.

Chigier, N.A. 1981. Energy, combustion and the environment. McGraw Hill, New York.

Chmielniak, T., Sciazko, M. 2003. Co-gasification of biomass and coal for methanol synthesis. Applied Energy 74:393–403.

Dandik, L., Aksoy, H.A. 1998. Pyrolysis of used sunflower oil in the presence of sodium carbon-ate by using fractionating pyrolysis reactor. Fuel Proc Technol 57:81–92.

Darnoko, D., Cheryan, M. 2000. Kinetics of palm oil transesterification in a batch reactor. JAOCS 77:1263–1267.

Das, M., Das, S.K., Suthar, S.H. 2002. Composition of Seed and Characteristics of Oil from Karingda [(Citrullus lanatus Thumb) Man of]. Int J Food Sci Technol 37:893–896.

Dasari, M.A., Kiatsimkul, P.P., Sutterlin, W.R., Suppes, G.J. 2005. Low-pressure hydrogenolysis of glycerol to propylene glycol. Applied Catalysis A: General 281:225–231.

Demirbas, A. 1998. Fuel properties and calculation of higher heating values of vegetable oils. Fuel 77:1117–1120.

Demirbas, A. 2000. Mechanism of liquifaction and pyrolysis reactions of biomass. Energy Convers Mgmt 41:633–46.

Demirbas, A. 2002. Diesel fuel from vegetable oil via transesterification and soap pyrolysis. Energy Sources 24:835–841.

Demirbas, A. 2003a. Biodiesel fuels from vegetable oils via catalytic and non-catalytic supercritical alcohol transesterifications and other methods: a survey. Energy Convers Mgmt 44: 2093–2109.

Demirbas, A. 2003b. Fuel conversional aspects of palm oil and sunflower oil. Energy Sources 25:457–466.

Demirbas, A. 2003c. Fuel conversional aspects of palm oil and sunflower oil. Energy Sources25:457–466.

Demirbas, A. 2004a. Ethanol from cellulosic biomass resources. Int J Green Energy 1:79–87.

Demirbas, A. 2004b. Pyrolysis of municipal plastic wastes for recovery of gasoline-range hydrocarbons. J Anal Appl Pyrolysis 2004;72:97–102.

Demirbas, A. 2005a. Potential applications of renewable energy sources, biomass combustion problems in boiler power systems and combustion related environmental issues. Progress Energy Combus Sci 31:171–192.

Demirbas, 2005b. Bioethanol from cellulosic materials: A renewable motor fuel from biomass. Energy Sources 27:327–337.

Demirbas, A. 2007. The influence of temperature on the yields of compounds existing in bio-oils obtaining from biomass samples via pyrolysis. Fuel Proc Technol 88:591–597.

Demirbas, A. 2008. The Importance of bioethanol and biodiesel from biomass. Energy Sources Part B 3:177–185.

Demirbas, A., Güllü, D. 1998. Acetic acid, methanol and acetone from lignocellulosics by pyrpolysis. Energy Edu. Sci. Technol. 1 (2):111–115.

Encinar, J.M., Gonzalez, J.F., Rodriguez, J.J., Tejedor, A. 2002. Biodiesel fuels from vegetable oils: Transesterification of Cynara cardunculus L. oils with ethanol. Energy Fuels 16:443–450.

Erickson, D.R., Pryde, E.H., Brekke, O.L., Mounts, T.L., Falb, R.A. 1980. Handbook of soy oil processing and utilization. American Soybean Association and the American Oil Chemistís Society. St. Louis, Missouri and Champaign, IL.

EPA (Environmental Protection Agency). 2002. A comprehensive analysis of biodiesel impacts on exhaust emissions, EPA Draft Technical Report No: 420-P-02-001.

Foidl, N., Foidl, G., Sanchez, M., Mittelbach, M., Hackel S. 1996. Jatropha curcas L. As a source for the production of biofuel in nicaragua. Biores Technol 58:77–82.

Freitas, L., Perez, V.H., Santos, J.C., de Castro, H.F. 2007. Enzymatic synthesis of glyceride esters in solvent-free system: Influence of the molar ratio, lipase source and functional activating agent of the support. J Braz Chem Soc 18:1360–1366.

Geyer, S.M., Jacobus, M.J., Lestz, S.S. 1984. Comparison of diesel engine performance and emissions from neat and transesterified vegetable oils. Transactions of the ASAE 27:375–381.

Giannelos, P.N., Zannikos, F., Stournas, S., Lois, E., Anastopoulos, G. 2002. Tobacco seed oil as an alternative diesel fuel: Physical and chemical properties. Ind Crop Prod 16:1–9.

Goering, E., Schwab, W., Daugherty, J., Pryde, H., Heakin, J., 1982. Fuel properties of eleven vegetable oils. Trans ASAE 25, 1472–1483.

Goering, C.E. 1984. Final report for project on effect of nonpetroleum fuels on durability of direct-injection diesel engines under contract 59-2171-1-6-057-0, USDA, ARS, Peoria, IL.

Grassi, G. 1999. Modern bioenergy in the European Union. Renewable Energy 16:985–990.

Gubitz, G.M., Mittelbach, M., Trabi, M. 1999. Exploitation of the tropical seed plant *Jatropha Curcas* L. Biores Technol 67:73–82.

Güllü, D., Demirbas, A. 2001. Biomass to methanol via pyrolysis process. Energy Convers Mgmt 42:1349–1356.

Gunstone, F.D., Hamilton, R.J (ed.). 2001. Oleochemicals manufacture and applications. Sheffield Academic Press/CRC Press, Sheffield, UK/Boca Raton, FL.

Gupta, V.P. 1995. Glycerine ditertiary butyl ether preparation, U.S. Patent KO. 5476971.

Gusakov, A.V., Sinitsyn, A.P., Salanovich, T.N., Bukhtojarov, F.E., Markov, A.V., Ustinov, B.B., van Zeijl, C., Punt, P., Burlingame, R. 2005. Purification, cloning and characterisation of two forms of thermostable and highly active cellobiohydrolase I (Cel7A) produced by the industrial strain of Chrysosporium lucknowense. Enzyme Microbial Technol 36:57–69.

Gusmão, J., Brodzki, D., Djéga-Mariadassou, G., Frety, R. 1989.Utilization of vegetable oils as an alternative source for diesel-type fuel: hydrocracking on reduced Ni/SiO_2 and sulphided $Ni-Mo/\gamma-Al_2O_3$. Catal Today 5:533–544.

Hall, D.O., Rosillo-Calle, F., Williams, R.H., Woods, J. 1993. Biomass for energy: Supply prospects. In: Johansson, T.B., Kelly, H., Reddy, A.K.N., Williams, R.H. (eds.), Renewable energy – Sources for fuels and electricity. Island Press, Washington, DC.

Hao, X.H., Guo, L.J. 2002. A review on investigation of hydrogen production by biomass catalytic gasification in supercritical water. Huagong Xuebao 53:221–228 [in Chinese].

He, Y., Bao, Y. D. 2003. Study on rapeseed oil as alternative fuel for a single-cylinder diesel engine. Renev Energy 28:1447–1453.

Isigigur, A., Karaosmonoglu, F., Aksoy, H.A. 1994. Methyl ester from safflower seed oil of Turkish origin as a biofuel for diesel engines. Appl Biochem Biotechnol 45/46:103–112.

Islam, M.N., Beg, M.R.A. 2004. The fuel properties of pyrolysis liquid derived from urban solid wastes in Bangladesh. Bioresource Technology 92:181–186.

Iwasa, N., Kudo, S., Takahashi, H., Masuda, S., Takezawa, N. 1993. Highly selective supported Pd catalysts for steam reforming of methanol. Catal Lett 19:211–216.

Jamróz, M.E., Jarosz, M., Witowska-Jarosz, J., Bednarek, E., Tęcza, W., Jamróz, M.H., Dobrowolski, J.C., Kijeński, J. 2007. Mono-, di-, and tri-tert-butyl ethers of glycerol. A molecular spectroscopic study. Specrochim Acta Part A: Mol Biomaol Spectroscopy 67:980–988.

Kadiman O.K. 2005. Crops: Beyond Foods, In: Proceedings of the 1st International Conferenceon Crop Security, Malang, Indonesia, September 20–23, 2005.

Kaplan, C., Arslan, R., Surmen, A. 2006. Performance characteristics of sunflower methyl esters as biodiesel. Energy Sources, Part A 28:751–755.

Karaosmonoglu, F. 1999. Vegetable oil fuels: a review. Energy Sources 21:221–31.

Karmee, S.K., Chadha, A. 2005. Preparation of biodiesel from crude oil of *Pongamia pinnata*. Biores Technol 96:1425–1429.

Kerschbaum, S., Rinke, G. 2004. Measurement of the temperature dependent viscosity of biodiesel fuels. Fuel 83:287–291.

Katikaneni, S.P.R., Adjaye, J.D., Bakhshi, N.N. 1995. Catalytic conversion of canola oil to fuels and chemicals over various cracking catalysts. Can J Chem Eng 73:484–497.

Kesling Jr., H.S., Karas, L.J., Liotta Jr., F.J. 1994. Diesel fuel, U.S. Patent 5,308,365, May 3, 1994.

Klepacova, K., Mravec, D., Hajekova, E., Bajus, M. 2003. Etherification of glycerol. Petroleum Coal 45:54–57.

Klepacova, K., Mravec, D., Bajus, M. 2005. tert-Butylation of glycerol catalysed by ion-exchange resins. Appl Catal A: General 294:141–147.

Klepacova, K., Mravec, D., and Bajus, M. 2006. Etherification of glycerol with tert-butyl alcohol catalysed by ion-exchange resins. Chem. Papers 60:224–230.

Komers, K., Stloukal, R., Machek, J., Skopal, F. 2001. Biodiesel from rapeseed oil, methanol andKOH 3. Analysis of composition of actual reaction mixture. Eur J Lipid Sci Technol 103:363–3471.

Knothe, G., Krahl, J., Van Gerpen, J., eds. 2005. The biodiesel handbook. AOCS Press, Champaign, IL.

Knothe, G., Sharp, C.A., Ryan, T.W. 2006. Exhaust emissions of biodiesel, petrodiesel, neat methyl esters, and alkanes in a new technology engine. Energy Fuels 20:403–408.

Kuhlmann, B., Arnett, E.M., Siskin, M. 1994. Classical organic reactions in pure superheated water. J Org Chem 59:3098–3101.

Kusdiana, D., Saka, S. 2001. Kinetics of transesterification in rapeseed oil to biodiesel fuels as treated in supercritical methanol. Fuel 80:693–698.

Kusdiana, D., Saka, S. 2004. Effects of water on biodiesel fuel production by supercritical methanol treatment. Biores Technol 91:289–295.

Leng, T.Y., Mohamed, A.R., Bhatia, S. 1999. Catalytic conversion of palm oil to fuels and chemicals. Can J Chem Eng 77:156–162.

Lima, D.G., Soares, V.C D., Ribeiro, E.B., Carvalho, D.A., Cardoso, E.C.V., Rassi, F.C., Mundim, K.C., Rubim, J.C., Suarez, P. A.Z. 2004. Diesel-like fuel obtained by pyrolysis of vegetable oils. J Anal Appl Pyrolysis 71:987–996.

Ma, F., Hanna, M.A. 1999. Biodiesel production: a review. Biores. Technol. 70:1–15.

Machacon, H.T.C., Matsumoto, Y., Ohkawara, C., Shiga, S., Karasawa, T., Nakamura, H. 2001. The effect of coconut oil and diesel fuel blends on diesel engine performance and exhaust emissions. JSAE Review 22:349–355.

Madras, G., Kolluru, C., Kumar, R. 2004. Synthesis of biodiesel in supercritical fluids. Fuel 83:2029–2033.

Matsumura, Y., Minowa, T. 2004. Fundamental design of a continuous biomass gasification process using a supercritical water fluidized bed. Int J Hydrogen Energy 29(7):701–7.

Meher, L.C. Kulkarni, M.G., Dalai, A.K., Naik, S.N. 2006a. Transesterification of karanja (*Pongamia pinnata*) oil by solid basic catalysts. European J Lipid Sci Technol 108:389–397.

Meher, L.C., Dharmagadda, V.S.S., Naik, S.N. 2006b. Optimization of alkali-catalyzed transesterification of *Pongamia pinnata* oil for production of biodiesel. Biores Technol 97:1392–1397.

Meher, L.C., Vidya Sagar, D., Naik, S.N. 2006c. Technical aspects of biodiesel production by transesterification – A review. Renew Sustain Energy Rev 10:248–268.

Mills, G.A., Howard, A.G. 1983. A preliminary investigation of polynuclear aromatic hydrocarbon emissions from a diesel engine operating on vegetable oil-based alternative fuels. J Inst Energy 56:131–137.

Mislavskaya, V.S., Leonow, V.E., Mislavskii, N.O., Ryzhak, I. A. 1982. Conditions of phase stability in a gasoline-methanol-cyclohexanol-water system. Soviet Chem Ind 14:270–276.

Mittelbach, M., Gangl, S. 2001.Long storage stability of biodiesel made from rapeseed and used frying oil. JAOCS 78:573–577.

Mohan, D., Pittman Jr,. C.U., Steele, P.H. 2006. Pyrolysis of wood/biomass for bio-oil: a critical review. Energy Fuels 2006;20:848–889.

Morrison, R.T., Boyd, R.N. 1983. Organic Chemistry, 4th edn. Allyn and Bacon Inc., Singapore.

Mulkins-Phillips, G.J., Stewart J.E. 1974. Effect of environmental parameters on bacterial degradation of bunker C oil, crude oils, and hydrocarbons. Appl Microbiol 28:915–922.

Najafi, G., Ghobadian, B., Yusaf, T.F., Rahimi, H. 2007. Combustion analysis of a CI engine performance using waste cooking biodiesel fuel with an artificial neural network aid. Am J Appl Sci 4:756–764.

Nelson, C.R., Courter, M.L. 1954. Ethanol by hydration of ethylene. Chem Eng Prog 50:526–32.

Nitschke, W.R., Wilson, C.M. 1965. Rudolph Diesel, pioneer of the age of power. The University of Oklahoma Press, Norman, OK.

Noureddini, H., Dailey, W.R., Hunt, B.A. 1998. Production of ethers of glycerol from grude glycerol. University of Nebraska, Lincoln. http://digitalcommons.unl.edu/chemeng_biomaterials/18/.

Noureddini, H., Gao, X., Philkana, R.S. 2005. Immobilized Pseudomonas cepacia lipase for biodiesel fuel production from soybean oil. Biores Technol 96:769–777.

Osten, D.W., Sell, N.J. 1983. Methanol-gasoline blends: Blending agents to prevent phase separation. Fuel 62:268–270.

Ouellette, N., Rogner, H.-H., Scott, D.S. 1997. Hydrogen-based industry from remote excess hydroelectricity. Int.J. Hydrogm Energy 22:397–403.

Ozkan, M., Ergenc, A.T., Deniz, O. 2005. Experimental performance analysis of biodiesel, traditional diesel and biodiesel with glycerine. Turkish J Eng Env Sci 29:89–94.

Pachauri, N., He, B. 2006. Value-added utilization of crude glycerol from biodiesel production: A survey of current research activities. ASABE Annual International Meeting, Portland, Oregon, 9–12 July 2006, pp. 1–16.

Pagliaro, M., Ciriminna, R., Kimura, H., Rossi, M., Pina, C.D. 2007. From glycerol to value-added products. Angew Chem Int Ed 46:4434–440.

Perry, R.H., Gren, D.W. 1997. Perry's chemical engineers' handbook. McGraw-Hill, New York, pp. 2–39.

Philippidis, G.P., Smith, T.K. 1995. Limiting factors in the simultaneous saccharification and fermentation process for conversion of cellulosic biomass to fuel ethanol. Appl Biochem Biotechnol 51/52:117–124.

Phillips, V.D., Kinoshita, C.M., Neill, D.R., Takashi, P.K. 1990. Thermochemical production of methanol from biomass in Hawaii. Appl. Energy 35:167–175.

Piel, W.J. 2001. Transportation fuels of the future? Fuel Proces Technol 71:167–179.

Pioch, D., Lozawa, M., Rasoanatoandro, C., Graile, S, Geneste P, Guida A. Biofuels from catalytic cracking of tropical vegetable oils. Oleagineaux 1993;48:289–291.

Pinto, A.C., Guarieiro, L.L.N., Rezende, M.J.C., Ribeiro, N.M., Torres, E.A., Lopes, W.A., Pereira, P. A.P., Andrade, J.B. 2005. Biodiesel: An overview. J Brazilian Chem Soc 16:1313–1330.

Pischinger, G.M., Falcon, A.M, Siekmann, W., Fernandes, F.R. 1982. Methyl esters of plant oils as diesels fuels, either straight or in blends. Vegetable Oil Fuels, ASAE Publication 4–82, Amer Soc Agric Engrs, St. Joseph, MI, USA.

Prakash, C.B.1998. A critical review of biodiesel as a transportation fuel in Canada. A Technical Report. GCSI – Global Change Strategies International Inc., Canada.

Pryde, E.H. 1983. Vegetable oil as diesel fuel: Overview. JAOCS 60, 1557–1558.

Pryor, R.W., Hanna, M.A., Schinstock, J.L., Bashford, L.L. 1982. Soybean oil fuel in a small diesel engine. Trans ASAE 26:333–338.

Puhan, S., Vedaraman, N., Sankaranarayanan, G., Bharat Ram B. V. 2005. Performance and emission study of Mahua oil (Madhuca indica oil) ethyl ester in a 4-stroke natural aspirated direct injection diesel engine. Renew Energy 30:1269–78.

Radovanovic, M., Venderbosch, R.H., Prins, W., van Swaaij, W.P.M. 2000. Some remarks on the viscosity measurement of pyrolysis liquids. Biomass Bioenergy 18:209–222.

Ramadhas, A.S., Jayaraj, S., Muraleedharan, C. 2004. Biodiesel production from high FFA rubber seed oil. Fuel 84:335–340.

Roels, J.A.N., Kossen, W.F. 1978. On the modeling of microbial metabolism. Progress Ind Microbiol 14:95–108.

Rowell, R.M., Hokanson, A.E. 1979. Methanol from wood: a critical assessment. In: Sarkanen, K.V., Tillman, D.A. (eds) Progress in biomass conversion, Vol. 1.Academic Press, New York.

Saka, S., Kusdiana, D. 2001. Biodiesel fuel from rapeseed oil as prepared in supercritical methanol. Fuel 80:225–231.

Sang, O.Y., Twaiq, F., Zakaria, R., Mohamed, A., Bhatia, S. 2003. Biofuel production from catalytic cracking of palm oil. Energy Sources 25:859–869.

Sasaki, M., Kabyemela, B.M., Malaluan, R.M., Hirose, S., Takeda, N., Adschiri, T., Arai, K. 1998.Cellulose hydrolysis in subcritical and supercritical water. J. Supercritical Fluids 13:261–268.

Saucedo, E. 2001. Biodiesel. Ingeniera Quimica 2001:20:19–29.

Saunders, B. 1997, Major overcapacity woes loom for oxygenates despite rising demand. Oil Gas J 95:21–22.

Schelenk, H., Gellerman, J.L. 1960. Esterification of fatty acids with diazomethane on a small scale. Anal. Chem. 32:1412–1414.

Schwab, A.W., Bagby, M.O., Freedman B. 1987. Preparation and properties of diesel fuels from vegetable oils. Fuel 66:1372–1378.

Sensoz, S., Angin, D., Yorgun, S. 2000. Influence of particle size on the pyrolysis of rapeseed (*Brassica napus L.*): fuel properties of bio-oil. Biomass Bioenergy 2000;19:271–279.

Shah, S., Sharma, A., Gupta, M.N. 2004. Extraction of oil from *Jatropha cursas* L. seed kernels by enzyme assisted three phase partitioning. J Crops Prod 20:275–279.

Shah, S., Sharma, A., Gupta, M.N. 2005. Extraction of oil from *Jatropha cursas* L. seed kernels by combination of ultrasonication and aqueous enzymatic oil extraction. Biores Technol 96:121–123.

Shah, S., Gupta, M.N. 2007. Lipase catalyzed preparation of biodiesel from *Jatropha* oil in a solvent free system. Proc Biochem 42:409–414.

Shay, E.G. 1993. Diesel fuel from vegetable oils: status and opportunities. Biomass Bioenergy 4:227–242.

Sheehan, J., Cambreco, V., Duffield, J., Garboski, M., Shapouri, H. 1998. An overview of biodiesel and petroleum diesel life cycles. A report by US Department of Agriculture and Energy, pp: 1–35.

Song, C. 2000. Introduction to chemistry of diesel fuels. In: Song, C., Hsu, C.S., Moshida, I (eds.), Chemistry of Diesel Fuels, p. 13. Taylor and Francis, London.

Sonntag, N.O.V. 1979. Reactions of fats and fatty acids. Bailey's industrial oil and fat products, vol. 1, 4th edition, ed. Swern, D., p. 99. Wiley, New York.

Sorensen, H.A. 1983. Energy conversion systems. Wiley., New York.

Spooner-Wyman, J., Appleby, D.B. 2003. Heavy-duty diesel emissions characteristics of glycerol ethers.25th Symposium on Biotechnology for Fuels and Chemicals, Breckenridge, CO: http://www.nrel.gov/biotechsymp25/session5_pp.html.

Srivastava, A., Prasad, R. 2000. Triglycerides-based diesel fuels. Renew. Sustain Energy Rev 4:111–133.

Sun, Y. 2002. Enzymatic hydrolysis of rye straw and bermudagrass for ethanol production, PhD Thesis, Biological and Agricultural Engineering, North Carolina State University, August 12, 2002.

Tabka, M.G., Herpoël-Gimbert, I., Monod, F., Asther, M., Sigoillot, J.C. 2006. Enzymatic saccharification of wheat straw for bioethanol production by a combined cellulase xylanase and feruloyl esterase treatment. Enzyme Microbial Technol 39:897–902.

Taherzadeh, M.J. 1999. Ethanol from lignocellulose: Physiological effects of inhibitors and fermentation dtrategies. Thesis for the Degree of Doctor of Philosophy, Department of Chemical Reaction Engineering, Chalmers University of Technology, Göteborg, Sweden.

Takezawa, N., Shimokawabe, M., Hiramatsu, H., Sugiura, H., Asakawa, T., Kobayashi, H. 1987. Steam reforming of methanol over Cu/ZrO_2. Role of ZrO_2 support. React Kinet Catal Lett 33:191–196.

Timell, T.E. 1967. Recent progress in the chemistry of wood hemicelluloses. Wood Sci Technol1:45–70.

Valliyappan, T., Bakhshi, N.N., Dalai, A.K.2007. Pyrolysis of glycerol for the production of hydrogen or syn gas. Biores Technol 99:4476–4483.

Väljamäe, P., Pettersson, G., Johansson, G. 2001. Mechanism of substrate inhibition in cellulose synergistic degradation. European J Biochem 268:4520–4526.

Van Gerpen, J., Shanks, B., Pruszko, R., Clemens, D., Knothe, G. 2004. Biodiesel production technology. National Renewable Energy Laboratory, NREL/SR-510-36240, Colorado, USA.

Vasudevan, P., Sharma, S., Kumar, A. 2005. Liquid fuel from biomass: An overview. J Sci Ind Res 64:822–831.

Vicente, G., Martinez, M., Aracil, J. 2004. Integrated biodiesel production: a comparison of different homogeneous catalysts systems. Biores. Technol. 92:297–305.

Walker, D., Petrakis, L., Colwell, R.R. 1976. Comparison of biodegradability of crude and fuel oils. Can J Microbiology 22:598–602.

Wang, D., Czernik, S., Chornet, E. 1998. Production of hydrogen from biomass by catalytic steam reforming of fast pyrolysis oils. Energy Fuels 12:19–24.

Wikiepedia Encyclopedia, 2008. Neem oil characteristics, web: http://en.wikipedia.org/wiki/Neem_oil.

Yarmo, M.A., Alimuniar, A., Ghani, R.A., Suliaman, A.R., Ghani, M., Omar, H., Malek, A. 1992. Transesterification products from the metathesis reaction of palm oil. J Mol Cat 76:373–379.

Yazdani, S.S., Gonzalez, R. 2007. Anaerobic fermentation of glycerol: a path to economic viability for the biofuels industry. Current Opinion Biotechnol.18: 213–219.

Yoshida, T., Oshima, Y., Matsumura, Y. 2004. Gasification of biomass model compounds and real biomass in supercritical water. Biomass Bioenergy 26:71–8.

Zhang, X. 1996. Biodegradability of biodiesel in the aquatic and soil environments. Ph. D. dissertation, Dept. of Biol. and Agr. Engr., University of Idaho, Moscow, ID.

Zhang, X., Peterson, C., Reece, D., Haws, R., Moller, G., 1998. Biodegradability of biodiesel in the aquatic environment. Transactions of the ASAE 41:423–1430.

Zhang, Y., Dub, M. A., McLean, D. D., Kates, M. 2003. Biodiesel production from waste cooking oil: 2. Economic assessment and sensitivity analysis. Biores Technol 90:229–240.

Zhang, Y.H.P., Himmel, M.E., Mielenz, J.R. 2006. Outlook for cellulase improvement: Screening and selection strategies. Biotechnol Advances 24:452–481.

Ziejewski, M., Goettler, H., Pratt, G. L. 1986. Paper No. 860301, International Congress and Exposition, Detroit, MI, 24–28 February.

Zhenyi, C., Xing, J., Shuyuan, L., Li, L. 2004. Thermodynamics calculation of the pyrolysis of vegetable oils. Energy Sources 26:849–856.

Chapter 5
Biorenewable Gaseous Fuels

5.1 Introduction to Biorenewable Gaseous Fuels

Main biorenewable gaseous fuels are biogas, landfill gas, gaseous fuels from pyrolysis and gasification of biomass, gaseous fuels from Fischer–Tropsch synthesis and biohydrogen. There are a number of processes for converting of biomass into gaseous fuels such as methane or hydrogen. One pathway uses plant and animal wastes in a fermentation process leading to biogas from which the desired fuels can be isolated. This technology is established and in widespread use for waste treatment. Anaerobic digestion of biowastes occurs in the absence of air, the resulting gas called biogas is a mixture consisting mainly of methane and carbon dioxide. Biogas is a valuable fuel that is produced in digesters filled with the feedstock like dung or sewage. The digestion is allowed to continue for a period of from ten days to a few weeks. A second pathway uses algae and bacteria that have been genetically modified to produce hydrogen directly instead of the conventional biological energy carriers. Finally, high-temperature gasification supplies a crude gas, which may be transformed into hydrogen by a second reaction step. This pathway may offer the highest overall efficiency.

Biobased products have plant and animal materials as their main ingredients. Sustainable and economically advantageous biobased products can be grown and processed close to their point of use. Biological treatment of organic waste is an old method. Biological waste treatment can be carried out in two principally different ways: by anaerobic digestion, *i.e.*, biogas production or by composting.

A xenobiotic compound is a (foreign = xeno) synthetic compound not normally found in nature. Examples include: (1) pesticides, (2) detergents, and (3) plastics and various other synthetic polymers. A wide range of xenobiotic compounds are metabolized by cytochrome P450 (CYP) enzymes, and the genes that encode these enzymes are often induced in the presence of such compounds. Biological treatment of organic waste by aerobic composting and anaerobic digestion can be compared with respect to a number of environmental effects and sustainability

A. Demirbas, *Biofuels*,
© Springer 2009

criteria including energy balance, nutrient recycling, global warming mitigation potential, emission of xenobiotic compounds, and economy.

5.2 Biogas

Anaerobic digestion (AD) is the conversion of organic material directly to a gas, termed biogas, a mixture of mainly methane and carbon dioxide with small quantities of other gases such as hydrogen sulfide. Methane is the major component of the biogas used in many homes for cooking and heating. Biogas has a chemical composition close to that of natural gas. The biodigester, or a biogas plant, is a physical structure used to provide an anaerobic condition that stimulates various chemical and microbiological reactions resulting in the decomposition of input slurries and the production of biogas – mainly methane (Demirbas and Ozturk, 2004).

Biogas can be used after appropriate gas clean-up as a fuel for engines, gas turbines, fuel cells, boilers, industrial heaters, other processes, or for the manufacturing of chemicals. Before landfilling, treatment or stabilization of biodegradable materials can be accomplished by a combination of anaerobic digestion followed by aerobic composting.

The same types of anaerobic bacteria that produced natural gas also produce methane today. Anaerobic bacteria are some of the oldest forms of life on Earth. They evolved before the photosynthesis of green plants released large quantities of oxygen into the atmosphere. Anaerobic bacteria break down or digest organic material in the absence of oxygen and produce biogas as a waste product.

The first methane digester plant was built at a leper colony in Bombay, India, in 1859 (Meynell, 1976). Most of the biogas plants utilize animal dung or sewage. The schematic of biogas plant utilizing cow dung is illustrated in Fig. 5.1 (Balat, 2008). Anaerobic digestion is a commercially proven technology and is widely

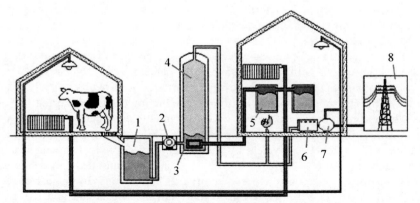

Fig. 5.1 Schematic of a biogas plant utilizing cow dung. 1: Compost storage, 2: pump, 3: internal heater, 4: digester, 5: combustor, 6–8: power generators

used for treating high moisture content organic wastes including +80%–90% moisture. Biogas can be used directly in spark ignition gas engines (SIGEs) and gas turbines. Used as a fuel in SIGE to produce electricity only, the overall conversion efficiency from biomass to electricity is about 10%–16% (Demirbas, 2006a).

5.2.1 Aerobic Conversion Processes

Aerobic conversion includes most commercial composting and activated sludge wastewater treatment processes. Aerobic conversion uses air or oxygen to support the metabolism of the aerobic microorganisms degrading the substrate. Nutritional considerations are also important for the proper functioning of aerobic processes. Aerobic processes operate at much higher reaction rates than anaerobic processes and produce more cell mass, but generally do not produce useful fuel gases. Aerobic decomposition can occur from as low as near freezing to about 344 K.

Respiration refers to those biochemical processes in which organisms oxidize organic matter and extract the stored chemical energy needed for growth and reproduction. Respiration patterns may be subdivided into two major groups, based on the nature of the ultimate election acceptor. Although alternative pathways exist for the oxidation of various organic substrates, it is convenient to consider only the degradation of glucose. The breakdown of glucose is *via* the Embden–Meyerof–Parnas glycolytic pathway, which yields 2 moles each of pyruvate, ATP, and reduced nicotinamide adenine dinucleotide (NAD) *per* mole of glucose.

Under aerobic conditions, the pyruvate is oxidized to CO_2 and H_2O *via* the tricarboxylic acid or Krebs cycle and the electron transport system. The net yield for glycolysis followed by complete oxidation is 38 moles ATP *per* mole glucose, although there is evidence that the yield for bacteria is 16 moles ATP *per* mole glucose (Aiba *et al.*, 1973). Thus, 673 kcal are liberated *per* mole glucose, much of which is stored as ATP.

5.2.2 Anaerobic Conversion Processes

Anaerobic digestion (AD) is a bacterial fermentation process that is sometimes employed in wastewater treatment for sludge degradation and stabilization. This is also the principal process occurring in the decomposition of food wastes and other biomass in landfills. AD operates without free oxygen and results in a fuel gas called biogas, containing mostly CH_4 and CO_2, but frequently carrying other substances such as moisture, hydrogen sulfide (H_2S), and particulate matter that are generally removed prior to use of the biogas. AD is a biochemical process for converting biogenic solid waste into a stable, humus-like product. Aerobic conversion uses air or oxygen to support the metabolism of the aerobic microorganisms degrading the substrate. Aerobic conversion includes composting and activated sludge wastewater treatment processes. Composting produces useful materials, such as mulch, soil additives and amendments, and fertilizers.

Digestion is a term usually applied to anaerobic mixed bacterial culture systems employed in many wastewater treatment facilities for sludge degradation and stabilization. Anaerobic digestion is also becoming more widely used in on-farm animal manure management systems, and is the principal process occurring in landfills that creates landfill gas (LFG). Anaerobic digestion operates without free oxygen and results in a fuel gas called biogas containing mostly methane (CH_4) and carbon dioxide (CO_2), but frequently carrying impurities such as moisture, hydrogen sulfide (H_2S), and particulate matter.

AD is known to occur over a wide temperature range from 10–71°C. Anaerobic digestion requires attention to the nutritional needs and the maintenance of reasonable temperatures for the facultative and methanogenic bacteria degrading the waste substrates. The carbon/nitrogen (C/N) ratio of the feedstock is especially important. Biogas can be used after appropriate gas clean-up as a fuel for engines, gas turbines, fuel cells, boilers, industrial heaters, other processes, and the manufacturing of chemicals. Anaerobic digestion is also being explored as a route for direct conversion to hydrogen.

Cellulose and hemicelluloses can be hydrolyzed to simple sugars and amino acids that are consumed and transformed by the fermentive bacteria. The lignin is refractory to hydrolysis and generally exits the process undigested. In fact, lignin may be the most recalcitrant naturally produced organic chemical. Lignin polymers are cross-linked carbohydrate structures with molecular weights on the order of 10,000 atomic mass units. As such, lignin can bind with or encapsulate some cellulose making that cellulose unavailable to hydrolysis and digestion. Lignin degradation (or delignification of lignocellulosis) in nature is due principally to aerobic filamentous fungi that decompose the lignin in order to gain access to the cellulose and hemicelluloses.

For anaerobic systems, methane gas is an important product. Depending on the type and nature of the biological components, different yields can be obtained for different biodegradable wastes. For pure cellulose, for example, the biogas product is 50% methane and 50% carbon dioxide. Mixed waste feedstocks yield biogas with methane concentrations of 40–60% (by volume). Fats and oils can yield biogas with 70% methane content.

Anaerobic digestion functions over a wide temperature range from the so-called psychrophilic temperature near 283 K to extreme thermophilic temperatures above 344 K. The temperature of the reaction has a very strong influence on the anaerobic activity, but there are two optimal temperature ranges in which microbial activity and the biogas production rate are highest, the so-called mesophilic and thermophilic ranges. The mesophilic regime is associated with temperatures of about 308 K, the thermophilic regime of about 328 K. Operation at thermophilic temperature allows for shorter retention time and a higher biogas production rate, however, maintaining the high temperature generally requires an outside heat source because anaerobic bacteria do not generate sufficient heat. Aerobic composting can achieve relatively high temperatures (up to 344 K) without heat addition because reaction rates for aerobic systems are much higher than those for anaerobic systems. If heat is not conducted away from the hot center of a compost pile, then thermochemical reactions can initiate, which can lead to spontaneous combustion if sufficient oxygen reaches the hot areas. Managed compost operations use aeration to provide oxygen to the bacteria but also to transport heat out of the pile. The anaerobic digestion of lignocellulosic waste occurs in a three-step process often termed hydrolysis, acetogenesis, and methanogenesis. The molecular structure of the biodegradable portion of the waste that contains proteins and carbohydrates is first broken down through hydrolysis. The lipids are converted to volatile fatty acids and amino acids. Carbohydrates and proteins are hydrolyzed to sugars and amino acids. In acetogenesis, acid forming bacteria use these byproducts to generate intermediary products such as propionate and butyrate. Further microbial action results in the degradation of these intermediary products into hydrogen and acetate. Methanogenic bacteria consume the hydrogen and acetate to produce methane and carbon dioxide.

Under anaerobic conditions, various pathways exist for pyruvate metabolism, which serve to reoxidize the reduced hydrogen carriers formed during glycolysis. The ultimate acceptor builds up as a waste product in the culture medium. The end products of the pathways are: (1) CO_2, ATP, and acetate; (2) CO_2 and ethanol; (3) H_2 and CO_2; (4) CO_2 and 2,3-butylene glycol; (5) CO_2, H_2, acetone, ATP, and butanol; (6) succinate; and (7) lactate. The pathway that occurs depends on the microorganism cultivated and the culture.

5.2.2.1 Experimental Considerations for Biogas Production

The experiments for biomass production are carried out using four 1800-ml working volume bottle reactors. The bottles are closed with butyl rubber stoppers maintained at the optimal mesophilic temperature range (308 ± 1.0 K). The methane yield was determined in the batch experimental set-up depicted in Fig. 5.2.

Total solids (TS) content in the slurry is determined by drying it in an oven at 378 K until a constant weight is obtained. The dried solid samples from the TS determination are ignited at 1225 K in a furnace for 7 min. The loss in weight is taken as the volatile solids of the substrate slurry.

After the first 6 days of digestion, methane production from manure increased exponentially, after 16 days it reaches a plateau value, and at the end of the 20th day, the digestion reached the stationary phase. For wheat straw and mixtures of manure/straw the rates of digestion are lower than that of manure.

The maximum daily biogas productions are between 4 and 6 days. During a 30-day digestion period, ~80–85% of the biogas is produced in the first 15–18 days. This implies that the digester retention time can be designed to 15–18 days instead of 30 days,

For the first 3 days, methane yield is almost 0%, and carbon dioxide generation is almost 100%. In this period, digestion occurs as fermentation to carbon dioxide. The yields of methane and carbon dioxide gases are 50–50 on the 11th

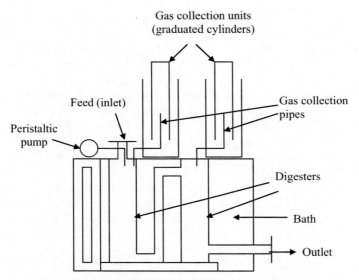

Fig. 5.2 Experimental set-up for batch anaerobic digestion of manure and/or straw slurries

day. At the end of the 20th day, the digestion reaches the stationary phase. The methane content of the biogas is in the range of 73–79% for the runs, the remainder being principally carbon dioxide. During digestion, the volatile fatty acid concentration is lower and the pH higher. The pH of the slurry with manure increased from 6.4 initially, to 6.9–7.0 at the maximum methane production rate. The pH of the slurry with wheat straw is around 7.0–7.1 at the maximum methane production rate.

5.2.3 Biogas Processing

A methane digester system, commonly referred to as an AD is a device that promotes the decomposition of manure or digestion of the organics in manure to simple organics and gaseous biogas products. There are three types of continuous digesters: vertical tank systems, horizontal tank or plug-flow systems, and multiple tank systems. Proper design, operation, and maintenance of continuous digesters produce a steady and predictable supply of usable biogas.

Biogas, a clean and renewable form of energy could very well substitute (especially in the rural sector) conventional sources of energy (fossil fuels, oil, *etc.*), which are causing ecological–environmental problems and at the same time depleting at a faster rate. Biogas is a modern form of bioenergy that is derived from the anaerobic digestion of organic matter, such as manure, sewage sludge, municipal solid waste, biodegradable waste, and agricultural slurry under anaerobic conditions.

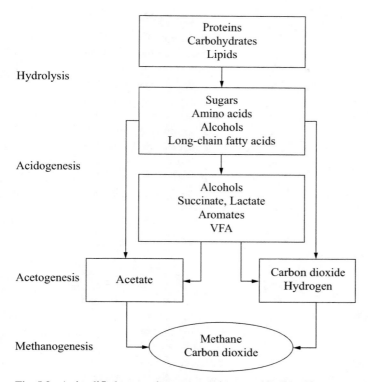

Fig. 5.3 A simplified conversion processes in anaerobic digestion

Anaerobic digestion of organic compounds is a complex process, involving several different types of microorganisms. This is the natural breakdown of organic matter, such as biomass, by bacterial populations in the absence of air into biogas, *i.e,* a mixture of methane (40–75% v/v) and carbon dioxide. The end products of anaerobic digestion are biogas and digestate, a moist solid, which is normally dewatered to produce a liquid stream and a drier solid. During anaerobic digestion, typically 30–60% of the input solids are converted to biogas; byproducts consist of undigested fiber and various water-soluble substances.

The anaerobic digestion process occurs in the following four basic steps: (1) hydrolysis, (2) acidogenesis, (3) acetogenesis, and (4) methanogenesis. A simplified model of anaerobic digestion process, showing the main steps, is shown in Fig. 5.3 (Asplund, 2005).

5.2.3.1   Hydrolysis

The first step in anaerobic degradation is the hydrolysis of complex organic compounds such as carbohydrates, proteins, and lipids. Organic compounds are hydrolyzed into smaller units, such as sugars, amino acids, alcohols, and long-chain fatty acids. During the hydrolysis step both solubilization of insoluble particulate matter and biological decomposition of organic polymers to monomers or dimers take place. Thermal, mechanic and chemical treatment have been investigated as a possible pretreatment step to accelerate sludge hydrolysis. The aim of the pretreatment is disintegrate sludge solids to facilitate the release of cell components and other organic matter.

Extracellular hydrolysis is often considered the rate-limiting step in the anaerobic digestion of organic wastes. Under anaerobic conditions, the hydrolysis rate of protein is generally slower than the hydrolysis of carbohydrates. Yu *et al.* (2003) have investigated the hydrolysis and acidogenesis of sewage sludge in an upflow reactor with an agitator and a gas-liquid-solid separator. They showed that 31–65% of carbohydrates, 20–45% of protein and 14–24% of lipid were acidified in this reactor.

5.2.3.2 Acidogenesis

In the second step, acidogenesis, another group of microorganisms ferments the break-down products to acetic acid, hydrogen, carbon dioxide, and other lower weight simple volatile organic acids like propionic acid and butyric acid, which are in turn converted to acetic acid. Acetate, carbon dioxide, and molecular hydrogen can be directly utilized as a substrate by another group of anaerobic microorganisms called methanogens. These organisms comprise a wide variety of different bacterial genera representing both obligate and facultative anaerobes.

During acidogenesis, the freshly fed biomass on the top produced a lot of volatile fatty acids (VFAs) intermediates that accumulated in the bed. A daily sprinkling of the bed once on one side introduced acidogenic organisms onto the upper regions of the biomass bed and it also carried down VFA intermediates to lower regions of the bed. This degradation pathway is often the fastest step and also gives a high-energy yield for the microorganisms, and the products can be used directly as substrates by the methanogenic microorganism.

5.2.3.3 Acetogenesis

The third step is the biological process of acetogenesis where the products of acidogenesis are further digested to produce carbon dioxide, hydrogen, and mainly acetic acid, although higher-molecular-weight organic acids (*e.g.*, propionic, butyric, valeric) are also produced. The products formed during acetogenesis are due to a number of different microbes, *e.g.*, s*yntrophobacter wolinii*, a propionate decomposer and s*ytrophomonos wolfei*, a butyrate decomposer. Other acid formers are *clostridium spp., peptococcus anerobus, lactobacillus*, and *actinomyces* (Verma, 2002). Acetogens are slow-growing microorganisms that are sensitive to environmental changes such as changes in the organic loading rate and flow rate. The microorganisms in the step are usually facultative heterotrophs that function best in a range of pH from 4.0 to 6.5 (Zhang *et al.*, 2005). Acetogenic reactions are represented as follows (Parawira, 2004).

$$CH_3CH_2COOH + 2H_2O \rightarrow CH_3COOH + CO_2 + 3H_2$$
$$\Delta G^0 = +76.1\,kJ \tag{5.1}$$

$$CH_3CH_2CH_2COOH + 2H_2O \rightarrow 2CH_3COOH + 2H_2$$
$$\Delta G^0 = +48.1\,kJ \tag{5.2}$$

$$C_2H_5OH + H_2O \rightarrow CH_3COOH + 2H_2$$
$$\Delta G^0 = +9.6\,kJ \tag{5.3}$$

$$CH_3CHOHCOOH + 2H_2O \rightarrow CH_3COOH + CO_2 + 2H_2 + H_2O$$
$$\Delta G^0 = -4.2\,kJ \tag{5.4}$$

5.2.3.4 Methanogenesis

Methanogens belong to *Archae*, a unique group of microorganisms, phylogenetically different from the main group of prokaryotic microorganisms. Only a limited number of compounds can act as substrates in methanogenesis, among these are acetate, H_2/CO_2, methanol, and formate. The most important methanogenic transformations in anaerobic digestion are the acetoclastic reaction and the reduction of carbon dioxide. Around 70% of methane is formed from VFA, and 30% from hydrogen and CO_2 by methanogenic bacteria (Garcia, 2005). Almost all known methanogens convert H_2/CO_2 to methane, whilst aceticlastic methanogenesis has been documented for only two methanogenic genera: *Methanosarcina* and *Methanosaeta*. *Methanosarcina* sp. has faster growth rates, higher apparent substrate affinity constants (K_M) for acetate use, and higher acetate threshold values (Aiyuk *et al.*, 2006). Species of *Methanosaeta* grow very slowly, with doubling times of 4 to 9 days. The methanogenesis reactions can be expressed as follows (Parawira, 2004):

$$CH_3COOH + H_2O \rightarrow CH_4 + CO_2 + H_2O \qquad \Delta G^0 = -31.0\,kJ \tag{5.5}$$

$$CO_2 + 4H_2 \rightarrow CH_4 + 2H_2O \qquad\qquad \Delta G^0 = -135.6\,kJ \tag{5.6}$$

Methanogens, sulfate-reducing bacteria (SRB), and acetogens are believed to be responsible for the removal of hydrogen in most anaerobic systems. SRB actually can out-compete methanogens during the anaerobic digestion process. Therefore, sulfide production generally proceeds to completion before methanogenesis occurs. The energetics of sulfate reduction with H_2 is favorable to the reduction of CO_2 with H_2, forming either CH_4 or acetate (McKinsey, 2003).

5.2.3.5 The Effect of Operational Parameters

Anaerobic digestion can occur under a wide range of environmental conditions, although narrower ranges are needed for optimum operation. The key factors to successfully control the stability and efficiency of the process are reactor configurations, temperature, pH, HRT, OLR, inhibitor concentrations, concentrations of total volatile fatty acid (TVFA), and substrate composition. In order to avoid a process failure and/or low efficiency, these parameters require an investigation so that they can be maintained at or near to optimum conditions. A variety of factors affect the rate of digestion and biogas production. The most important is temperature. Anaerobic bacteria communities can endure temperatures ranging from below freezing to above 330.4 K, but they thrive best at temperatures of about 309.9 K (mesophilic) and 329.6 K (thermophilic). Bacterial activity, and thus biogas production, falls off significantly between about 312.4 K and 324.9 K, and gradually from 328.2 K to 273.2 K.

Temperature significantly influences anaerobic digestion process, especially in methanogenesis wherein the degradation rate is increases with temperature. It has been found that the optimum temperature ranges for anaerobic digestion are mesophilic (303–313 K), and thermophilic (323–333 K) (Braun, 2007). Based on the temperature chosen, the duration of the process and effectiveness in destroying pathogens will vary. In mesophilic digestion, the digester is heated to 308 K and the typical time of retention in the digester is 15–30 days, whereas in thermophilic digestion the digester is heated to 328 K, and the time of retention is typically 12–14 days (Erickson *et al.*, 2004). It has been observed that higher temperatures in the thermophilic range reduce the required retention time (Verma, 2002). However, anaerobes are most active in the mesophilic and thermophilic temperature range. The activities of microorganisms increase with the increase of temperature, reflecting stable degradation of the substrate.

A change from mesophilic to thermophilic conditions has been shown to result in an immediate shift in the methanogenic population due to the rapid death of mesophilic organisms. Thermophilic digestion has many advantages such as a higher metabolic rate and higher consequent specific growth rate compared with mesophilic digestion, although the death rate of thermophilic bacteria is higher. The disadvantage of thermophilic digestion is the often-found high effluent VFA concentrations (Parawira, 2004).

Temperature is a universal process variable. It influences the rate of bacterial action as well as the quantity of moisture in the biogas. The biogas moisture content increases exponentially with temperature. Temperature also influences the quantity of gas and volatile organic substances dissolved in solution as well as the concentration of ammonia and hydrogen sulfide gas.

pH is a major variable to be monitored and controlled. The range of acceptable pH in digestion is theoretically between 5.5–8.5. It may be necessary to use a base or buffer to maintain the pH in the biodigester. The VFAs and LCFAs produced by the degradation of fat are inhibitors of methanogenic activity because they decrease the pH. For example, calcium carbonate ($CaCO_3$) can be used as a buffer and calcium hydroxide ($Ca(OH)_2$) can be used to precipitate LCFAs that are toxic to methanogenic bacteria (Erickson *et al.*, 2004). In general, pH has been reported to be one of the most important parameters for the inhibition of the methanogenic activity in an acid phase reactor, operating in the pH range 4.5–6.5. Maximum propionate degradation and methane production were observed at pH 7.5 and pH 7.0, respectively. Efficient methanogenesis from a digester operating in a steady state should not require pH control, but at other times, for example, during start-up or with unusually high feed loads, pH control may be necessary. pH can only be used as a process indicator when treating waste with low buffering capacity, such as carbohydrate-rich waste (Parawira, 2004).

It is generally accepted that alkalinity is very important to assess the anaerobic digestion stability. The main alkalinity components in a digester are bicarbonate and VFA, which are consumed and produced through the process steps. Bicarbonate buffers the system in the optimum pH range for the process to run efficiently. VFA buffers the system at low pH that is inhibitory to the biomass matrix in the digester. Bicarbonate is maintained by CO_2 production in the digester. Alkalinity is not an absolute value but depends upon the choice of pH endpoint for the titration. pH values often used are 4.2, 4.3, and 5.8.

The relationship between the amount of carbon and nitrogen present in organic materials is represented by the carbon-to-nitrogen (C/N) ratio (Verma, 2002). For instance, microorganisms utilize carbon during anaerobic digestion 20 to 30 times faster than nitrogen (Pesta, 2007). Thus to meet this requirement, microbes need a 20–30:1 ratio of C to N with the largest percentage of the carbon being readily degradable. A high C/N ratio will lead to a rapid consumption of nitrogen by the methanogenic bacteria and lower gas production rates. In raw materials with a high C/N ratio, such as rice straw (C/N:60.3), paper mill sludge (C/N:140), and sawdust (C/N:242), the level of nitrogen in the grass fiber-filter paper bags increased (Shiga, 1997). On the other hand, a lower C/N ratio causes ammonia accumulation and pH values exceeding 8.5, which is toxic to methanogenic bacteria. Additionally, the quality of the compost resulting from the digestate decreases with ammonia production.

The organic loading rate (OLR) is the quantity of organic matter fed *per* unit volume of the digester *per* unit time, (*e.g.,* kg VS m^{-3} d^{-1}). OLR plays an important role in anaerobic wastewater treatment in continuous systems and is a useful criterion for assessing performance of the reactors (Parawira, 2004). A higher OLR feed rate may cause crashing of anaerobic digestion if the acidogenic bacteria multiply and produce acids rapidly. Maximum OLR for an anaerobic digester depends on a number of parameters, such as reactor design, wastewater characteristics, the ability of the biomass to settle, and activity, *etc.*

Hydraulic retention time (HRT) is an important control parameter in many wastewater treatment processes. HRT exerts a profound influence on the hydraulic conditions and the contact time among different reactants within the reactor. To optimize process performance, a proper the HRT should be judiciously selected and carefully maintained. In activated sludge systems, typical values of the HRT range from 4–8 h for aeration basins treating domestic wastewater.

In tropical countries like India, HRT varies from 30–50 days, while in countries with colder climate it may go up to 100 days. Shorter retention time is likely to face the risk of washout of active bacterial population, while longer retention time requires a large volume of the digester and hence more capital cost. Hence there is a need to reduce the HRT for domestic biogas plants based on solid substrates. It is possible to carry out methanogenic fermentation at low HRTs without stressing the fermentation process at mesophilic and thermophilic temperature ranges. A high solids reactor operating in the thermophilic range has a retention time of 14 days (Verma, 2002).

If anaerobic digestion is to compete with other MSW disposal options, the retention time must be lower than the current standard of 20 days. HRT is determined by the average time it takes for organic material to digest completely, as measured by the COD and BOD of exiting effluents. COD removal efficiency at 20% in the anaerobic completely stirred tank reactor and up to 86% in the anaerobic filter operating at 10–12 days HRT was observed by Glass et al. (2005).

An average of 68% of the cultivated land produces grains with wheat ranking first, barley second, and corn third in developing countries. Agricultural solid residues are potential renewable energy resources. Wheat straw wastes represent a potential energy resource if they can be properly and biologically converted to methane. They are renewable and their net CO_2 contribution to the atmosphere is zero.

In a process of manure and straw mixture digestion, for the first 3 days, the methane yield was almost 0%, and carbon dioxide generation was almost 100%. In this period, digestion occurred as aerobic fermentation to carbon dioxide. The yields of methane and carbon dioxide gases were 50–50 on the 11th day. At the end of the 20th day, the digestion reached the stationary phase. The methane content of the biogas was in the range of 73–79% for the runs, the remainder being principally carbon dioxide. During a 30-day digestion period, ~80–85% of the biogas was produced in the first 15–18 days. This implies that the digester retention time can be designed to 15–18 days instead of 30 days.

Agricultural residues contain low nitrogen and have carbon-to-nitrogen ratios (C/N) of around 60–90. The proper C/N ratio for anaerobic digestion is 25–35 (Hills and Roberts, 1981); therefore, nitrogen needs to be supplemented to enhance the anaerobic digestion of agricultural solid residues. Nitrogen can be added in inorganic form such as ammonia or in organic form such as livestock manure, urea, or food wastes. Once nitrogen is released from the organic matter, it becomes ammonium, which is water soluble. Recycling nitrogen in the digested liquid reduces the amount of nitrogen needed.

5.2.4   Reactor Technology for Anaerobic Digestion

In the last two decades, anaerobic digestion technology has been significantly improved by the development of sludge bed digesters, based on granular biomass. The most widely employed systems are granular sludge-based bioreactors, such as the upflow anaerobic sludge blanket (UASB), the expanded granular sludge bed (EGSB), and the anaerobic hybrid reactor (AHR), which consists of a granular sludge bed and an upper fixed bed section.

Anaerobic batch digestion is useful because it can be performed with simple, inexpensive equipment and with waste with total solids concentration as high as 90%, *e.g.*, straw. The major disadvantages of batch systems are their large footprint, a possible need for a bulking agent, and a lower biogas yield caused by impairment of the percolation process due to channeling or clogging due to compaction (Parawira, 2004)

The UASB reactor has been widely used to treat many types of wastewater because it exhibits positive features such as high organic loadings, low energy demand, short HRT, long sludge retention time, and little sludge production. The sludge bed is a layer of biomass settled at the bottom of the reactor. The sludge blanket is a suspension of sludge particles mixed with gases produced in the process. When the UASB system is seeded only with non-granular anaerobic sludge, it can take several months before a highly effective granular bed can be cultivated. This clearly restricts the general application in countries where granules from operating the UASB systems are not readily available, unless the granulation reaction can be induced in other treatment systems.

In UASB systems, the sludge bed acts as a filter to the suspended solids (SS), thereby increasing their specific residence time. This way, the UASB reactor may achieve high COD and SS removals at very short HRTs. Currently, the UASB reactors represent more than 65% of all anaerobic digesters installed for treating industrial wastewater. However, in spite of the existence of more than 900 UASB units operating all over the world, it is recognized that some basic mechanisms underlying granulation are still unclear. They are seldom applied to treat low-strength wastewater with COD concentration lower than 1,500–2,000 mg/L because the development of granules in the UASB reactors is very difficult when treating such wastewaters. Bench-scale and pilot-scale studies indicate that it is possible to operate this type of reactor at an organic loading rate (OLR) of 40 kg COD m^{-3} d^{-1} at HRTs of 4–24 h with a COD reduction of more than 80% (Parawira, 2004).

In addition, the amount and activity of methanogenic populations are very important to improve the process capacity of UASB reactors. Retention of an adequate level of methanogens in the UASB reactor will give not only a good digester performance in terms of COD removal and methane yield, but also a better quality effluent. The methane producing intensity in UASB reactors will be very violent with an increase of OLR. As high methane intensity can make anaerobic sludge leak out from UASBs, it is necessary to modify the three-phase (gas–liquid–solid) separator for maintaining a high concentration of biomass in UASBs and meet the demand of high OLR.

The EGSB reactor is quite a promising version of UASB reactors operated at high superficial upflow velocities, obtained by means of high recycling rates, biogas production, and elevated height/diameter ratios. EGSB reactors are gaining more popularity and gradually replacing UASB applications, which is most likely due to the EGSB higher loading rates favored by hydrodynamics. Compared with conventional UASB reactors (0.5–2 m/h), the advantage of the EGSB system (>4 m/h) is the significantly better contact between sludge and wastewater (Liu *et al.*, 2006). EGSB reactors can be operated as ultra-high-loaded anaerobic reactors (up to 30 kg COD m^{-3} d^{-1}) to treat effluents from the chemical, biochemical, and biotechnological industries, and EGSB systems have been shown to be suited to low temperatures (10 °C) and low strengths (<1g$_{COD}$/L), and for the treatment of recalcitrant toxic substrates (Nicolella *et al.*, 2000).

The anaerobic hybrid reactor (AHR) has been developed to combine the advantages of AFR and UASB reactors, where the UASB allocated in the bottom part of reactor and the region of attached biomass on support media is in the upper part of reactor. AHRs can be used for a wide variety of industrial effluents, and it is possible to maintain the desired pH conditions for both the acidogens and the methanogens. The performance of AHRs depends on contact of the wastewater with both the suspended growth in the sludge layer and the attached biofilm in the material matrix. So, AHR configurations generally have better operating characteristics than fully packed reactors (Kara, 2007). If an AHR has to maintain a high solid retention, a key factor to be attended to is having support materials to form a filter inside the reactor. The significance of the media is arguably comparable to granular sludge in an upflow-sludge bed-type reactor. Considering the system's efficiency, an AHR has often been compared with other anaerobic digestion systems.

A significant limitation of UASB reactors is the interference of suspended solids in the incoming wastewater with granulation and reactor performance. Hence, other high-rate systems, such as the anaerobic sequencing batch reactor (ASBR), have been developed to better handle high-suspended solids in wastewater. ASBRs are single-vessel bioreactors that operate in a four-step cycle: (1) wastewater is fed into the reactor with settled biomass, (2) wastewater and biomass are mixed intermittently, (3) biomass is settled, and (4) the effluent is withdrawn from the reactor. ASBRs are particularly useful for agricultural waste and have recently been scaled up for on-farm treatment of dilute swine waste. This design does not require feed distribution and gas–solids separation systems, which simplifies its configuration. A disadvantage of the ASBR is the non-continuous operating mode. Figure 5.4 shows a schematic diagram of the ASBR, UASB, and AMBR reactor (Angenent and Sung, 2001).

The anaerobic baffled reactor (ABR) is a high-rate reactor that contains between three to eight compartments in which the liquid flow is alternately upwards and downwards between compartment partitions. This reactor consists of a series of baffled compartments where the wastewater flows upward through a bed of anaerobic sludge. The ABR does not require the sludge to granulate in order to perform effectively, although granulation does occur over time. The compartmentalized design allows operation without a gas–solids-separation system, which simplifies the process, and biomass retention during shock-load conditions has improved.

The most common reactor type used for anaerobic digestion of wastewaters is the continuously stirred tank reactor (CSTR). Animal manures typically have a low solids content (<10% TS), and thus, the anaerobic digestion technology applied in manure processing is mostly based on wet processes, mainly on the use of CSTRs. However, for dimensioning the fermenter size of CSTRs both the OLR

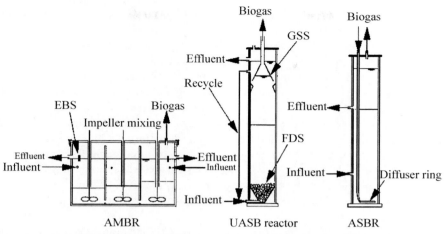

Fig. 5.4 Schematic diagram of the AMBR, UASB, and ASBR reactor. EBS = effluent-baffle system, GSS: gas–solids separator, FDS: feed-distribution system

and the HRT are the parameters that are applied most frequently in practice. Stable CSTR operation requires HRTs of 15–30 days. HRTs are relatively long (*ca.* 30 days) and a comparable solid content waste can be used as compared to the CSTR. Because of the slow growth rates of syntrophic and methanogenic bacteria, reduction of the HRT in CSTRs risks causing washout of the active biomass, with consequent process failure.

5.3 Landfill Gas

Biogas can be obtained from several sources. It is obtained from decomposing organic material. Contents of domestic solid waste are given in Table 5.1. Biogas is composed by methane (CH_4), carbon dioxide (CO_2), air, ammonia, carbon monoxide, hydrogen, sulfur gases, nitrogen, and oxygen. Among its components, methane is the most important, particularly for the combustion process in vehicle engines (Kuwahara *et al.*, 1999). A typical analysis of raw landfill gas is given in Table 5.2. CH_4 and CO_2 make up around 90% of the gas volume produced. The main constituents of landfill gas are methane and carbon dioxide, both of which are major contributors to global warming. Because of the widely varying nature of the contents of landfill sites, the constituents of landfill gases vary widely.

Landfill leachate treatment has received significant attention in recent years, especially in municipal areas (Uygur and Kargi, 2004). The generation of municipal solid wastes (MSW) has increased parallel to rapid industrialization. Approximately 16% of all discarded MSW is incinerated (EPA, 1994); the remainder is disposed of in landfills. Effective management of these wastes has become a major social and environmental concern (Erses and Onay, 2003). Disposal of MSW in sanitary landfills is usually associated with soil, surface water, and groundwater contamination when the landfill is not properly constructed. The flow rate and composition of leachate vary from site to site, seasonally at each site, and depending on the age of the landfill. Young leachate normally contains high amounts of volatile fatty acids (Timur and Ozturk, 1999). MSW statistics and management practices including waste recovery and recycling initiatives have been evaluated (Metin *et al.*, 2003). The organic MSW has been chemically and biologically characterized, in order to study its behavior during anaerobic digestion, and its pH, biogas production, alkalinity, and volatile fatty acid production has been determined by Plaza *et al.* (1996). Anaerobic digestion of the organic food fraction of MSW, on its own or co-digested with primary sewage sludge, produces high quality biogas, suitable as renewable energy (Kiely *et al.*, 1997). The processing of MSW (*i.e.*, landfill, incineration, aerobic composting) secures many advantages and limitations (Braber, 1995). Greenhouse gas emissions can be reduced by the uncontrolled release of methane from improperly disposed organic waste in a large landfill (Al-Dabbas, 1998)

Decomposition in landfills occurs in a series of stages, each of which is characterized by the increase or decrease of specific bacterial populations and the formation and utilization of certain metabolic products. The first stage of decomposition,

Table 5.1 Contents of domestic solid waste (wt% of total)

Component	Lower limit	Upper limit
Paper waste	33.2	50.7
Food waste	18.3	21.2
Plastic matter	7.8	11.2
Metal	7.3	10.5
Glass	8.6	10.2
Textile	2.0	2.8
Wood	1.8	2.9
Leather and rubber	0.6	1.0
Miscellaneous	1.2	1.8

Source: Demirbas, 2006b

Table 5.2 Typical analysis of raw landfill gas

Component	Chemical formula	Content
Methane	CH_4	40–60 (% by vol.)
Carbon dioxide	CO_2	20–40 (% by vol.)
Nitrogen	N_2	2–20 (% by vol.)
Oxygen	O_2	<1 (% by vol.)
Heavier hydrocarbons	C_nH_{2n+2}	<1 (% by vol.)
Hydrogen sulfide	H_2S	40–100 ppm
Complex organics	–	1000–2000 ppm

Source: Demirbas, 2006b

which usually lasts less than a week, is characterized by the removal of oxygen from the waste by aerobic bacteria (Augenstein and Pacey, 1991). In the second stage, which is termed the anaerobic acid stage, a diverse population of hydrolytic and fermentative bacteria hydrolyzes polymers, such as cellulose, hemicellulose, proteins, and lipids, into soluble sugars, amino acids, long-chain carboxylic acids, and glycerol (Micales and Skog, 1997). Figure 5.5 shows the behavior of biogas production with time, in terms of the biogas components. Figure 5.5 indicates that the economic exploitation of CH_4 is worthwhile after one year from the start of the landfill operation. The main components of landfill gas are byproducts of the decomposition of organic material, usually in the form of domestic waste, by the action of naturally occurring bacteria under anaerobic conditions.

Methods developed for treatment of landfill leachates can be classified as physical, chemical, and biological, and are usually used in combinations in order to improve the treatment efficiency. Biological leachate treatment methods can be classified as aerobic, anaerobic, and anoxic processes and are widely used for the removal of biodegradable compounds (Kargi and Pamukoglu, 2004a). Biological treatment of landfill leachate usually results in low nutrient removals because of high chemical oxygen demand (COD), high ammonium-N content, and the presence of toxic compounds such as heavy metals (Uygur and Kargi, 2004). Landfill leachate obtained from the solid waste landfill area contains high COD and ammonium ions, which results in low COD and ammonium removals by direct biological treatment (Kargi and Pamukoglu, 2003a). Several anaerobic and aerobic treatment systems have been studied in landfill leachate (Ozturk et al., 2003).

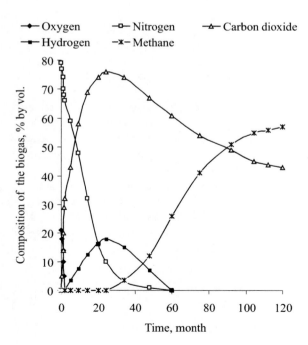

Fig. 5.5 Production of biogas components with time in landfill
Source: Demirbas, 2006b

Leachates contain non-biodegradable substrates that are not removed by biological treatment alone, and an increase of leachate input may cause reduction in substrate removal (Cecen *et al.*, 2003). Raw landfill leachate has been subjected to pretreatment by coagulation-flocculation and air stripping of ammonia before biological treatment (Kargi and Pamokoglu, 2004b). In order to improve biological treatability of the leachate, coagulation-flocculation and air stripping of ammonia have been used as pretreatment (Kargi and Pamukoglu, 2003b). Natural zeolite and bentonite can be utilized as a novel landfill liner material (Kayabali, 1997).

5.4 Crude Gases from Pyrolysis and Gasification of Biomass

A large number of research projects in the field of thermochemical conversion of biomass, mainly on pyrolysis, carbonization, and gasification, have been carried out. The pyrolysis of carbonaceous materials refers to incomplete thermal degradation resulting in char, condensable liquid or tar, and gaseous products. In its strictest definition, pyrolysis is carried out in the absence of air. Some solids, liquids, and gases are produced in every thermal degradation process, including gasification. However, pyrolysis differs from gasification in that the products of interest are the char and liquids, which as a result of the incomplete nature of the process retain much of the structure, complexity, and signature of the raw material undergoing pyrolysis.

Gasification is a form of thermal decomposition, carried out at high temperatures in order to optimize the gas production. The resulting gas, known as the producer gas, is a mixture of carbon monoxide, hydrogen, and methane, together with carbon dioxide and nitrogen. The gas is more versatile than the original solid biomass (usually wood or charcoal): it can be burnt to produce process heat and steam, or used in gas turbines to produce electricity. Biomass gasification technologies are expected to be an important part of the effort to meet the goals of expanding the use of biomass. Gasification technologies provide the opportunity to convert renewable biomass feedstocks into clean fuel gases or synthesis gases. These gaseous products can be burned to generate heat or electricity, or they can potentially be used in the synthesis of liquid transportation fuels, hydrogen, or chemicals (Demirbas, 2006c).

Assuming a gasification process using biomass as a feedstock, the first step of the process is a thermochemical decomposition of the lignocellulosic compounds with production of char and volatiles. Further the gasification of char and some other equilibrium reactions occur as shown in Eqs. 5.7–5.10.

$$C + H_2O = CO + H_2 \tag{5.7}$$

$$C + CO_2 = 2CO \tag{5.8}$$

$$CO + H_2O = H_2 + CO_2 \tag{5.9}$$

$$CH_4 + H_2O = CO + 3H_2 \tag{5.10}$$

The main gaseous products from biomass are:

pyrolysis of biomass $\rightarrow$ H_2 + CO_2 + CO + hydrocarbon gases $\qquad$ (5.11)

catalytic steam reforming of biomass $\rightarrow$ H_2 + CO_2 + CO $\qquad$ (5.12)

gasification of biomass $\rightarrow$ H_2 + CO_2 + CO + N_2 $\qquad$ (5.13)

Hydrogen gas has been produced on a pilot scale by steam gasification of charred cellulosic waste material. The gas was freed from moisture and carbon dioxide. The beneficial effect of some inorganic salts such as chlorides, carbonates, and chromates on the reaction rate and production cost of the hydrogen gas has been investigated (Rabah and Eldighidy, 1989). Steam reforming C_1–C_5 hydrocarbons, naphtha, gas oils, and simple aromatics are commercially practiced through well-known processes. Steam reforming of hydrocarbons; partial oxidation of heavy oil residues, selected steam reforming of aromatic compounds, and gasification of coals and solid wastes to yield a mixture of H_2 and CO (syngas), followed by water–gas shift conversion to produce H_2 and CO_2, these are well-established processes (Duprez, 1992). When the objective is to maximize the production of H_2, the stoichiometry describing the overall process is

$$C_nH_m + 2nH_2O \rightarrow nCO_2 + [2n + (m/2)]H_2 \qquad (5.14)$$

The simplicity of Eq. 5.14 hides the fact that, in a hydrocarbon reformer, the following reactions take place concurrently:

$$C_nH_m + nH_2O = nCO + [2n + (m/2)]H_2 \qquad (5.15)$$

Under normal reforming conditions, steam reforming of higher hydrocarbons (C_nH_m) is irreversible (Eq. 5.14), whereas the methane reforming (Eq. 5.15) and the shift conversion (Eq. 5.15) reactions approach equilibrium. A large molar ratio of steam to hydrocarbon will ensure that the equilibrium for Eqs. 5.14 and 5.15 is shifted toward H_2 production.

5.5 Biohydrogen from Biorenewable Feedstocks

As a sustainable energy source, hydrogen is a promising alternative to fossil fuels. It is a clean and environmentally friendly fuel (Han and Shin, 2004). Hydrogen is the fuel of the future mainly due to its high conversion efficiency, recyclability, and non-polluting nature.

Hydrogen produced from water, biorenewable feedstocks, either biologically (biophotolysis and fermentation) or photobiologically (photodecomposition), is termed "biohydrogen". Biohydrogen technology will play a major role in future because it can utilize the renewable sources of energy. Hydrogen is currently more expensive than conventional energy sources. There are different technologies presently being applied to produce hydrogen economically from biomass (Nath and Das, 2003).

5.5.1 Hydrogen from Biorenewable Feedstocks via Thermochemical Conversion Processes

Hydrogen can be produced from biomass *via* two thermochemical processes: (1) gasification followed by reforming of the syngas, and (2) fast pyrolysis followed by reforming of the carbohydrate fraction of the bio-oil. In each process, water–gas shift is used to convert the reformed gas into hydrogen, and pressure swing adsorption is used to purify the product. Gasification technologies provide the opportunity to convert biorenewable feedstocks into clean fuel gases or synthesis gases. The synthesis gas includes mainly hydrogen and carbon monoxide ($H_2 + CO$), which is also called syngas. Biosyngas is a gas rich in CO and H_2 obtained by gasification of biomass.

Hydrogen can be produced from biomass by pyrolysis, gasification, steam gasification, steam-reforming of bio-oils, and enzymatic decomposition of sugars. Hydrogen is produced from pyroligneous oils produced from the pyrolysis of lignocellulosic biomass. The yield of hydrogen that can be produced from biomass is relatively low, 16–18% based on dry biomass weight (Demirbas, 2001).

The strategy is based on producing hydrogen from biomass pyrolysis using a co-product strategy to reduce the cost of hydrogen, and it is concluded that only this strategy can compete with the cost of the commercial hydrocarbon-based technologies (Wang *et al.*, 1998). This strategy will demonstrate how hydrogen and biofuel are economically feasible and can foster the development of rural areas when practiced on a larger scale. The process of biomass to activated carbon is an alternative route to hydrogen with a valuable co-product that is practiced commercially. The yield of hydrogen that can be produced from biomass is relatively low, 12–14% based on the biomass weight (Demirbas, 2005). In the proposed second process, fast pyrolysis of biomass is used to generate bio-oil and catalytic steam reforming of the bio-oil to hydrogen and carbon dioxide.

Gasification of solid wastes and sewage is a recent innovation. Hydrogen can be generated from biomass, but this technology urgently needs further development. The production of hydrogen from biomass is already economically competitive today. Hydrogen from biomass has many advantages (Tetzlaff, 2001):

- Independence from oil imports
- Net product remains within the country
- Stable pricing level
- Peace keeping
- The carbon dioxide balance can be improved by around 30%.

Hydrogen can be generated from water by electrolysis, photolysis, direct thermal decomposition or thermolysis, and biological processes (Das and Veziroglu, 2001; Momirlan and Veziroglu, 2002). Many studies have reported on biohydrogen production by photocatalytic (Hwang *et al.*, 2004) or enzymatic (de Vrije *et al.*, 2002; Han and Shin, 2004) processes. Figure 5.6 shows main alternative processes of hydrogen production and hydrogen use.

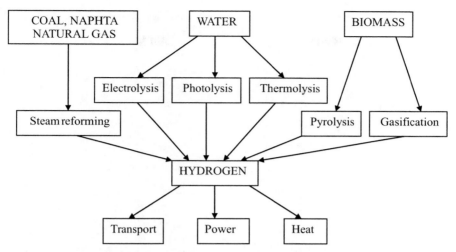

Fig. 5.6 Main alternative processes of hydrogen production and hydrogen use

In the pyrolysis and gasification processes, water–gas shift is used to convert the reformed gas into hydrogen, and pressure swing adsorption is used to purify the product. The cost of hydrogen production from supercritical water gasification of wet biomass was several times higher than the current price of hydrogen from steam methane reforming (Demirbas, 2005).

The yield from steam gasification increases with increasing water-to-sample ratio. The yields of hydrogen from the pyrolysis and the steam gasification increase with increasing temperature. In general, the gasification temperature is higher than that of pyrolysis, and the yield of hydrogen from the gasification is higher than that of the pyrolysis. The highest yields (% dry and ash free basis) were obtained from the pyrolysis (46%) and steam gasification (55%) of wheat straw, and the lowest yields from olive waste. The yield of hydrogen from supercritical water extraction was considerably high (49% by volume) at lower temperatures.

The pyrolysis was carried out at the moderate temperatures and steam gasification at the highest temperatures. The pyrolysis-based technology, in particular because it has co-product opportunities, has the most favorable economics.

Catalytic aqueous-phase reforming might prove useful for the generation of hydrogen-rich gas from carbohydrates extracted from renewable biomass and biomass waste streams. The biomass-derived hydrocarbons are suitable for hydrogen generation from biomass, as well as for reforming.

It is believed that in the future biomass can become an important sustainable source of hydrogen. Biomass has the advantage of low environmental impact compared with that for fossil fuels. The price of hydrogen obtained by direct gasification of lignocellulosic biomass, however, is about three times higher than that for hydrogen produced by steam reforming of natural gas (Spath *et al.*, 2000).

Figure 5.7 shows the yields of hydrogen and carbon monoxide obtained from pyrolysis of tallow (beef) at different temperatures. As seen from Fig. 5.7, the

Fig. 5.7 Yields of hydrogen and carbon monoxide (% by volume of total gas products) obtained from pyrolysis of tallow (beef) at different temperatures

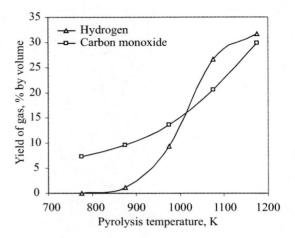

yields hydrogen and carbon monoxide from pyrolysis of the tallow increases with increasing temperature. The yield of hydrogen from pyrolysis of the tallow sharply increases from 9.4 to 31.7% by volume of total gaseous products with increasing of temperature from 975 K to 1175 K. The yield of carbon monoxide from the pyrolysis increases from 20.6 to 26.7% by volume of total gaseous products with increasing of temperature from 1075 K to 1175 K.

Hydrogen gas can be produced from the biomass material by direct and catalytic pyrolysis when the final pyrolysis temperature is generally increased from 775 K to 1025 K (Demirbas, 2001). Hydrogen and carbon monoxide-rich gas products can be obtained from triglycerides by pyrolysis. The total yield of combustible gases (mainly H_2 and CO) for the triglyceride samples increase with increasing pyrolysis temperature from 775 K to 1175 K. The most important reaction parameters are temperature and resistance time.

5.5.1.1 The Steam Reforming Process

In the steam-reforming reaction, steam reacts with hydrocarbons in the feed to predominantly produce carbon monoxide and hydrogen, commonly called synthesis gas. Steam reforming can be applied various solid waste materials including, municipal organic waste, waste oil, sewage sludge, paper mill sludge, black liquor, refuse-derived fuel, and agricultural waste. Steam reforming of natural gas, sometimes referred to as steam methane reforming, is the most common method of producing commercial bulk hydrogen. Steam reforming of natural gas is currently the least expensive method of producing hydrogen and is used for about half of the world's production of hydrogen.

Hydrogen production from carbonaceous solid wastes requires multiple catalytic reaction steps: For the production of high purity hydrogen, the reforming of fuels is followed by two water–gas shift reaction steps, a final carbon monoxide purification and carbon dioxide removal. Steam reforming, partial oxidation and autothermal reforming of methane are well-developed processes for the production of hydrogen. Stepwise steam reforming of methane for production of carbon monoxide-free hydrogen has been investigated at various process conditions by Choudhary and Goodman (2000). The process consists of two steps involving the decomposition of methane to carbon monoxide-free hydrogen and surface carbon in the first step, followed by steam gasification of this surface carbon in the second step. The amount of carbon monoxide-free hydrogen formed in the first step hydrogen is produced in the second step of the reaction. The mixture of gases can be separated and methane-rich gas mixture returned to the first step (Choudhary and Goodman, 2000). Steam, at high temperatures (975–1375 K) is mixed with methane gas in a reactor with a Ni-based catalyst at 3–25 bar pressure to yield carbon monoxide (CO) and hydrogen (H_2). Steam reforming is the process by which methane and other hydrocarbons in natural gas are converted into hydrogen and carbon monoxide by reaction with steam over a nickel catalyst on a ceramic support. The hydrogen and carbon monoxide are used as initial material for other industrial processes.

$$CH_4 + H_2O \leftrightarrows CO + 3H_2 \qquad \Delta H = +251 \, kJ/mol \qquad (5.16)$$

It is usually followed by the shift reaction:

$$CO + H_2O \leftrightarrows CO_2 + H_2 \qquad \Delta H = -42 \, kJ/mol \qquad (5.17)$$

The theoretical percentage of hydrogen to water is 50%. The further chemical reactions for most hydrocarbons that take place are:

$$C_nH_m + n \, H_2O \leftrightarrows n \, CO + (m/2 + n) \, H_2 \qquad (5.18)$$

It is possible to increase the efficiency to over 85% with an economic profit at higher thermal integration. There are two types of steam reformers for small-scale hydrogen production: Conventional reduced-scale reformers and specially designed reformers for fuel cells.

Commercial catalysts consist essentially of Ni supported on a-alumina. Mg-promoted catalysts showed a greater difficulty for Ni precursor's reduction besides different probe molecules (H_2 and CO) adsorbed states. In the conversion of cyclohexane, Mg inhibited the formation of hydrogenolysis products. Nonetheless, the presence of Ca did not influence the metallic phase. The impregnated Ni/MgO-catalyst performed better than the other types (Santos et al., 2004).

In comparison with other biomass thermochemical gasification such as air gasification or steam gasification, supercritical water gasification can directly deal with the wet biomass without drying, and there is high gasification efficiency in lower temperature. The cost of hydrogen production from supercritical water gasification of wet biomass was several times higher than the current price of hydrogen from steam methane reforming. Biomass was gasified in supercritical water at a series of temperature and pressure during different resident times to form a product gas composed of H_2, CO_2, CO, CH_4, and a small amount of C_2H_4 and C_2H_6 (Demirbas, 2004).

The yield of hydrogen from conventional pyrolysis of corncob increases from 33% to 40% with an increase in temperature from 775 K to 1025 K. The yields of hydrogen from steam gasification increase from 29% to 45% for (water/solid) = 1 and from 29% to 47% for (water/solid) = 2 with an increase in temperature from 975 K to 1225 K (Demirbas, 2006d). The pyrolysis is carried out at the moderate temperatures and steam gasification at the highest temperatures.

5.5.2   Biohydrogen from Biorenewable Feedstocks

Biological generation of hydrogen (biohydrogen) technologies provide a wide range of approaches to generate hydrogen, including direct biophotolysis, indirect biophotolysis, photo-fermentations, and dark-fermentation (Levin *et al.*, 2004). Biological hydrogen production processes are found to be more environmentally friendly and less energy intensive as compared to thermochemical and electrochemical processes (Das and Veziroglu, 2001). Researchers have been investigating hydrogen production with anaerobic bacteria since the 1980s (Nandi and Sengupta, 1998; Chang *et al.*, 2002)

There are three types of microorganisms of hydrogen generation: cyanobacteria, anaerobic bacteria, and fermentative bacteria. The cyano-bacteria directly decompose water to hydrogen and oxygen in the presence of light energy by photosynthesis. Photosynthetic bacteria use organic substrates like organic acids. Anaerobic bacteria use organic substances as the sole source of electrons and energy, converting them into hydrogen. Biohydrogen can be generated using bacteria such as *Clostridia* by temperature, pH control, reactor hydraulic retention time (HRT), and other factors of the treatment system.

Biological hydrogen can be generated from plants by biophotolysis of water using microalgae (green algae and cyano-bacteria), fermentation of organic compounds, and photo-decomposition of organic compounds by photo-synthetic bacteria. To produce hydrogen by fermentation of biomass, a continuous process using a non-sterile substrate with a readily available mixed microflora is desirable (Hussy *et al.*, 2005). A successful biological conversion of biomass to hydrogen depends strongly on the processing of raw materials to produce feedstock, which can be fermented by the microorganisms (de Vrije *et al.*, 2002).

Hydrogen production from the bacterial fermentation of sugars has been examined in a variety of reactor systems. Hexose concentration has a greater effect on H_2 yields than HRT. Flocculation also was an important factor in the performance of the reactor (Van Ginkel and Logan, 2005).

Hydrogen gas is a product of the mixed acid fermentation of *Escherichia coli*, the butylene glycol fermentation of *Aerobacter*, and the butyric acid fermentations of *Clostridium* spp. (Aiba *et al.*, 1973). A study was conducted to improve hydrogen fermentation of food waste in a leaching-bed reactor by heat-shocked anaerobic sludge, and also to investigate the effect of the dilution rate on the production of hydrogen and metabolites in hydrogen fermentation (Han and Shin, 2004).

5.6 Gaseous Fuels from Fischer–Tropsch Synthesis of Biomass

Syngas (a mixture of carbon monoxide and hydrogen) can be produced by gasification of biorenewable feedstocks, also called biosyngas. Biosyngas can be converted into a large number of organic compounds that are useful as chemical feedstocks, fuels, and solvents. Many of the conversion technologies were developed for coal gasification, but process economics have resulted in a shift to natural-gas-derived syngas. These conversion technologies successively apply similarly to biomass-derived biosyngas. Franz Fischer and Hans Tropsch first studied conversion of syngas into larger, useful organic compounds in 1923 (Balat, 2006). The fundamental reactions of synthesis gas chemistry are methanol synthesis, Fischer–Tropsch Synthesis (FTS), oxo synthesis (hydroformylation), and methane synthesis (Prins *et al.*, 2004).

To produce biosyngas from biorenewable feedstocks the following procedures are necessary: (a) gasification of the fuel, (b) cleaning of the product gas, (c) usage of the synthesis gas as energy carrier in fuel cells, and (d) usage of the synthesis gas to produce chemicals.

Biorenewable feedstocks can be converted to biosyngas by non-catalytic, catalytic, and steam gasification processes. The main aim of FTS is synthesis of long-chain hydrocarbons from CO and H_2 gas mixture. The FTS is described by the set of equations (Anderson, 1984; Schulz, 1999; Sie and Krishna, 1999):

$$nCO + (n + m/2)\,H_2 \rightarrow C_nH_m + nH_2O \tag{5.19}$$

where n is the average length of the hydrocarbon chain and m is the number of hydrogen atoms *per* carbon. All reactions are exothermic, and the product is a mixture of different hydrocarbons where paraffin and olefins are the main parts.

In FTS one mole of CO reacts with two moles of H_2 in the presence of a cobalt (Co)-based catalyst to afford a hydrocarbon chain extension ($-CH_2-$). The reaction of synthesis is exothermic ($\Delta H = -165\,kJ/mol$):

$$CO + 2H_2 \rightarrow -CH_2- + H_2O \qquad \Delta H = -165\,kJ/mol \tag{5.20}$$

$-CH_2-$ is a building stone for longer hydrocarbons. A main characteristic regarding the performance of FTS is the liquid selectivity of the process (Tijmensen *et al.*, 2002). For this reaction given with Eq. 5.20 is necessary a H_2/CO ratio of at least 2 for the synthesis of the hydrocarbons. The reaction of synthesis is exothermic ($\Delta H = -42\,kJ/mol$). When the ratio is lower it can be adjusted in the reactor with the catalytic water-gas shift reaction according to Eq. 5.21:

$$CO + H_2O \rightarrow CO_2 + H_2 \qquad \Delta H = -42\,kJ/mol \tag{5.21}$$

When iron (Fe)-based catalysts are used with water–gas shift reaction activity the water produced in the reaction equation 5.21 can react with CO to form additional H_2. The reaction of synthesis is exothermic (ΔH = –204 kJ/mol). In this case, a minimal H_2/CO ratio of 0.7 is required:

$$2CO + H_2 \rightarrow -CH_2- + CO_2 \qquad \Delta H = -204\,kJ/mol \qquad (5.22)$$

Typical operation conditions for the FTS are a temperature range of 475–625 K and pressures of 15–40 bar, depending on the process. All over reactions are exothermic. The kind and quantity of liquid product obtained is determined by the reaction temperature, pressure and residence time, the type of reactor, and the catalyst used. Iron catalysts have a higher tolerance for sulfur, are cheaper, and produce more olefin products and alcohols. However, the lifetime of the Fe catalysts is short and in commercial installations generally limited to 8 weeks. Co catalysts have the advantage of a higher conversion rate and a longer life (over 5 years). Co catalysts are in general more reactive for hydrogenation and, therefore, produce less unsaturated hydrocarbons and alcohols compared to iron catalysts.

The products from FTS are mainly aliphatic straight-chain hydrocarbons (C_xH_y). Besides the C_xH_y also branched hydrocarbons, unsaturated hydrocarbons, and primary alcohols are formed in minor quantities. The product distribution obtained from FTS includes the light hydrocarbons methane (CH_4), ethene (C_2H_4) and ethane (C_2H_6), LPG ($C_3–C_4$, propane and butane), gasoline ($C_5–C_{12}$), diesel fuel ($C_{13}–C_{22}$), and light and waxes ($C_{23}–C_{33}$). Any raw biosyngas contains trace contaminants like NH_3, H_2S, HCl, dust, and alkalis in ash. The distribution of the products depends on the catalyst and the process parameters such as temperature, pressure, and residence time. The distribution of products is described by the so-called Schulz–Flory equation (Anderson, 1984):

$$X_n = \alpha n^{1-\alpha} \qquad (5.23)$$

where X_n is the mole fraction of the product n. The composition of the synthesis gas, temperature, pressure, and the composition of the catalyst affect on the value of the parameter α. The effect of the parameter α on the composition of the FTS products is given in Table 5.3. The catalyst activation affects the reaction rate and synthesis gas conversion (Bukur et al., 1995). Table 5.4 shows the higher heating values of fuel gases.

Figure 5.8 shows the production of diesel fuel from biosyngas by FTS. The design of a biomass gasifier integrated with a FTS reactor must be aimed at achieving a high yield of liquid hydrocarbons. For the gasifier, it is important to avoid methane formation as much as possible, and convert all carbon in the biomass to mainly carbon monoxide and carbon dioxide (Prins et al., 2004).

Gas cleaning is an important process before FTS. It is even more important for the integration of a biomass gasifier and a catalytic reactor. To avoid poisoning of the FTS catalyst, tar, hydrogen sulfide, carbonyl sulfide, ammonia, hydrogen cyanide, alkali, and dust particles must be removed thoroughly (Tijmensen et al., 2002).

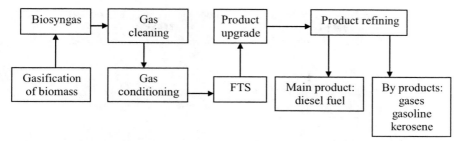

Fig. 5.8 Production of diesel fuel from biosyngas by the Fischer–Tropsch synthesis (FTS)

Table 5.3 Effect of the parameter α on the composition of the FTS products (% by mole)

Carbon number	Range of the value of the parameter α				
	0.5–0.6	0.6–0.7	0.7–0.8	0.8–0.9	0.9–1.0
C_2–C_4	51.0–59.4	59.5–64.8	64.9–78.6	79.7–91.3	91.4–98.4
C_5–C_{10}	7.8–13.5	13.6–25.7	25.8–41.4	41.5–61.8	61.9–91.6
C_{11}–C_{20}	0.4–0.9	1.0–2.9	3.0–10.6	10.7–34.6	34.7–79.8
C_{21}–C_{33}	0	0–0.2	0.3–0.6	0.7–12.6	12.6–66.8

Sources: Balat, 2006; Demirbas, 2007

Table 5.4 Higher heating values of fuel gases

Gas	Higher heating value (MJ/Nm^3)
Hydrogen	12.8
Methane	39.9
Ethane	70.4
Propane	101.7
Butane	132.4
Carbon monoxide	12.7

References

Al-Dabbas, M.A.F. 1998. Reduction of methane emissions and utilization of municipal waste for energy in Amman. Renewable Energy 14:427–434.

Anderson, R.B. 1984. The Fischer–Tropsch synthesis. Academic Press, New York.

Augenstein, D., Pacey, J. 1991. Landfill methane models. Proceedings from the Technical Sessions of SWANA's 29th Annual International Solid Waste Exposition, SWANA, Silver Springs, MD.

Aiba, S., Humphrey, A.E., Milis, N.F. 1973. Biochemical engineering, 2d ed, Academic Press, New York.

Aiyuk, S., Forrez, I., Lieven, D.K., van Haandel, A., Verstraete, W. (2006). Anaerobic and complementary treatment of domestic sewage in regions with hot climates – A review. Biores Technol 97:2225–2241.

Angenent, L.T., Sung, S. 2001. Development of anaerobic migrating blanket teactor (AMBR), a novel anaerobic treatment system. Water Research 35:1739–1747.

Asplund, S. 2005. The biogas production plant at Umea dairy evaluation of design and startup. Degree thesis, Linköping University, Sweden, October 5.

Balat, M. 2006. Sustainable transportation fuels from biomass materials Energy Edu Sci Technol 17:83–103.

Balat, M. 2008. Progress in biogas production processes. Energy Edu Sci Tecnol 22:15–35.

Braber, K. 1995. Anaerobic digestion of municipal solid waste: A modern waste disposal option on the verge of breakthrough. Biomass Bioenergy 9:365–376.

Braun, R. 2007. Anaerobic digestion: a multi-faceted process for energy, environmentalmanagement and rural development. In: Improvement of Crop Plants for Industrial End Uses, Springer, Netherlands, p.335–416.

Bukur, D.B., Nowicki, L., Manne, R.V., Lang, X. 1995. Activation studies with a precipitated iron catalysts for the Fischer–Tropsch synthesis. J. Catalysis 155:366–75.

Cecen, F., Erdincler, A., Kilic, E. 2003. Effect of powdered activated carbon addition on sludge dewaterability and substrate removal in landfill leachate treatment. Advances Environ Res 7:707–713.

Chang, J.-S., Lee, K.-S., Lin, P.-J. 2002. Biohydrogen production with fixed-bed bioreactors. Int J Hydrogen Energy 27:1167–1174.

Choudhary, T.V., Goodman, D.W. 2000. CO-free production of hydrogen via stepwise steam reforming of methane. J Catal 192:316–312.

Das, D., Veziroglu, T.N. 2001. Hydrogen production by biological processes: a survey of literature. Int J Hydrogen Energy 26:13–28.

Demirbas, A. 2001. Yields of hydrogen of gaseous products *via* pyrolysis from selected biomass samples. Fuel 80:1885–1891.

Demirbas, A. 2004. Hydrogen rich gas from fruit shells *via* supercritical water extraction. Int J Hydrogen Energy 29:1237–1243.

Demirbas, A. 2005. Hydrogen production from biomass *via* supercritical water extraction. Energy Sources 27:1409–1417.

Demirbas, A. 2006a. Biogas potential of manure and straw mixtures. Energy Sources Part A 28:71–78.

Demirbas, A. 2006b. Biogas production from the organic fraction of municipal solid waste. Energy Sources Part A 28:1127–1134.

Demirbas, A. 2006c. Biomass gasification for power generation in Turkey. Energy Sources Part A 28:433–445.

Demirbas, M.F. 2006d. Hydrogen from various biomass species *via* pyrolysis and steam gasification processes. Energy Sources, Part A 28:245–252.

Demirbas A. 2007. Progress and recent trends in biofuels. Prog Energy Combus Sci 33: 1–18.

Demirbas, A., Ozturk, T. 2004. Anaerobic digestion of agricultural solid residues. Int J Gren Energy 1:483–494.

de Vrije, T., de Haas, G.G., Tan, G.B., Keijsers, E.R.P., Claassen, P. A.M. 2002. Pretreatment of Miscanthus for hydrogen production by Thermotoga elfii. Int J Hydrogen Energy 27:1381–1390.

Duprez, D. 1992. Selective steam reforming of aromatic compounds on metal catalysts. Appl Catal A 82:111–157.

EPA. 1994. Characterization of potential of municipal solid waste (MSW) components. Municipal Solid Waste in the United States: 1992 Update. EPA/530-R-94-042, NTS #PB 95-147690. Solid Waste and Emergency Response (5305), Washington. D.C.: U.S. Environmental Protection Agency (EPA).

Erickson, L.E., Fayet, E., Kakumanu, B.K., Davis, L.C. 2004. Anaerobic digestion. In: Carcass disposal: A comprehensive review, Chapter 7, pp. 1–19, National Agricultural Biosecurity Center, Kansas State University, Manhattan, Kansas, August.

Erses, A.S., Onay., T.T. 2003. In situ heavy metal attenuation in landfills under methanogenic conditions. J Hazard Mat B99:159–175.

Garcia, S.G. 2005. Farm scale anaerobic digestion integrated in an organic farming system. JTI (Institutet för Jordbruks- och Miljöteknik) Report.

Glass, C.C., Chirwa, E.M.N., Bozzi, D. 2005. Biogas production from steam-treated municipal solid waste wastewater. Environ Eng Sci 22:510–524.

Han, S.-K., Shin, H.-S. 2004. Biohydrogen production by anaerobic fermentation of food waste. International Journal of Hydrogen Energy 29:569–577.

Hills, D.J., Roberts, D.W. 1981. Anaerobic digestion of dairy manure and field crop residues. Agricultural Wastes 3:179–189.

Hussy, I., Hawkes, F.R., Dinsdale, R., Hawkes, D.L. 2005. Continuous fermentative hydrogen production from sucrose and sugarbeet. International Journal of Hydrogen Energy 30:471–483.

Hwang, D.W., Kim, H.GH., Jang, J.S., Bae, S.W., Ji, S.M., Lee, J.S. 2004. Photocatalytic decomposition of water–methanol solution over metal-doped layered perovskites under visible light irradiation. Catalysis Today 93:845–850.

Kara, M. 2007. Anaerobic fitler performance at different conditions. Master's thesis, Graduate School of Natural and Applied Sciences of Dokuz Eylül University, Izmir, Turkey, June.

Kargi, F., Pamukoglu, M.Y. 2003a. Aerobic biological treatment of pre-treated landfill leachate by fed-batch operation. Enzyme Microbial Technol. 33:588–595.

Kargi, F., Pamukoglu, M.Y. 2003b. Simultaneous adsorption and biological treatment of pre-treated landfill leachate by fed-batch operation. Process Biochem. 38:1413–1420.

Kargi, F., Pamukoglu, M.Y. 2004a. Adsorbent supplemented biological treatment of pre-treated landfill leachate by fed-batch operation. Biores. Technol. 94:285–291.

Kargi, F., Pamukoglu, M.Y. 2004b. Repeated fed-batch biological treatment of pre-treated landfill leachate by powdered activated carbon addition. Enzyme Microbial Technol. 34:422–428.

Kiely, G., Tayfur, G., Dolan, C., Tanji, K. 1997. Physical and mathematical modelling of anaerobic digestion of organic wastes. Water Res 31:534–540.

Kuwahara, N., Berni, M.D., Bajay, S.V. 1999. Energy supply from municipal wastes: The potential of biogas-fuelled buses in Brazil. Renewable Energy 16:1000–1003.

Levin, D.B., Pitt, L., Love, M. 2004. Biohydrogen production: prospectsand limitationsto practical application. Int J Hydrogen Energy 29:173–185.

Liu, Y.H., He, Y.L., Yang, S.C., An, C.J. 2006. Studies on the expansion characteristics of the granular bed present in EGSB bioreactors. Water SA 32:555–560.

McKinsey, Z.S. 2003. Removal of hydrogen sulfide from biogas using cow manure compost.Master Thesis, Faculty of the Graduate School of Cornell University, USA, January.

Meynell, P.-J. 1976. Methane: Planning a digester. Schocken Books, New York.

Metin, E., Erozturk, A., Neyim, C. 2003. Solid waste management practices and review of recovery and recycling operations in Turkey. Waste Mgmt 23:425–432.

Micales, J.A, Skog, K.E. 1997. The Decomposition of forest products in landfills. Int Biodeterioration Biodegradation 39:145–158.

Momirlan, M., Veziroglu, T. 2002. Current status of hydrogen energy. Renew Sustain Energy Rev 6:141–79.

Nandi, R., Sengupta, S. 1998. Microbial production of hydrogen – An overview. Critical Review Microbiology 24:61–84.

Nath, K., Das, D. 2003. Hydrogen from biomass. Current Sci 85:265–271.

Nicolella, C., van Loosdrecht, M.C.M., Heijnen, S.J. 2000. Particle-based biofilm reactor technology. Trends Biotechnol 18:312–320.

Ozturk, I., Altinbas, M., Koyuncu, I., Arikan, O., Gomec-Yangin, C. 2003. Advanced physicochemical treatment experiences on young municipal landfill leachates. Waste Mgmt 23: 441–446.

Parawira, W. 2004. Anaerobic treatment of agricultural residues and wastewater – Application of high-rate reactors. PhD thesis, Department of Biotechnology, Lund University, Sweden, December 14.

Pesta, G. 2007. Utilization of by-products and treatment of waste in the food industry. In Anaerobic digestion of organic residues and wastes, pp. 53–72. Springer, Netherlands.

Plaza, G., Robredo, P., Pacheco, O., Toledo, A.S. 1996. Anaerobic treatment of municipal solid waste. Water Sci. Technol. 33:169–175.

Prins, M.J., Ptasinski, K.J., Janssen, F.J.J.G. 2004. Exergetic optimisation of a production process of Fischer–Tropsch fuels from biomass. Fuel Proc Technol 86:375–389.

Rabah, M.A., Eldighidy, S.M. 1989. Low cost hydrogen production from waste. Int J Hydrogen Energy 1989;14:221–227.

Santos, D.C.R.M., Lisboa, J.S., Passos, F.B., Noronha, F.B. 2004. Characterization of steam-reforming catalysts. Braz J Chem Eng 21:203–209.

Schulz, H. 1999. Short history and present trends of FT synthesis. Applied Catalysis A: General 186:1–16.

Shiga H. 1997. The decomposition of fresh and composted organic materials in soil. Food Research and Development Center for Dairy Farming, Kiyota-Ku, Sapporo, Japan, <www.agnet.org/library/eb/447/>.

Sie, S.T., Krishna, R. 1999. Fundamentals and selection of advanced FT-reactors. Appl Catal A: General 186:55–70.

Spath, P., Lane, J., Mann, M., Amos, W. 2000. Update of hydrogen from biomass: Determination of the delivered cost of hydrogen. NREL Milestone Report, April.

Tetzlaff, K.-H. 2001. HYPOTHESIS IV, Hydrogen Power – Theoretical and Engineering Solutions, International Symposium, 9–14 September, Stralsund – Germany; Proceedings Vol. 1, pp.118–122.

Tijmensen, M.J.A., Faaij, A.P.C., Hamelinck, C.N., van Hardeveld, M.R.M. 2002. Exploration of the possibilities for production of Fischer–Tropsch liquids and power *via* biomass gasification. Biomass Bioenergy 23:129–152.

Timur, H., Ozturk, I. 1999. Anaerobic sequencing batch reactor treatment of landfill leachate. Wat Res 33:3225–3230.

Uygur, A., Kargi F. 2004. Biological nutrient removal from pre-treated landfill leachate in a sequencing batch reactor. J Environ Mgmt 71: 9–14.

Van Ginkel, S.W., Logan., B. 2005. Increased biological hydrogen production with reduced organic loading. Water Research 39:3819–3826.

Verma, S. 2002. Anaerobic digestion of biodegradable organics in municipal solid wastes. Master's thesis, Applied Science Columbia University, New York, May.

Wang, D., Czernik, S., Chornet, E. 1998. Production of hydrogen from biomass by catalytic steam reforming of fast pyrolysis oils. Energy Fuels 12:19–24.

Yu, H.Q., Zheng, X.J., Hu, Z.H., Gu, G.W. 2003. High-rate anaerobic hydrolysis and acidogenesis of sewage sludge in a modified upflow reactor. Water Science and Technology 48:69–75.

Zhang, X., Ylikorpi, T., Pepe, G. 2005. Biomass-based fuel cells for manned space exploration.European Space Agency Report: Ariadna AO/1-4532/03/NL/MV, October 31.

Chapter 6
Thermochemical Conversion Processes

6.1 Introduction to Thermochemical Conversion Processes

Thermochemical biomass conversion includes a number of possible roots to produce from the initial biorenewable feedstock useful fuels and chemicals. Biorenewable feedstocks can be used as a solid fuel, or converted into liquid or gaseous forms for the production of electric power, heat, chemicals, or gaseous and liquid fuels. Thermochemical conversion processes include three subcategories: pyrolysis, gasification, and liquefaction. Figure 6.1 shows biomass thermal conversion processes. A variety of biomass resources can be used to convert to liquid, solid, and gaseous fuels with the help of some physical, thermochemical, biochemical, and biological conversion processes. Main biomass conversion processes are direct liquefaction, indirect liquefaction, physical extraction, thermochemical conversion, biochemical conversion, and electrochemical conversion. Figure 6.2 shows the types and classification of biomass conversion processes. The conver-

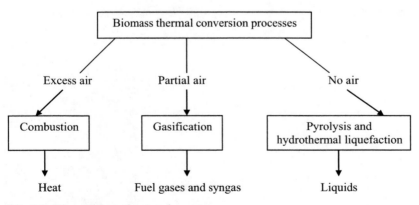

Fig. 6.1 Biomass thermal conversion processes

A. Demirbas, *Biofuels*,
© Springer 2009

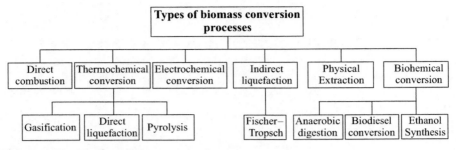

Fig. 6.2   Classification of biomass conversion processes

sion of biomass materials has the precise objective to transform a carbonaceous solid material, which is originally difficult to handle, bulky and of low energy concentration, into fuels having physico-chemical characteristics that permit economic storage and transferability through pumping systems.

Gasification of biomass for use in internal combustion engines for power generation provides an important alternate renewable energy resource. Gasification is partial combustion of biomass to produce gas and char at the first stage and subsequent reduction of the product gases, chiefly CO_2 and H_2O, by the charcoal into CO and H_2. The process also generates some methane and other higher hydrocarbons depending on the design and operating conditions of the reactor.

Pyrolysis is the fundamental chemical reaction process that is the precursor of both gasification and combustion of solid fuels, and is simply defined as the chemical changes occurring when heat is applied to a material in the absence of oxygen. Flash pyrolysis of biomass is the thermochemical process that converts small dried biomass particles into a liquid fuel (biocrude) with a yield of almost 75%, and char and non-condensable gases by heating the biomass to 775 K in the absence of oxygen. Char in the vapor phase catalyzes secondary cracking. Figure 6.3 shows the bio-oil from flash pyrolysis of the biomass pyrolysis process.

Plasma arc and radio frequency (or microwave) heating refer to specific devices providing heat from electricity for gasification, pyrolysis, or combustion depending on the amount of reactive oxygen, hydrogen, steam, or other reactant

Fig. 6.3   Bio-oil from flash pyrolysis of biomass pyrolysis

fed to the reactor. Very high temperatures are created in the ionized plasma. Plasma arc processes use electricity passing through electrodes to produce a discharge converting the surrounding gas to an ionized gas or plasma. Gases heated in plasmas typically reach temperatures of 377 K and higher.

Thermochemical conversion (TCC) technologies were studied as early as the 17th century with the first patent issued in 1788 by Robert Gardner for his work in the gasification area. However, during the time span 1800–1970 TCC technologies were forgotten due to an abundance of oil. When TCC research continued, it began to focus on sources outside of wood and coal. TCC technologies include, but are not limited to, gasification, liquefaction, pyrolysis, direct combustion, and supercritical fluid extraction. Gasification and liquefaction continue to be heavily researched and used commercially throughout the world. Researchers are focusing efforts to attempt to understand the complex reaction mechanisms that occur during these processes.

Thermochemical conversion is characterized by higher temperatures and conversion rates than most other processes. Thermochemical conversion includes a continuum of processes ranging from thermal decomposition in a primarily nonreactive environment (commonly called pyrolysis) to decomposition in a chemically reactive environment (usually called gasification if the products are primarily fuel gases). Pyrolysis can be considered an incomplete gasification process, in which a mixture of gaseous, liquid, and solid products is produced, each of which may have some immediate use to sustain the process. The characteristics of each of these processes can also vary depending on the oxidizing or reducing media, process temperature, and process pressure.

Two biomass conversion processes using water have been studied: hydrothermal upgrading (HTU) under subcritical water and supercritical water gasification (SCWG) in supercritical water conditions. For the design of the both biomass conversion processes, the following contributions of thermodynamics have been presented: phase behavior and phase equilibria in the reactor and separators, and an indication of favorable operation conditions and the trends in product distribution for the conversion reactions. A wide variety of fluids have been dealt with, from small molecules to large molecules, including non-polar and polar substances (Feng *et al.*, 2004).

HTU or direct liquefaction is a promising technology to treat waste streams from various sources and produce valuable bio-products such as biocrudes. HTU is a thermochemical process for the conversion of wet biomass material under sub (near)-critical water conditions and produces a hydrophobic oil layer (biocrude), aqueous liquid products, gasses, and some solid remains. Biocrude is obtained in the yield of typically 40–50% and has reduced oxygen content (average 12%). The resulting bio-oil can be used for electricity production, but also upgraded into transportation fuel. The project involves the visualization of the process and the development of new reactors. SCWG provides a direct conversion of glucose to hydrogen, CO, CO_2 and CH_4 in water of 875 K and 30 MPa.

Processes relating to the liquefaction of biomass are based on the early research of Appell *et al.* (1971). These workers reported that a variety of biomass such as

Table 6.1 Comparison of liquefaction and pyrolysis

Process	Temperature (K)	Pressure (MPa)	Drying
Liquefaction	525–600	5–20	Unnecessary
Pyrolysis	650–800	0.1–0.5	Necessary

agricultural and civic wastes can be converted, partially, into a heavy oil-like product by reaction with water and carbon monoxide/hydrogen in the presence of sodium carbonate.

The pyrolysis and direct liquefaction with water processes are sometimes confused with each other, and a simplified comparison of the two follows. Both are thermochemical processes in which feedstock organic compounds are converted into liquid products. In the case of liquefaction, feedstock macromolecule compounds are decomposed into fragments of light molecules in the presence of a suitable catalyst (Balat, 2008). At the same time, these fragments, which are unstable and reactive, repolymerize into oily compounds having appropriate molecular weights (Molten *et al.*, 1983). With pyrolysis, on the other hand, a catalyst is usually unnecessary, and the light decomposed fragments are converted to oily compounds through homogeneous reactions in the gas phase. The differences in operating conditions for liquefaction and pyrolysis are shown in Table 6.1.

6.2 Thermal Decomposition Mechanisms of Biorenewables

In previous studies, the mechanisms of thermal depolymerization of biomass were not extensively identified. Mechanisms have been poorly studied in the past due to difficulties in sampling the slurry during the process. These proposed reactions are general and the type of biomass will dictate the type of processes or reactions required to breakdown and rearrange molecules. The more complex the raw biomass is chemically, the more complex the reaction mechanisms required, and thus the increased difficulty in determining them.

The hemicelluloses, which are present in deciduous woods chiefly as pentosans and in coniferous woods almost entirely as hexosanes, undergo thermal decomposition very readily. It was therefore to be expected that furan derivatives would readily be found among the decomposition products. The hemicelluloses decomposed more readily than cellulose during heating. The thermal degradation of hemicelluloses begins above 475 K.

Thermal degradation of cellulose proceeds through a gradual degradation, decomposition, and charring on heating at lower temperatures, and a rapid volatilization accompanied by the formation of levoglucosan on pyrolysis at higher temperatures. The glucose chains in cellulose are first cleaved to glucose and, in a second stage, glucosan is formed by the splitting off of one molecule of water. Initial degradation reactions include depolymerization, hydrolysis, oxidation, dehydration, and decarboxylation (Demirbas. 2000). Cellulose decomposes more readily than lignin by heating.

Lignin decomposes over a wider temperature range. As the resistance time increases the quantity of unstable substances decrease. For example, quantitatively, 1-hydroxy-2-propanone and 1-hydroxy-2-butanone present high concentrations in the liquid products from fast pyrolysis. These two alcohols are partly esterified by acetic acid. In conventional slow pyrolysis, these two products are not found in so great a quantity because of their low stability (Beaumont, 1985; Demirbas, 2007). Many unstable fragments and radicals form from degradation of lignin. As a result, it is believed that as the reaction progresses the remaining mass becomes less reactive and forms stable chemical structures, and consequently the activation energy increases as the conversion level of biomass increases.

It is difficult to determine exactly what types of reactions occur during liquefaction processes. Liquefaction of carbonaceous materials takes place through a sequence of structural and chemical changes (Chornet and Overend, 1985; Demirbas, 2000).

The reaction mechanisms may not follow the exact order described above, and the reactions in biomass depolymerization process are much more complex. Many researchers have tried to investigate and propose reaction mechanisms, but no definitive study has been conducted.

A summary of some the proposed mechanisms is given below. After "cracking" reactions a hydrolyzing of the polymers and monomers, such as glucose, are then further reduced with the presence of reductive compounds. The oxygen element is eliminated, and high hydrogen and carbon containing compounds are yielded. To increase the conversion rate of organic matter to oil, a high hydrogen content in the feedstock is desirable. The use of hydrogen gas has been studied by Datta and McAuliffe (1993) and Kranich (1984). No difference in yields between CO and H used as a process gas were noted. Appell *et al.* (1980) concluded that the use of CO was more than efficient for conversion to oil products.

It has been stated that the water-gas shift reaction under high temperatures would be responsible for the increase in the conversion rate from organic matter to oil. He *et al.* (2008) determined that any process gas can produce bio-oil, and that the initial gas prevented complete vaporization of the slurry during processing. A water–gas shift reaction has been proposed to be responsible for the increase in the conversion rates of organic matter to oil (Appell *et al.*, 1980). According to the water-gas shift reaction, the hydrogen radicals then react with other oxygen containing functional groups to eliminate oxygen elements and yield hydrocarbon like compounds. Carbon monoxide, as a highly reductive compound, participates in the redox reactions directly. It combines with oxygen in the hydroxyl and carboxyl groups to form CO_2. The hydrogen radical that is released is then ready to combine with carbon. Carbon dioxide forms through numerous routes and can form through decarboxylation reactions as well.

It is difficult to determine exactly what types of reactions occur during liquefaction processes. Liquefaction of carbonaceous materials takes place through a sequence of structural and chemical changes, which involve at least the following steps (Chornet and Overend, 1985; Demirbas, 2000).

1. Cracking and reduction of polymers such as lignin and lipids.
2. Hydrolysis of cellulose and hemicelluloses to glucose.
3. Hydrogenolysis in the presence of hydrogen.
4. Reduction of amino acids.
5. New molecular rearrangements through dehydration and decarboxylation.
6. Hydrogenation of functional groups.

The reaction mechanisms may not follow the exact order described above, and the reactions in biomass depolymerization process are much more complex. Many researchers have tried to investigate and propose reaction mechanisms, but no definitive study has been conducted.

The hydrogen radicals then react with other oxygen containing functional groups to eliminate oxygen elements and yield hydrocarbon like compounds. Carbon monoxide, as a highly reductive compound, participates in the redox reactions directly. It combines with oxygen in the hydroxyl and carboxyl groups to form CO_2. The hydrogen radical that is released is then ready to combine with carbon. Carbon dioxide forms through numerous routes and can form through decarboxylation reactions as well.

6.3   Hydrothermal Liquefaction of Biorenewable Feedstocks

Liquefaction was developed for coal conversion over a century ago. Liquefaction used for biomass conversions to bio-oils is grouped under the TCC area of energy conversion methods along with gasification and pyrolysis.

Liquefaction can be accomplished directly or indirectly. Direct liquefaction involves hydrothermal liquefaction and rapid pyrolysis to produce liquid tars and oils and/or condensable organic vapors. Indirect liquefaction involves the use of catalysts to convert non-condensable, gaseous products of pyrolysis or gasification into liquid products. The liquefaction of biomass has been investigated in the presence of solutions of alkalis (Eager et al., 1982), propanol and butanol (Ogi and Yokoyama, 1993), and glycerine (Demirbas, 1985), or by direct liquefaction (Ogi et al., 1985; Minowa et al., 1994).

HTL or direct liquefaction is a promising technology to treat waste streams from various sources and produce valuable bio-products such as biocrudes. A major problem with commercializing the HTL processes for biomass conversion today is that it remains uneconomical when compared to the costs of diesel or gasoline production. High transportation costs of large quantities of biomass increase production costs, and poor conversion efficiency coupled with a lack of understanding complex reaction mechanisms inhibits growth of the process commercially.

In the HTU process, biomass is reacted in liquid water at elevated temperature and pressure. The phase equilibria in the HTU process are very complicated due to the presence of water, supercritical carbon dioxide, alcohols, as well as the so-called biocrude. The biocrude is a mixture with a wide molecular weight distribution and consists of various kinds of molecules. Biocrude contains 10–13% oxygen. The biocrude is upgraded by catalytic hydrodeoxygenation in a central facility.

Biomass, such as wood, with a lower energy density is converted to biocrude with a higher energy density, organic compounds including mainly alcohols and acids, gases mainly including CO_2. Water is also a byproduct. In the products, CO_2, the main component of the gas product, can be used to represent all gas produced, and methanol and ethanol represent organic compounds. In Table 6.2, the weight fraction of each component is assigned on the basis of the data of the vacuum flash of biocrude and the data of a pilot plant (Feng *et al.*, 2004). Figure 6.4 shows the block scheme of commercial HTU plant. The feedstocks, reaction conditions, and the products for the HTU process are given in Table 6.3.

Table 6.2 Representatives for the products from the HTU process

Product	Component	Weight fraction (%)
Biocrude	Polycarbonates	47.5
	Methyl-*n*-propyl ether	2.5
Gas	Carbon dioxide	25.0
Organic compounds	Methanol	5.0
	Ethanol	3.5
Water	Water	16.5

Table 6.3 Feedstocks, reaction conditions, and products for the HTU process

Biomass feedstocks	Wood and forest wastes
	Agricultural and domestic residues
	Municipal solid wastes
	Organic industrial residues
	Sewage sludge
Reaction conditions	Temperature: 300–350°C
	Pressure: 12–18 MPa
	Resistance time: 5–20 min
	Medium: liquid water
Main chemical reactions	Depolymerization
	Decarboxylation
	Dehydration
	Oxygen removed as CO_2 and H_2O
	Hydrodeoxygenation
	Hydrogenation
Products (%w on feedstock)	Biocrude: 45
	Water soluble organics: 10
	Gas (>90% CO_2): 25
	Process water: 20
Thermal efficiency	70–90%

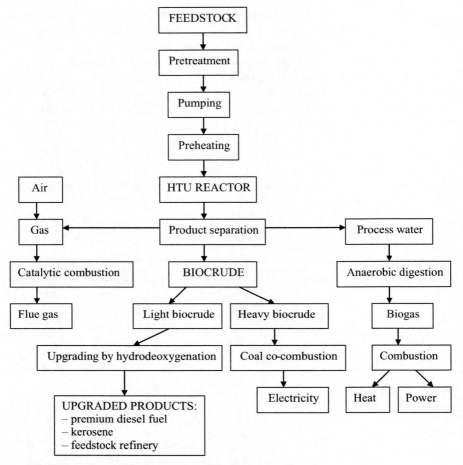

Fig. 6.4 Block scheme of a commercial HTU plant

One of the first HTL studies was conducted by Kranich (1984) using municipal waste materials (MSW) as a source to produce oil. Three different types of materials from a MSW plant were used: primary sewage sludge, settled digester sludge, and digester effluent. Using a magnetically stirred batch autoclave with a hydrogen-feed system, a slurry feed device, a pressure and temperature recorder, and a wet-test meter for measuring gas product, Kranich processed the waste sources. The feedstock was first dried then powdered. The wastes were also separated into different oil and water slurries and processed separately. Temperatures ranged from 570–720 K with pressures up to 14 MPa. Retention times also varied between 20–90 minutes. Hydrogen was used as the reducing gas with initial pressures up to 8.3 MPa. Three types of catalyst were studied: sodium carbonate, nickel carbonate, and sodium molybdate. The slurry feedstock was injected into the reactor through a pressurized injector, and the oil product was extracted by pentane and toluene. Results showed that organic conversion rates varied from 45–99%, and oil production rates were reported from 35.0–63.3%. Gas products were found to contain H_2, CO_2, and C_1–C_4 hydrocarbons. The experimental results showed no significant differences between the applications of the three different catalysts. Kranich recommended that the water slurry system was not feasible for scale-up, and considerations of a commercial scale process were confined to only the oil slurry system. It was also concluded that no further development work on hydroliquefaction of sewage sludge to oil was necessary. Kranich's recommendation did not hold, mainly due to increases in crude oil prices and the need to find new technologies for energy procurement, and thus many studies on liquefaction of sewage sludge have since been conducted. Research has indicated that liquefaction is a feasible method for the treatment of sewage sludge wastes and has a high oil producing potential (Suzuki *et al.*, 1986; Itoh *et al.*, 1994; Inoue *et al.*, 1997). Today, HTL research is still being conducted with sewage sludge; however, focus has shifted to include many varieties of biomass materials.

Several technologies have been developed to convert biomass into a liquid biofuel with a higher heating value, such as gasification, fast pyrolysis and the HTU. In the HTU, the biomass is treated during 5–20 min with water under subcritical conditions (575–625 K, 10–18 MPa) to give a heavy organic liquid (biocrude) with a heating value of 30–35 MJ/kg. During this process, the oxygen content of the organic material is reduced from about 40% to between 10% and 15%. The removed oxygen ends up in CO_2, H_2O and CO. After 1.6 s at 320°C and 25MPa, 47% conversion of cellulose in water was obtained yielding hydrolysis products (cellobiose, glucose, *etc.*, 44%) and decomposition products of glucose (erythrose, 1,6-anhydroglucose, 5-hydroxymethylfurfural, 3%. Furthermore, it has been shown that cellobiose decomposes *via* hydrolysis to glucose, and *via* pyrolysis to glycosylerythrose and glycosylglycolaldehyde, which are further hydrolyzed into glucose, erythrose, and glycolaldehyde. Hydrolysis refers to splitting up of the organic particles into smaller organic fragments in water. Hydrothermal decomposition also acts on the large organic molecules reducing them into smaller fragments, some of which dissolve in water.

In the HTU process, biomass chips are pressurized and digested at 200–250°C with recycled water from the process. Subsequently the digested mass is pressurized to 12–18 MPa and reacted in liquid water at 300–400°C for 5–15 min. Under these conditions decarboxylation and depolymerization take place and a biocrude is formed, which separates from the water phase. Part of the process water is recycled. Obviously, the process is very simple high efficiency.

Hydrothermal reaction involves applying heat under pressure to achieve reaction in an aqueous medium. The treatment of organic wastes by SCW reaction in a homogeneous phase undergoes that interface mass transfer limitations are avoided, and reaction efficiencies of 99.9% can be achieved at residence times lower than 1 min. Because of the distinctive characteristics of water described above, hydrothermal reaction is an effective method for the treatment of organic wastes. The reaction can be performed under subcritical or supercritical conditions. It can also be classified into two broad categories: (a) oxidative, *i.e.*, involving the use of oxidants, and (b) non-oxidative, *i.e.*, excluding the use of oxidants.

The process at subcritical temperatures results in the production of abundant quantities of dissolved organic matter. The dissolved organic matter contains significant quantities of volatile fatty acids, especially acetic acid, which remains in the liquid phase and can be significantly removed at supercritical temperatures.

6.3.1   The Role of Water During the HTL Process

Most substrates are not soluble in water under normal conditions, but salvation can occur between the hydroxyl groups and water under high temperatures and pressures. Water is a medium for intermediate hydrolysis of cellulose and other high-molecular weight carbohydrates to water-soluble sugars. The primary reactions in the conversion to oil are likely to involve the formation of low-molecular weight, water soluble compounds such as glucose.

Alkaline catalysts are water soluble as well, facilitating their dispersion throughout the process vessel in a readily available form. Water is also used to mix reactants, and to diminish condensations to chars by diluting the reaction intermediates.

Water is a reactant at high temperatures. Hydrogen may be added to the substrate through the water-gas shift reactions, which consumes carbon monoxide from carbon dioxide and hydrogen.

6.3.2   HTU Applications

The conversions of biomass wastes into biocrude or hydrogen have been successfully performed by the HTU processes. The application of hydrothermal reaction is transfer from organic wastes to useful materials, such as liquid fuel, hydrogen, glucose, organic acid, *etc.*

Hydrothermal reaction is a prominent method for the treatment of organic wastes and has been attracting worldwide attention. During the process, various reactions such as oxidation, hydrolysis, dehydration, and thermal decomposition can be carried out energetically, so that the reaction can be successfully used for oxidizing organic wastes to CO_2 and other innocuous end products, as well as for conversion of organic wastes to fuels or useful materials, such as biocrude, hydrogen, glucose, lactic acid, acetic acid, amino acids, *etc.* (He *et al.*, 2008).

In the HTU process, biomass chips are pressurized and digested at 473–523 K with recycled water from the process. Subsequently the digested mass is pressurized to 12–18 MPa and reacted in liquid water at 573–673 K for 5–15 min. Under these conditions decarboxylation and depolymerization take place and a biocrude is formed, which separates from the water phase. Part of the process water is recycled. Obviously, the process is very simple with high efficiency (Zhong *et al.*, 2002).

Hydrothermal reaction involves applying heat under pressure to achieve reaction in an aqueous medium. The treatment of organic wastes by SCW reaction in a homogeneous phase undergoes that interface mass transfer limitations are avoided and reaction efficiencies of 99.9% can be achieved at residence times lower than 1 min (Tester and Cline, 1999). Because of the distinctive characteristics of water described above, hydrothermal reaction is an effective method for the treatment of organic wastes. The reaction can be performed under near-critical or supercritical conditions. It can also be classified into two broad categories (Shanableh and Jomaa, 1998): (a) oxidative, *i.e.*, involving the use of oxidants, and (b) non-oxidative, *i.e.*, excluding the use of oxidants.

The process at subcritical temperatures results in the production of abundant quantities of dissolved organic matter (Jomaa, 2001). The dissolved organic matter contains significant quantities of volatile fatty acids, especially acetic acid, which remains in the liquid phase and can be significantly removed at supercritical temperatures (Jomaa *et al.*, 2003).

6.4   Direct Combustion of Biomass

Direct combustion is the old way of using biomass. Biomass thermochemical conversion technologies such as pyrolysis and gasification are certainly not the most important options at present; combustion is responsible for over 97% of the world's bioenergy production.

Research on direct liquefaction has been widely studied in the past, especially in the late 1970s and early 1980s for the purpose of alternative energy production. The feedstocks mainly consisted of wood and municipal solid wastes (MSW). Many aspects of the process are still being studied: the type and condition of various feedstocks, the operating carrier media, and reducing reagents. More specifically, researchers are focusing on various operating conditions, such as pH, processing gas, temperature, pressure, catalyst, retention time, solid content, gas to volatile solid ratio, and solvents for extraction or processing. In addition to studying these conditions researchers are still focusing their efforts on understanding the complex reactions that occur during the process. Biomass is complex by nature and varies by location. Developing a process that will handle many biomass sources, and one that is flexible to handle variations of biomass, is desirable to increase the potential impact that the process may have. Economics currently limit large scale biomass liquefaction treatment facilities, and on-site treatment remains difficult and expensive. However, researchers continue to move forward with their studies, and many alternative organic feedstocks have been processed through this technology as a means of waste management, as well as renewable energy production.

Combustion is the oxidation of the fuel for the production of heat at elevated temperatures without generating commercially useful intermediate fuel gases, liquids, or solids. Combustion of MSW or other secondary materials is generally referred to as incineration. Particle temperatures in heterogeneous (*e.g.*, unsteady reactions between solid and gas phases) combustion can differ from the surrounding gas temperatures, depending on radiation heat transfer conditions.

Combustion of solids involves the simultaneous processes of heat and mass transport, progressive pyrolysis, gasification, ignition, and burning, with no intermediate steps and with an unsteady, sometimes turbulent, fluid flow. Normally, combustion employs an excess of oxidizer to ensure maximum fuel conversion, but it can also occur under fuel-rich conditions.

Combustion is a basic chemical process that releases energy from a fuel and air mixture. For combustion to occur, fuel, oxygen, and heat must be present together. Combustion is the chemical reaction of a particular substance with oxygen. Combustion represents a chemical reaction, during which from certain matters other simple matters are produced, this is a combination of inflammable matter with oxygen of the air accompanied by heat release. The quantity of heat evolved when one mole of a hydrocarbon is burned to carbon dioxide and water is called the heat of combustion. Combustion to carbon dioxide and water is characteristic of organic compounds; under special conditions it is used to determine their carbon and hydrogen content. During combustion the combustible part of fuel is subdivided into volatile part and solid residue. During heating it evaporates together with a part of carbon in the form of hydrocarbons combustible gases and carbon monoxide release by thermal degradation of the fuel. Carbon monoxide is mainly formed the following reactions: (a) from reduction of CO_2 with unreacted C,

$$CO_2 + C \rightarrow 2CO \tag{6.1}$$

and (b) from degradation of carbonyl fragments (–CO) in the fuel molecules at 600–750 K temperature.

The combustion process is started by heating the fuel above its ignition temperature in the presence of oxygen or air. Under the influence of heat, the chemical bonds of the fuel are cleaved. If complete combustion occurs, the combustible elements (C, H and S) react with the oxygen content of the air to form CO_2, H_2O and mainly SO_2.

If not enough oxygen is present, or the fuel and air mixture is insufficient, then the burning gases are partially cooled below the ignition temperature and the combustion process stays incomplete. The flue gases then still contain combustible components, mainly carbon monoxide (CO), unburned carbon (C), and various hydrocarbons (C_xH_y).

The standard measure of the energy content of a fuel is its heating value (HV), sometimes called the calorific value or heat of combustion. In fact, there are multiple values for HV, depending on whether it measures the enthalpy of combustion (ΔH) or the internal energy of combustion (ΔU), and whether for a fuel containing hydrogen product water is accounted for in the vapor phase or the condensed (liquid) phase. With water in the vapor phase, the lower heating value (LHV) at constant pressure measures the enthalpy change due to combustion. The heating value is obtained by the complete combustion of a unit quantity of solid fuel in an oxygen-bomb calorimeter under carefully defined conditions. The gross heat of combustion or higher heating value (GHC or HHV) is obtained by the oxygen-bomb colorimeter method as the latent heat of moisture in the combustion products is recovered.

6.4.1 Combustion Efficiency

In general, combustion efficiency as defined in flue gas analysis standards is simply reduced by the stack loss. Combustion efficiency is based on the flue gas temperature and inlet air temperature. Combustion efficiency calculations assume complete fuel combustion and are based on the three following factors: (1) the chemistry of the fuel, (2) the net temperature of the stack gases, and (3) the percentage of oxygen or CO_2 by volume after combustion.

Combustion efficiency relates to the part of the reactants that combine chemically. Combustion efficiency increases with increasing temperature of the reactants, the increasing time that the reactants are in contact, increasing vapor pressures, increasing surface areas, and increasing stored chemical energy.

Figure 6.5 shows a typical diagram for combustion efficiency. Without enough combustion air the combustion efficiency is low. The combustion efficiency increases with fuel/air ratio. The combustion efficiency is lower than maximum value in the stoichiometric fuel/air mixture. Figure 6.6 shows the typical combustion efficiency *vs.* excess of air for natural gas.

6.4.1.1   Combustion Efficiency of Biofuels

Ethanol is an oxygenated liquid fuel and its combustion heat is considerably lower than those of petroleum-based fuels. Its characteristics as a transportation fuel can be attributed to its chemical composition. The oxygen provides more efficient

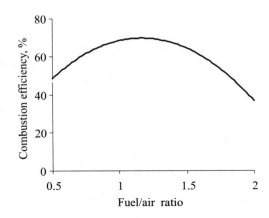

Fig. 6.5 A typical diagram for combustion efficiency

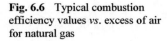

Fig. 6.6 Typical combustion efficiency values *vs.* excess of air for natural gas

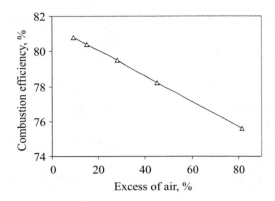

combustion and cleaner emissions. At a stoichiometric air/fuel ratio of 9:1 in comparison with gasoline's 14.7:1, it is obvious that more ethanol is required to produce the chemically correct products of CO_2 and water. Ethanol has a higher octane number (108), broader flammability limits, higher flame speeds, and higher heats of vaporization than gasoline. These properties allow for a higher compression ratio, shorter burn time, and leaner burn engine, which lead to theoretical efficiency advantages over gasoline in an internal combustion engine. The octane number of ethanol allows it to sustain significantly higher internal pressures than gasoline, before being subjected to predetonation. The disadvantages of ethanol include its lower energy density than gasoline, its corrosiveness, low flame luminosity, lower vapor pressure, miscibility with water, and toxicity to ecosystems.

Methyl alcohol has the lowest combustion energy of all engine fuels. However, it also has the lowest stoichiometric or chemically correct air-fuel ratio. Therefore, an engine burning methyl alcohol would produce the most power. It also is possible to take advantage of the higher octane number of methyl (114) alcohol and increase the engine compression ratio. This would increase the efficiency of converting the potential combustion energy to power. Finally, alcohols burn more completely, thus increasing combustion efficiency.

Biofuels, except biohydrogen, are oxygenated compounds. Oxygenates are just preused hydrocarbons having a structure that provides a reasonable antiknock value. Also, as they contain oxygen, fuel combustion is more efficient, reducing hydrocarbons in the exhaust gases. The only disadvantage is that oxygenated fuel has less energy content. For the same efficiency and power output, more fuel has to be burned.

The advantages of biofuels such as biodiesel, vegetable oil, bioethanol, biomethanol, and biomass pyrolysis oil as engine fuel are liquid nature-portability, ready availability, renewability, higher combustion efficiency, lower sulfur and aromatic content, and biodegradability. Full combustion of a fuel requires in existence the amount of stoichiometric oxygen. However, the amount of stoichiometric oxygen generally is not enough for full combustion so as not to oxygenate the fuel. The structural oxygen content of fuel increases combustion efficiency of the fuel due to an increase of the homogeneity of oxygen with the fuel during combustion. Because of this, the combustion efficiency and cetane numbers of vegetable oils and biodiesels are higher than diesel fuel; moreover the combustion efficiency of methanol/ethanol is higher than that of gasoline. The cetane number measures the readiness of a fuel to auto-ignite.

6.5 Direct Liquefaction

Liquefaction was developed for coal conversion over a century ago. Liquefaction used for biomass conversions to bio-oils is grouped under the TCC area of energy conversion methods along with gasification and pyrolysis. Liquefaction can be accomplished directly or indirectly. Direct liquefaction involves hydrothermal liquefaction and rapid pyrolysis to produce liquid tars and oils, and/or condensable organic vapors. Indirect liquefaction involves the use of catalysts to convert non-condensable, gaseous products of pyrolysis or gasification into liquid products.

Alkali salts, such as sodium carbonate and potassium carbonate, can act like the hydrolysis of cellulose and hemicelluloses, breaking them into smaller fragments. The degradation of biomass into smaller products mainly proceeds by depolymerization and deoxygenation. In the liquefaction process, the amount of solid residue increases in proportion to the lignin content. Lignin is a macromolecule, which consists of alkylphenols and has a complex three-dimensional structure. It is generally accepted that free phenoxyl radicals are formed by thermal decomposition of lignin above 525 K, and that the radicals have a random tendency to form a solid residue through condensation or repolymerization.

The changes during the liquefaction process involve all kinds of processes such as solvolysis, depolymerization, decarboxylation, hydrogenolysis, and hydrogenation. Solvolysis results in micellar-like substructures of the biomass. The depolymerization of biomass leads to smaller molecules. It also leads to new molecular rearrangements through dehydration and decarboxylation. When hydrogen is present, hydrogenolysis and hydrogenation of functional groups, such as hydroxyl groups, carboxyl groups, and keto groups, also occur.

Due to the high moisture content and low heating value, biomass is not suitable to be used as energy directly. Aqueous liquefaction of lignocellulosic materials involves disaggregation followed by partial depolymerization of the constitutive families (hemicelluloses, cellulose and lignin). Wood, bark, and sugar cane bagasse have been directly liquefied by Ogi and Yokoyama (1993) under the same conditions. They observed that the yields of heavy oil were 50% ± 5% for wood. As for bark, the yields of heavy oil ranged from 20–27%, which was much lower than those of the wood, while for sugar cane bagasse the yield was similar to those of wood.

Liquefaction of biomass and wastes is accomplished by natural, direct, and indirect thermal and fermentation methods. Natural liquefaction systems were the

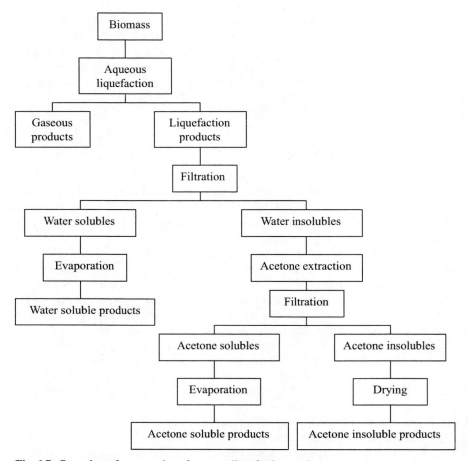

Fig. 6.7   Procedures for separation of aqueous liquefaction products

connection with certain arid-land plants and microalgae growth, and the resultant formation of lipids and hydrocarbons. In the case of liquefaction, feedstock macromolecule compounds are decomposed into fragments of light molecules in the presence of a suitable catalyst.

Direct liquefaction of biomass by thermochemical means has been studied as a process for fuel production for the last 20 years. Direct liquefaction is either reaction of biomass components with smaller molecules such as H_2 and CO, or short-term pyrolytic treatment, sometimes in the presence of gases such as H_2. Figure 6.7 shows the procedures for separation of aqueous liquefaction products

Indirect liquefaction involves successive production of an intermediate, such as synthesis gas or ethylene, and its chemical conversion to liquid fuels via Fischer–Tropsch processes.

6.6 Pyrolysis Processes

Pyrolysis dates back to at least ancient Egyptian times, when tar for caulking boats and certain embalming agents were made by pyrolysis. In the 1980s, researchers found that the pyrolysis liquid yield can be increased using fast pyrolysis where a biomass feedstock is heated at a rapid rate and the vapors produced are also condensed rapidly (Mohan *et al.*, 2006). Pyrolysis has been used since the dawn of civilization. If some means is applied to collect the off-gasses (smoke), the process is called wood distillation. The ancient Egyptians practiced wood distillation by collecting tars and pyroligneous acid for use in their embalming industry. Pyrolysis of wood to produce charcoal was a major industry in the 1800s, supplying the fuel for the industrial revolution, until it was replaced by coal. In the late 19th century and early 20th century wood distillation was still profitable for producing the soluble tar, pitch, creosote oil, chemicals, and noncondensable gasses often used to heat boilers at the facility. The wood distillation industry declined in the 1930s due to the advent of the petrochemical industry and its lower priced products.

Pyrolysis is the thermal decomposition of organic matter occurring in the absence of oxygen or when significantly less oxygen is present than required for complete combustion. Pyrolysis is the basic thermochemical process for converting biomass to a more useful fuel. Biomass is heated in the absence of oxygen, or partially combusted in a limited oxygen supply, to produce a hydrocarbon rich gas mixture, an oil-like liquid, and a carbon rich solid residue. The products of pyrolysis can be gaseous, liquid, and/or solid. Flash pyrolysis describes the rapid, moderate temperature (67–875 K) pyrolysis that produces liquids. Biomass is heated at rates of 100–10,000 K/s, and the vapor residence time is normally less than 2 seconds. The oil products are maximized at the expense of char and gas.

Pyrolysis is a process similar to gasification except generally optimized for the production of fuel liquids (pyrolysis oils) that can be used straight or refined for higher quality uses such as engine fuels, chemicals, adhesives, and other products. Pyrolysis typically occurs at temperatures in the range of 675–975 K. Pyrolysis and combustion of pyrolysis-derived fuel liquids and gases also produce the same categories of end products as direct combustion of solids. Like gasification, their pollution control and conversion efficiencies may be improved.

Pyrolysis and direct liquefaction processes are sometimes confused with each other, and a simplified comparison of the two follows. Both are thermochemical processes in which feedstock organic compounds are converted into liquid products. In the case of liquefaction, feedstock macromolecule compounds are decomposed into fragments of light molecules in the presence of a suitable catalyst. At the same time, these fragments, which are unstable and reactive, repolymerize into oily compounds having appropriate molecular weights (Demirbas, 2000). With pyrolysis, on the other hand, a catalyst is usually unnecessary, and the light decomposed fragments are converted to oily compounds through homogeneous reactions in the gas phase.

Fast pyrolysis utilizes biomass to produce a product that is used both as an energy source and a feedstock for chemical production. Considerable efforts have been made to convert wood biomass to liquid fuels and chemicals since the oil crisis in the mid-1970s. Most work has been performed on wood, because of its consistency and comparability between tests. However, nearly 100 types of biomass have been tested, ranging from agricultural wastes such as straw, olive pits, and nut shells, to energy crops such as sweet sorghum, sugar cane, sugar beet, saccharum, fiber sorghum, switchgrass, and miscanthus. The effect of pyrolysis reaction rate and the yield of the volatiles are the biomass composition and structure, heating rate, residence time, catalyst, and particle size. Although very fast and very slow pyrolyses of biomass produce markedly different products, the variety of heating rates, temperatures, residence times, and feedstock varieties found in the literature make generalizations difficult to define, in regard to trying to critically analyze the literature (Mohan *et al.*, 2006).

Catalytic cracking is a thermochemical process that employs catalysts using hydrogen-driven reducing reactions to accelerate the breakdown of high molecular weight compounds (*e.g.*, plastics) into smaller products for the purposes of improving selectivity and imparting certain desirable characteristics to the final product, such as volatility and flashpoint of liquid fuels. This cracking process is often employed in oil refinery operations to produce lower molecular weight hydrocarbon fuels from waste feedstocks. These include gasoline from heavier oils, distillation residuals, and waste plastic.

Rapid heating and rapid quenching produces intermediate pyrolysis liquid products, which condense before further reactions break down higher-molecular-weight species into gaseous products. High reaction rates minimize char formation. Under some conditions, no char is formed. At higher fast pyrolysis temperatures, the major product is gas. Many researchers have attempted to exploit the complex degradation mechanisms by conducting pyrolysis in unusual environments (Mohan *et al.*, 2006).

Pyrolysis is the simplest and almost certainly the oldest method of processing one fuel in order to produce a better one. Pyrolysis can also be carried out in the presence of a small quantity of oxygen ("gasification"), water ("steam gasification"), or hydrogen ("hydrogenation"). One of the most useful products is methane, which is a suitable fuel for electricity generation using high-efficiency gas turbines.

Cellulose and hemicelluloses form mainly volatile products on heating, due to the thermal cleavage of the sugar units. The lignin forms mainly char since it is not readily cleaved to lower molecular weight fragments. The progressive increase in the pyrolysis temperature of wood leads to the release of volatiles thus forming a solid residue that is different chemically from the original starting material (Demirbas, 2000). Cellulose and hemicelluloses initially break into compounds of lower molecular weight. This forms an "activated cellulose", which decomposes by two competitive reactions; one forming volatiles (anhydrosugars) and the other char and gases. The thermal degradation of the activated cellulose and hemicelluloses to form volatiles and char can be divided into categories depending on the reaction temperature. Within a fire all these reactions take place concurrently and consecutively. Gaseous emissions are predominantly a product of pyrolytic cracking of the fuel. If flames are present, fire temperatures are high, and more oxygen is available from thermally induced convection.

Biomass pyrolysis is attractive because solid biomass and wastes can be readily converted into liquid products. These liquids, such as crude bio-oil or slurry of charcoal of water or oil, have advantages in transport, storage, combustion, retrofitting, and flexibility in production and marketing. Among the liquid products, methanol is one of the most valuable products. The liquid fraction of the pyrolysis products consists of two phases: an aqueous phase containing a wide variety of organo-oxygen compounds of low molecular weight and a non-aqueous phase containing insoluble organics of high molecular weight. This phase is called tar and is the product of greatest interest. The ratios of acetic acid, methanol, and acetone of the aqueous phase are higher than those of the non-aqueous phase. The point where the cost of producing energy from fossil fuels exceeds the cost of biomass fuels has been reached. With a few exceptions, energy from fossil fuels will cost more money than the same amount of energy supplied through biomass conversion.

The kinematic viscosity of pyrolysis oil varies from as low as 11 cSt to as high as $115 \, \text{mm}^2/\text{s}$ (measured at 313 K) depending on the nature of the feedstock, the temperature of the pyrolysis process, the thermal degradation degree and catalytic cracking, the water content of the pyrolysis oil, the amount of light ends that have collected, and the pyrolysis process used. The pyrolysis oils have water contents of typically 15–30 wt% of the oil mass, which cannot be removed by conventional methods like distillation. Phase separation may partially occur above certain water contents. The water content of pyrolysis oils contributes to their low energy density, lowers the flame temperature of the oils, leads to ignition difficulties, and, when preheating the oil, can lead to premature evaporation of the oil and resultant injection difficulties. The higher heating value (HHV) of pyrolysis oils is below 26 MJ/kg (compared to 42–45 MJ/kg for conventional petroleum fuel oils). In contrast to petroleum oils, which are non-polar and in which water is insoluble, biomass oils are highly polar and can readily absorb over 35% water (Demirbas, 2007).

The pyrolysis oil (bio-oil) from wood is typically a liquid, almost black through dark red brown. The density of the liquid is about $1200 \, \text{kg/m}^3$, which is higher than that of fuel oil and significantly higher than that of the original biomass. The bio-oils have a water contents of typically 14–33 wt.%, which cannot be removed by conventional methods like distillation. Phase separation may occur above a certain water content. The higher heating value (HHV) is below 27 MJ/kg (compared to 43–46 MJ/kg for conventional fuel oils).

The bio-oil formed at 725 K contained high concentrations of compounds such as acetic acid, 1-hydroxy-2-butanone, 1-hydroxy-2-propanone, methanol, 2,6-dimethoxyphenol, 4-methyl-2,6-dimethoxyphenol, and 2-cyclopenten-1-one, *etc*. A significant characteristic of the bio-oils is the high percentage of alkylated compounds especially methyl derivatives. As the temperature increases, some of these compounds are transformed *via* hydrolysis. The formation of unsaturated compounds from biomass materials generally involves a variety of reaction pathways such as dehydration, cyclization, Diels–Alder cycloaddition reactions, and ring rearrangement. For example, 2,5-hexanedione can undergo cyclization under hydrothermal conditions to produce 3-methyl-2-cyclopenten-1-one with very high selectivity of up to 81% (Demirbas, 2008).

The influence of temperature on the compounds existing in liquid products obtained from biomass samples *via* pyrolysis has been examined in relation to the yield and composition of the product bio-oils. The product liquids were analyzed by a gas chromatography mass spectrometry combined system. The bio-oils were composed of a range of cyclopentanone, methoxyphenol, acetic acid, methanol, acetone, furfural, phenol, formic acid, levoglucosan, guaiocol, and their alkylated phenol derivatives. Thermal depolymerization and decomposition of biomass structural components, such as cellulose, hemicelluloses, and lignin form liquids and gas products, as well as a solid residue of charcoal. The structural components of the biomass samples mainly affect pyrolytic degradation products. A reaction mechanism is proposed that describes a possible reaction route for the formation of the characteristic compounds found in the oils. The supercritical water extraction and liquefaction partial reactions also occur during the pyrolysis. Acetic acid is formed in the thermal decomposition of all three main components of biomass. In the pyrolysis reactions of biomass water is formed by dehydration, acetic acid comes from the elimination of acetyl groups originally linked to the xylose unit, furfural is formed by dehydration of the xylose unit, formic acid proceeds from carboxylic groups of uronic acid, and methanol arises from methoxyl groups of uronic acid (Demirbas, 2007).

Pyrolysis of wood has been studied as a zonal process with zone A (easily degrading zone) occurring at temperatures up to 475 K. The surface of the wood becomes dehydrates at this temperature, and along with water vapor, carbon dioxide, formic acid, acetic acid, and glyoxal are given off. When temperatures of 475–535 K are attained, the wood is said to be in zone B and is evolving water vapor, carbon dioxide, formic acid, acetic acid, glyoxal, and some carbon monoxide. The reactions to this point are mostly endothermic, the products are largely non-condensable, and the wood is becoming charred. Pyrolysis actually begins between 535 K and 775 K, which is called zone C. The reactions are exothermic, and unless heat is dissipated, the temperature will rise rapidly. Combustible gases such as carbon monoxide from cleaving of carbonyl group, methane, formaldehyde, formic acid, acetic acid, methanol, and hydrogen are being liberated and charcoal is being formed. The primary products are beginning to react with each other before they can escape the reaction zone. If the temperature continues to rise above 775 K, a layer of charcoal will be formed, which is the site of vigorous secondary reactions and is classified as zone D. Carbonization is said to be complete at temperatures of 675 K to 875 K. Thermal degradation properties of hemicelluloses, celluloses, and lignin can be summarized as follows (Demirbas, 2000):

Thermal degradation of hemicelluloses > of cellulose >> of lignin

The liquid fraction of pyrolysis products consists of two phases: an aqueous phase containing a wide variety of organo-oxygen compounds of low molecular weight and a non-aqueous phase containing insoluble organics (mainly aromatics) of high molecular weight. This phase is called bio-oil or tar and is the product of greatest interest. The ratios of acetic acid, methanol, and acetone of aqueous phase are higher than those of non-aqueous phase.

If the purpose is to maximize the yield of liquid products resulting from biomass pyrolysis, a low temperature, high heating rate, and short gas residence time process are required. For a high char production, a low temperature, low heating rate process are chosen. If the purpose is to maximize the yield of fuel gas resulting from pyrolysis, a high temperature, low heating rate, long gas residence time process should be preferred.

6.6.1 Reaction Mechanism of Pyrolysis

Earlier kinetic studies have been conducted under a variety of experimental conditions, resulting in conflicting data with a wide range of kinetic parameters (Stamm, 1956; Hofmann and Antal, 1984; Desrosiers and Lin, 1984; Demirbas, 1998; Koullas et al., 1998). Thermogravimetric analysis (TGA) is the general approach applied to determine the weight loss of pyrolyzed samples at various reaction temperatures. A comparison of kinetic data from literature is given in Table 6.4.

The pyrolysis process is always initially endothermic, and almost linear in mass loss. At high heating rates, this character is maintained throughout the process. At low heating rates, the well-known exothermic char forming processes at some point begin to complete with the basic endothermic nature of the process, and drive it back towards thermoneutrality. There is, however, no evidence of thermo-

Table 6.4 Comparison of kinetic data from the literature

Biomass	Frequency factor (min^{-1})	Activation energy (kJ/mole)	Temperature range (K)	Researcher
Douglas fir sawdust	1.1×10^{11}	105	368–525	Stamm (1956)
Wood	1.4×10^{6}	84.2	>605	Barooah and Long (1976)
Douglas fir bark	1.3×10^{10}	100–201	450–850	Tran and Rai (1978)
Missouri oak sawdust	1.5×10^{8}	106	595–665	Thurner and Mann (1981)
Sawdust	1.9×10^{7}	95.9	565–665	Koullas et al. (1998)
Hazelnut shell sawdust	4.7×10^{13}	92–170	450–750	Demirbas (1998)

dynamically different pathways being followed at high and low heating rates during the initial stages of pyrolysis. This alone does not assure that a change of mechanism does not occur, but there is also no other evidence to suggest such a change in mechanism with the heating rate.

It is believed that as the pyrolysis reaction progresses the carbon residue (semi-char) becomes less reactive and forms stable chemical structures, and consequently the activation energy increases as the conversion level of biomass increases (Tran and Charanjit, 1978).

The general changes that occur during pyrolysis are enumerated below (Babu and Chaurasia, 2003; Mohan *et al.*, 2006):

1. Heat transfer from a heat source, to increase the temperature inside the fuel.
2. The initiation of primary pyrolysis reactions at this higher temperature releases volatiles and forms char.
3. The flow of hot volatiles toward cooler solids results in heat transfer between hot volatiles and cooler unpyrolyzed fuel.
4. Condensation of some of the volatiles in the cooler parts of the fuel, followed by secondary reactions, can produce tar;
5. Autocatalytic secondary pyrolysis reactions proceed while primary pyrolytic reactions simultaneously occur in competition.
6. Further thermal decomposition, reforming, water gas shift reactions, radicals recombination, and dehydrations can also occur, which are a function of the residence time/temperature/pressure profile of the process.

A comparison of pyrolysis, ignition, and combustion of coal and biomass particles reveals the following:

1. Pyrolysis starts earlier for biomass as compared with coal.
2. The VM content of biomass is higher compared with that of coal.
3. The fractional heat contribution by VM in biomass is of the order of 70% compared with 36% for coal.
4. Biomass char has more O_2 compared with coal. The fractional heat contribution by biomass is of the order of 30% compared with 70% for coal.
5. The heating value of volatiles is lower for biomass as compared with that of coal.
6. Pyrolysis of biomass chars mostly releases CO, CO_2, and H_2O.
7. Biomass has ash that is more alkaline in nature, which may aggravate fouling problems.

The organic compounds from biomass pyrolysis are the following groups:

1. A gas fraction containing: CO, CO_2, some hydrocarbons and H_2.
2. A condensable fraction containing: H_2O and low molecular weight organic compounds (aldehydes, acids, ketones, and alcohols).
3. A tar fraction containing: higher molecular weight sugar residues, furan derivatives, phenolic compounds and airborne particles of tar and charred material which form smoke.

The mechanism of pyrolysis reactions of biomass has been extensively discussed in an earlier study (Demirbas, 2000). Water is formed by dehydration. In the pyrolysis reactions, methanol arises from the breakdown of methyl esters and/or ethers from decomposition of pectin-like plant materials. Methanol also arises from methoxyl groups of uronic acid. Acetic acid is formed in the thermal decomposition of all three main components of wood. When the yield of acetic acid originating from the cellulose, hemicelluloses, and lignin is taken into account, the total is considerably less than the yield from the wood itself. Acetic acid comes from the elimination of acetyl groups originally linked to the xylose unit.

In the pyrolysis processes, furfural is formed by dehydration of the xylose unit. Quantitatively, 1-hydroxy-2-propanone and 1-hydroxy-2-butanone present high concentrations in the liquid products. These two alcohols are partly esterified by acetic acid. In conventional slow pyrolysis, these two products are not found in such a great quantity because of their low stability. If wood is completely pyrolyzed, the resulting products are about what would be expected by pyrolyzing the three major components separately. The hemicelluloses would break down first, at temperatures of 470–530 K. Cellulose follows in the temperature range of 510–620 K, with lignin being the last component to pyrolyze at temperatures of 550–770 K. A wide spectrum of organic substances is contained in the pyrolytic liquid fractions given in the literature (Beaumont, 1985). Degradation of xylan yields eight main products: water, methanol, formic, acetic and propionic acids, 1-hydroxy-2-propanone, 1-hydroxy-2-butanone, and 2-furfuraldeyde. The methoxy phenol concentration decreased with increasing temperature, while phenols and alkylated phenols increased. The formation of both methoxy phenol and acetic acid was possibly as a result of the Diels–Alder cycloaddition of a conjugated diene and unsaturated furanone or butyrolactone.

Timell (1967) described the chemical structure of the xylan as the 4-methyl-3-acetylglucuoronoxylan. It has been reported that the first runs in the pyrolysis of the pyroligneous acid consist of about 50% methanol, 18% acetone, 7% esters, 6% aldehydes, 0.5% ethyl alcohol, 18.5% water, and small amounts of furfural (Demirbas, 2000). Pyroligneous acids disappear in high-temperature pyrolysis.

The composition of the water soluble products was not ascertained but it has been reported to be composed of hydrolysis and oxidation products of glucose such as acetic acid, acetone, simple alcohols, aldehydes, sugars, *etc*. Pyroligneous acids disappear in high-temperature pyrolysis. Levoglucosan is also sensitive to heat and decomposes to acetic acid, acetone, phenols, and water. Methanol arises from the methoxyl groups of aronic acid (Demirbas, 2000).

6.7 Gasification Research and Development

Gasification, one of thermochemical conversion routes, is widely recognized at present because its end product gas can find flexible application by industries or by home users, particularly in decentralized energy production coupled with microturbine/gas, turbine/engines, boiler, and even fuel cells (Chen *et al.*, 2004). Gasification of biomass is a well-known technology that can be classified depending on the gasifying agent: air, steam, steam-oxygen, air-steam, O_2-enriched air, *etc.*

Gasification describes the process in which oxygen-deficient thermal decomposition of organic matter primarily produces synthesis gas. Gasification is a combination of pyrolysis and combustion. Gasification typically refers to conversion of solid or liquid carbon-based materials by direct internal heating provided by partial oxidation using substoichiometric air or oxygen to produce fuel gases (synthesis gas, producer gas), principally CO, H_2, methane, and lighter hydrocarbons in association with CO_2 and N_2, depending on the process used. Alternative configurations using either indirect heating methods such as externally fired burners, or autothermal methods using exothermic reducing reactions have been demonstrated.

Gasification has more potential for near-term commercial application than other thermochemical processes. Benefits of gasification over combustion include: more flexibility in terms of energy applications, more economical and thermodynamic efficiency at smaller scales, and potentially lower environmental impact when combined with gas cleaning and refining technologies. An efficient gasifier will decompose high-molecular-weight organic compounds released during pyrolysis into low-molecular-weight, non-condensable compounds in a process referred to as tar cracking. Undesirable char that is produced during gasification will participate in a series of endothermic reactions at temperatures above 800°C, which converts carbon into a gaseous fuel. Typically gaseous products include: CO, H_2, and CH_4. Fischer–Tropsch processes may be used to upgrade gaseous products to liquid fuels through the use of catalysts. Gasification requires feedstocks that contain less than 10% moisture.

Gasification is a form of pyrolysis carried out in the presence of a small quantity of oxygen at high temperatures in order to optimize the gas production. The resulting gas, known as the producer gas, is a mixture of carbon monoxide, hydrogen and methane, together with carbon dioxide and nitrogen. The gas is more versatile than the original solid biomass (usually wood or charcoal); it can be burnt to produce process heat and steam, or used in gas turbines to produce electricity.

Gasification of solids with subsequent combustion of the gasification-derived fuel gases generates the same categories of products as direct combustion of solids, but pollution control and conversion efficiencies may be improved. Alternatively, the produced synthesis gases can be used directly for liquid fuel or chemical synthesis, eliminating or delaying the combustion process, and the emission of the resulting effluent.

Biomass gasification technologies are expected to be an important part of the effort to meet these goals of expanding the use of biomass. Gasification technologies provide the opportunity to convert renewable biomass feedstocks into clean fuel gases or synthesis gases. Biomass gasification is the latest generation of biomass energy conversion processes, and is being used to improve the efficiency, and to reduce the investment costs of biomass electricity generation through the use gas turbine technology. High efficiencies (up to about 50%) are achievable using combined-cycle gas turbine systems, where waste gases from the gas turbine are recovered to produce steam for use in a steam turbine. Economic studies show that biomass suffocation plants can be as economical as conventional coal-fired plants.

Commercial gasifiers are available in a range of size and types, and run on a variety of fuels, including wood, charcoal, coconut shells, and rice husks. Power output is determined by the economic supply of biomass, which is limited to 80 MW in most regions. The process of synthetic fuels (synfuels) from biomass will lower energy cost, improve waste management, and reduce harmful emissions. This triple assault on plant operating challenges is a proprietary technology that gasifies biomass by reacting it with steam at high temperatures to form a clean burning syngas. The molecules in the biomass (primarily carbon, hydrogen, and oxygen) and the molecules in the steam (hydrogen and oxygen) reorganize to form this syngas.

6.7.1 Biomass Gasification

Biomass gasification technologies have historically been based upon partial oxidation or partial combustion principles, resulting in the production of a hot, dirty, low heating value gas that must be directly ducted into boilers or dryers. In addition to limiting applications and often compounding environmental problems, these technologies are an inefficient source of usable energy.

Generating electricity and useful heat from the same power plant is called "cogeneration" in North America and "combined heat and power (CHP)" in Europe. Biomass integrated gasification combined cycle cogeneration technology is not yet commercially available. Gasification is an energy process producing a gas that can substitute fossil fuels in high efficiency power generation, heat and/or cogeneration applications, and can be used for the production of liquid fuels and chemicals *via* synthesis gas.

Biomass gasification has attracted the highest interest amongst the thermochemical conversion technologies as it offers higher efficiencies in relation to combustion, while flash pyrolysis is still in the development stage. The comparison of the environmental impact of biomass use in gasifiers and incinerators is very important when considering the effective use of biomass. However, the high alkali content in biomass can form compounds with low melting temperature during combustion. The low melting ash constituents can induce in-bed-agglomeration, in addition to fouling and corrosion problems.

The energy crisis of the 1970s brought a renewed interest. The technology was perceived as a relatively cheap indigenous alternative for small-scale industrial and utility power generation in those developing countries that suffered from high world market petroleum prices and had sufficient sustainable biomass resources. In the beginning of the 1980s at least ten (mainly European) manufacturers were offering small-scale wood and charcoal fired power plants (up to approximately 250 kW$_{el}$). At least four developing countries (The Philippines, Brazil, Indonesia, and India)

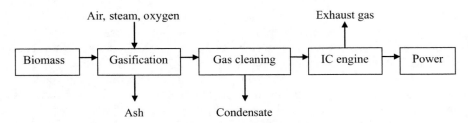

Fig. 6.8 System for power production by means of biomass gasification

started gasifier implementation programs based on locally developed technologies. Figure 6.8 shows the system for power production by means of biomass gasification. The gasification system of biomass in fixed-bed reactors provides the possibility of combined heat and power production in the power range of 100 kWe up to 5 MWe. A system for power production by means of fixed-bed gasification of biomass consists of the main unit gasifier, gas cleaning system, and engine.

The biomass gasification process is similar to processes used for many years by chemical and petrochemical manufacturers, including methanol, ammonia, and ethylene producers. The hydrogen and oxygen molecules in the steam are liberated, and a series of reactions result in a reorganization of the compounds to form synthesis gas (primarily H_2, CO and CO_2). This synthesis gas is then catalytically converted into methanol, ammonia, or another product.

Biomass gasification is the latest generation of biomass energy conversion processes, and is being used to improve the efficiency and to reduce the investment costs of biomass electricity generation through the use gas turbine technology. High efficiencies (up to about 50%) are achievable using combined-cycle gas turbine systems, where waste gases from the gas turbine are recovered to produce steam for use in a steam turbine. Economic studies show that biomass suffocation plants can be as economical as conventional coal-fired plants.

Various gasification technologies include gasifiers where the biomass is introduced at the top of the reactor and the gasifying medium is either directed co-currently (downdraft) or counter-currently up through the packed bed (updraft). Other gasifier designs incorporate circulating or bubbling fluidized beds. Tar yields can range from 0.1% (downdraft) to 20% (updraft) or greater in the product gases. Table 6.5 shows gasifiers and reactor types using gasification processes.

Table 6.5 Gasifiers and reactor types using gasification processes

Gasifier	Reactor type
Fixed bed	Downdraft, Updraft, Co-current, Counter-current, Cross current, Others
Fluidized bed	Single reactor, Fast fluid bed, Circulating bed
Entrained bed	–
Twin reactor	–
Moving bed	–
Others	–

Table 6.6 Composition of gaseous products from various biomass fuels by different gasification methods (% by volume)

H_2	CO_2	O_2	CH_4	CO	N_2
10–19	10–15	0.4–1.5	1–7	15–30	43–60

Commercial gasifiers are available in a range of size and types, and run on a variety of fuels, including wood, charcoal, coconut shells, and rice husks. Power output is determined by the economic supply of biomass, which is limited to 80 MW in most regions. The producer gas is affected by various gasification processes from various biomass feedstocks. Table 6.6 shows the composition of gaseous products from various biomass fuels by different gasification methods.

The relative simplicity of the gasification system enables its operation to be within the technical expertise of most operators experienced with conventional boilers and furnaces, and results in favorable project economics. Its modular design allows a wide range of scale-up or scale-down possibilities, so the systems can vary in size from about 1 ton *per* hour of residue to 20 tons *per* hour or more, with the size being limited only by biomass availability.

6.7.2   Biomass Gasification Systems

The system can gasify a wide variety of biomass wastes and other organic materials generated by many industries. It has gasified, and in most cases has data on expected fuel composition, char analyses, and emissions analyses from the syngas produced for the following feedstocks: hardwood and pine saw dust, bark/hogged fuel, sander/grinder dust from panel board mills, pulp and paper mill sludge, whole and ground rice hulls, sugar cane bagasse, sewage sludge, the cellulosic fraction of municipal solid waste, and several grades of lignite and subbituminous coal.

The main steps in the gasification process are:

- Step 1. Biomass is delivered to a metering bin from which it is conveyed with recycled syngas or steam, without air or oxygen into the gasifier.
- Step 2. The material is reformed into a hot syngas that contains the inorganic (ash) fraction of the biomass and a small amount of unreformed carbon.
- Step 3. The sensible heat in the hot syngas is recovered to produce heat for the reforming process.
- Step 4. The cool syngas passes through a filter and the particulate in the syngas is removed as a dry, innocuous waste. The clean syngas is then available for combustion in engines, turbines, or standard natural gas burners with minor modifications.

The major components of a gasification system are: (a) feed system, (b) primary heat exchanger, (c) primary reformer, (d) gas filter, and (e) final syngas cooler.

Depending upon the size, consistency and nature of the biomass, the material is often routed through a hammer-mill or tub grinder/classifier before entering the plant's metering bin located above the feed system. The material is fed by gravity into the metering bin where it enters a screw feed system. The material is then conveyed through a proprietary sealing mechanism that serves as the pressure seal on the front end of the system, keeping air out of the reformer and keeping syngas from backing up into the feed system. The material received from the screw feeder is then conveyed with recycled compressed syngas into the primary heat exchanger.

The primary heat exchanger serves two functions. First, biomass is conveyed with syngas into the convection section where preheating, devolitization, and evaporation of water occurs. In addition, after reforming, as the hot syngas leaves the primary reformer, it gives up its sensible heat energy to the primary heat exchanger and is cooled to the desired process temperature before it exits to the gas filter.

The preheated, partially reformed (gasified) biomass and conveying syngas pass from the convection section of the primary heat exchanger into the radiant coil section of the primary reformer where high temperature steam reforming takes place.

This unit receives syngas from the primary heat exchanger. The syngas and any char (inorganic solids and any unreformed carbon) are routed through barrier type filter elements where the char is collected and removed as a dry, innocuous residue. The char is delivered to a collection bin for alternative beneficial reuse or disposal. This air-cooled heat exchanger receives clean syngas from the gas filter and reduces the gas temperature to the desired level for supplying power generation equipment, or other fuel uses.

6.7.2.1 Types of Gasifiers

Numerous types of gasifiers have been developed and tested, and many industrial applications can use the technology. Gasifiers have been built and operated using a wide variety of configurations, including:

1. fixed bed (updraft or downdraft fixed beds) gasifiers,
2. fluidized bed (fluidized or entrained solids serve as the bed material) gasifiers, and
3. others including moving grate beds and molten salt reactors.

Schematic diagrams of updraft (counter current) and downdraft (co-current) fixed bed gasifiers are shown in Figs. 6.9 and 6.10, respectively. Other gasifier designs incorporate circulating or bubbling fluidized beds.

In the drying zone, feedstock descends into the gasifier and moisture is removed using the heat generated in the zones below by evaporation. In the distillation zone, pyrolysis and partial oxidation takes place using the thermal energy released by the partial oxidation of the pyrolysis products. Tar yields can range from 0.1% (downdraft) to 20% (updraft) or greater in the product gases. The oxidation reactions of the volatiles are very rapid, and the oxygen is consumed before it can diffuse to the surface of the char. In the reduction zone (often referred to as

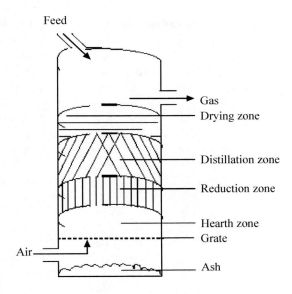

Fig. 6.9 Schematic diagram of an updraft fixed bed gasifier

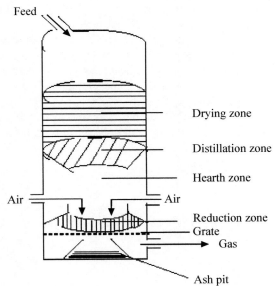

Fig. 6.10 Schematic diagram of a downdraft fixed bed gasifier

the gasification zone) the char is converted into product gas by reaction with the hot gases from the upper zones. Depending in the end use, it is necessary to cool and clean the gas in order to remove as much water vapor, dust, and pyrolytic products as possible from the gas, especially if it is to be used in an internal combustion engine.

Fixed-bed gasifiers are the most suitable for biomass gasification. Fixed-bed gasifiers are usually fed from the top of the reactor and can be designed in either updraft or downdraft configurations. The product gases from these two gasifier configurations vary significantly. At larger scale, fixed-bed gasifiers can encounter problems with bridging of the biomass feedstock. This leads to uneven gas flow. Achieving uniform temperatures throughout the gasifier at large scale can also be difficult due to the absence of mixing in the reaction zone. Most fixed-bed gasifiers are air-blown and produce low-energy gases (Stevens, 2001).

With fixed-bed updraft gasifiers, the air or oxygen passes upward through a hot reactive zone near the bottom of the gasifier in a direction counter-current to the flow of solid material. Exothermic reactions between air/oxygen and the charcoal in the bed drive the gasification process. Heat in the raw gas is transferred to the biomass feedstock as the hot gases pass upward, and biomass descending through the gasifier sequentially undergoes drying, pyrolysis, and finally gasification. Fixed-bed updraft gasifiers can be scaled up; however, they produce a product gas with very high tar concentrations. This tar should be removed for the major part from the gas, creating a gas-cleaning problem.

Fixed-bed downdraft gasifiers were widely used in World War II for operating vehicles and trucks. During operation, air is drawn downward through a fuel bed; the gas in this case contains relatively less tar compared with the other gasifier types. Fixed-bed downdraft gasifiers are limited in scale and require a well-defined fuel, making them not fuel-flexible.

Fluidized bed (FB) gasifiers are a more recent development that takes advantage of the excellent mixing characteristics and high reaction rates of this method of gas-solid contacting. Examples of FB gasifier systems are the bubbling fluidized bed (BFB) gasifiers, the entrained bed (EB) gasifiers, and the circulating fluidized bed (CFB) gasifiers.

The FB gasifiers are typically operated at 1050–1250 K (limited by the melting properties of the bed material) and are therefore not generally suitable for coal

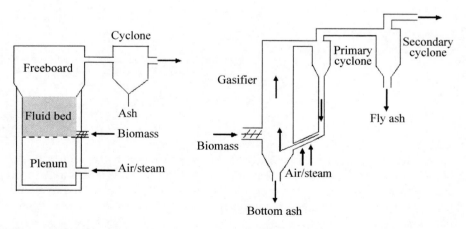

Fluidized bed gasifier   Circulating fluidized bed gasifier

Fig. 6.11   Fluidized bed and circulating fluidized bed gasifier systems

gasification, as due to the lower reactivity of coal compared to biomass, a higher temperature is required (>1550 K). Heat to drive the gasification reaction can be provided in a variety of ways in FB gasifiers. Direct heating occurs when air or oxygen in fluidizing gas partially oxidizes the biomass, and heat is released by the exothermic reactions that occur. Indirect heating methods such as internal heat exchangers, using preheated bed material, or other means can also be used to drive the gasification reactions (Stevens, 2001). The BFB gasifier tends to produce a gas with tar content between that of the updraft and downdraft gasifiers. Some pyrolysis products are swept out of the fluid bed by gasification products, but are then further converted by thermal cracking in the freeboard region (Warnecke, 2000). The CFB gasifiers employ a system where the bed material circulates between the gasifier and a secondary vessel. The CFB gasifiers are suitable for fuel capacity higher than 10 MWth. The FB gasifier and the CFB gasifier systems are shown in Fig 6.11.

6.7.2.2   Short Review of Biomass Gasification Chemistry

Gasification is a complex thermochemical process that consists of a number of elementary chemical reactions, beginning with the partial oxidation of a biomass fuel with a gasifying agent, usually air, oxygen, or steam. The chemical reactions involved in gasification include many reactants and many possible reaction paths. While the reactions that take place in a gasifier are complex, they can be categorized as follows: flash evaporation of inherent moisture, devolatization of higher organics, heavy hydrocarbon cracking, pyroylysis, and steam reforming. Most biomass gasification systems utilize air or oxygen in partial oxidation or combustion processes. Volatile matter partially oxidizes to yield the combustion products H_2O and CO_2, plus heat to continue the endothermic gasification process. Yields are product gases from thermal decomposition composed of CO, CO_2, H_2O, H_2, CH_4, other gaseous hydrocarbons (CHs), tars, char, inorganic constituents, and ash. Gas composition of the product from the biomass gasification depends heavily on the gasification process, the gasifying agent, and the feedstock composition. A generalized reaction describing biomass gasification is as follows:

$$\text{Biomass} + O_2 \;\rightarrow\; CO, CO_2, H_2O, H_2, CH_4 + \text{other (CHs)}$$
$$+ \text{tar} + \text{char} + \text{ash} \tag{6.2}$$

The relative amount of CO, CO_2, H_2O, H_2, and (CHs) depends on the stoichiometry of the gasification process. If air is used as the gasifying agent, then roughly half of the product gas is N_2. The air/fuel ratio in a gasification process generally ranges from 0.2–0.35 and if steam is the gasifying agent, the steam/biomass ratio is around 1. The actual amount of CO, CO_2, H_2O, H_2, tars, and (CHs) depends on the partial oxidation of the volatile products, as shown in Eq. 6.3.

$$C_nH_m + (n/2m/4) O_2 \leftrightarrows nCO + (m/2) H_2O \tag{6.3}$$

Most biomass gasification systems utilize air or oxygen in partial oxidation or combustion processes. These processes suffer from low thermal efficiencies and low Btu gas because of the energy required to evaporate the moisture typically inherent in the biomass and the oxidation of a portion of the feedstock to produce this energy.

In essence, the system embodies a fast, continuous process for pyrolyzing or thermally decomposing biomass and steam reforming the resulting constituents. The entire process occurs in a reducing environment, not an oxidizing environment like other biomass gasifiers. While the reactions that take place in the gasifier are complex, they can be categorized as follows: flash evaporation of inherent moisture, devolatization of higher organics, heavy hydrocarbon cracking, pyrolysis, and steam reforming. The major thermochemical reactions include the following:

Steam and methane:

$$CH_4 + H_2O \leftrightarrows CO + 3H_2 \tag{6.4}$$

Water–gas shift:

$$CO + H_2O \leftrightarrows CO_2 + H_2 \tag{6.5}$$

Carbon char to methane:

$$C + 2H_2 \leftrightarrows CH_4 \tag{6.6}$$

Carbon char oxides (Boudouard reaction):

$$C + CO_2 \leftrightarrows 2CO \tag{6.7}$$

The process is extremely efficient, achieving over 97% conversion of biomass carbon to useful syngas without producing troublesome tars, oils, or contaminated effluents. The process is also quite robust in its ability to handle feedstocks with varying degrees of inherent moisture, ranging from bone dry wood (in which case moisture is actually added to the feedstock) to organic sludge with moisture contents of over 60%. Unlike traditional partial oxidation systems where any moisture in the feedstock results in an energy efficiency penalty, the process utilizes a portion of the moisture to produce hydrogen and other combustible gases. At 40% moisture in the feedstock, the system achieves a cold gas efficiency of about 80%.

Char gasification is the rate-limiting step in the production of gaseous fuels from biomass. Arrhenius kinetic parameters have been determined for the reaction of chars prepared by pyrolysis of cottonwood at 1275 K with steam and carbon dioxide. Results indicate that both reactions are approximately zero order with respect to char; the overall reaction rate is fairly constant throughout and declines only when the char is nearly depleted. This suggests that the reaction rate depends on such factors as the total available active surface area or the interfacial area between the char and catalyst particles. These parameters would remain relatively constant during the gasification process. Sodium and potassium catalysts were equally effective for the gasification of wood char. The iron and nickel transition metals provided the highest initial catalytic activity, but lost their activity well before the char completely reacted. Softwood and hardwood chars exhibited similar gasification behavior. Results indicate that the mineral (ash) content and composition of the original biomass material, and pyrolysis conditions under which char is formed significantly influence the char gasification reactivity (Demirbas, 2000).

The char yield in a gasification process can be optimized to maximize carbon conversion or the char can be thermally oxidized to provide heat for the process. Char is partially oxidized or gasified according to the following reactions:

$$C + \frac{1}{2}O_2 \leftrightarrows CO \tag{6.8}$$

$$C + H_2O \leftrightarrows CO + H_2 \tag{6.9}$$

The gasification product gas composition, particularly the $H_2:CO$ ratio, can be further adjusted by reforming and shift chemistry. Additional hydrogen is formed when CO reacts with excess water vapor according to the water-gas shift reaction given in Eq. 6.5. Carbon chars can be converted to methane according to Eq. 6.6.

6.7.3 Electricity from Cogenerative Biomass Firing Power Plants

Biomass has historically been a dispersed, labor-intensive, and land-intensive source of energy. Biomass provides a clean, renewable energy source that may dramatically improve our environment, economy, and energy security. Biomass energy also creates thousands of jobs and helps revitalize rural communities. Therefore as industrial activity has increased in countries, more concentrated and convenient sources of energy have been substituted for biomass. Biomass accounts for 35% of primary energy consumption in developing countries, raising the world total to 14% of primary energy consumption.

In the future, biomass has the potential to provide a cost-effective and sustainable supply of energy, while at the same time aiding countries in meeting their greenhouse gas reduction targets. By the year 2050, it is estimated that 90% of the world population will live in developing countries. It is critical therefore that the biomass processes used in these countries be sustainable. Traditionally, biomass has been utilized through direct combustion, and this process is still widely used in many parts of the world. World production of biomass is estimated at 146 billion metric tons a year, mostly wild plant growth.

The future of biomass electricity generation lies in biomass integrated gasification/gas turbine technology, which offers high-energy conversion efficiencies. Electricity is produced by direct combustion of biomass, advanced gasification and pyrolysis technologies are almost ready for commercial scale use. Biomass power plants (BPPs) use technology that is very similar to that used in coal-fired power plants. For example, biomass plants use similar steam-turbine generators and fuel delivery systems. BPP efficiencies of is about 25%. Electricity costs are in the 6–8 c/kWh range. The average BPP is about 20 MW in size, with a few dedicated wood-fired plants in the 40–50 MW size range. As the biomass-to-electricity industry grows, it will be characterized by larger facilities of 50–150 MW capacity, with gas turbine/steam combined cycles. Biomass is burned to produce steam, and the steam turns a turbine and drives a generator, producing electricity. Because of potential ash build up, only certain types of biomass materials are used for direct combustion. Heat is used to thermochemically convert biomass into a pyrolysis oil. The oil, which is easier to store and transport than solid biomass material, is then burned like petroleum to generate electricity.

Biomass can be used as a primary energy source or as a secondary energy source to power gas turbines. As a secondary energy source, biomass is used to make a fuel, which can be used to fire a gas turbine. The heat produced from the electricity generating process is captured and utilized to produce domestic purposes and can be used in steam turbines to generate additional electricity. Cogeneration is the simultaneous production of electricity and useful thermal energy from a single source.

One alternative for producing electricity from biomass in a gas turbine is direct combustion of biomass as a primary energy source. Biomass is burned directly to produce steam, the steam turns a turbine, and the turbine derives a generator, producing electricity. Direct combustion usually involves reducing the biomass into fine pieces for fueling a close-coupled turbine system. In a close-coupled system, biomass is burned in a combustion chamber separated from the turbine by a filter.

The pulverized wood fuel can then be burned in a flame in the same way as oil or gas at the same high power output. Burners for coal or peat powder are suitable when adjusted for this proper mixture of fuel and air. Since the same type of boilers and control systems are used, it is easy to combine the different fuels.

Analysis and modeling of combustion in stoves, furnaces, boilers, and industrial processes require adequate knowledge of wood properties. Fuel properties for combustion analysis of wood can be conveniently grouped into physical, thermal, chemical, and mineral properties. Bark properties should be distinguished from wood properties. Thermal degradation products of wood consist of moisture, volatiles, char, and ash. Volatiles are further subdivided into gases and tars. Some properties vary with species, location within the tree, and growth conditions. Other properties depend on the combustion environment. Combustion systems using wood fuel may be generally grouped into fixed bed, suspension burning, and fluidized bed systems. The systems range from residential to commercial and industrial to utility scale. The fuel property data needed depend, of course, on the type of application and the details of the model.

A steam power plant is actually a two-fluid system, that is, energy is exchanged between the combustion gases and water. The feasibility of combining gas and steam expansion in a power cycle has been extensively explored. Because steam generation involves the flow of large volumes of combustion gases, gas expansion is most appropriately accomplished gases in a gas turbine.

Practically all biomass-based electricity generation plants employ steam turbine systems. Such electricity generation has been established in developed countries in order to upgrade lignocellulosic-based waste materials. Most systems are based on

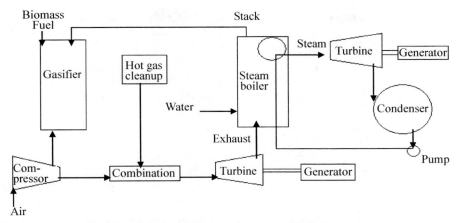

Fig. 6.12 Integrated biomass gas turbine/steam combined cycle power plant

low-pressure boilers (about 20–25 bar) with efficiencies slightly below 20%. Modern biomass powered high pressure (60–100 bar) boiler turbine systems produce electricity with efficiencies approaching 32%. Figure 6.12 shows the integrated biomass gas turbine/steam combined cycle power plant.

How the Gasifier Makes Electricity

1. Wood chips or other biomass materials are loaded into the fuel hopper. The wood handling system removes metal impurities and oversized fuel, and discharges the fuel into the gasifier.
2. Initially, dry wood will be used as fuel, but in later project phases a dryer will be installed to allow wetter fuels to be used. Sand that has been heated to 1275 K is added to the gasifier where the fuel is turned into a gas. The mixture is "fluidized" with steam and enters a cyclone separator.
3. The unburned char and cooled sand is removed and discharged to the combustor.
4. The char is burned by adding combustion air, which heats the sand to 1275 K. The heated sand is removed by the cyclone separator and returned to the gasifier. The fuel gas is cleaned in a scrubber and used to fuel a boiler or gas turbine.

Another alternative is to produce a fuel from biomass as a secondary energy source. Gasifiers are used to convert biomass into a combustible gas (biogas). The biogas is than used to drive a high efficiency, combined cycle gas turbine. Gaseous fuels consist of low-calorific-value and medium-calorific-value gases; the liquid is a primary-pyrolysis oil called biocrude. A number of gasifiers have been developed to produce biogases from biomass and peat. The biogas is then used to drive a high-efficiency, combined-cycle gas turbine.

Heat is used to chemically convert biomass into pyrolysis oil and biocrude. The oil, which is easier to store and transport than solid biomass material, is then burned like petroleum to generate electricity. Biocrude produced at about 775 K and 1 s residence time and containing water have about the same oxygen and energy content as the original feed. Entrained-flow and fluid-bed pyrolysis processes have been developed.

Gas turbines are used by themselves in a very wide range of applications, most notably for powering aircrafts of all types but also in industrial plants for driving mechanical equipment such as compressors, pumps, and electric generators in electrical utilities, and for producing electric power for peak loads, as well as for intermediate and some base-load duties. The compressor compresses the combustion air to desired pressure (typically 10–25 bar), which reacts with the fuel in the combustor. The hot pressurized flue gases expand in the turbine, which drives the compressor and the additional load (generator). The efficiency of the simple cycle gas turbine can be increased by recovering some heat of the exhaust gases by heating the compressed combustion air. An increase in efficiency of 30% is possible.

In a combined cycle, the heat contained in the exiting flue gases from a gas turbine without heat recovery can be used to produce steam. This steam can either be injected in the turbine of the gas turbine, the so-called steam injected gas turbine, or in a separate steam turbine in a so-called steam and gas turbine.

6.7.4   Fischer–Tropsch Synthesis (FTS)

FTS is a well established process for the production of synfuels. The process is used commercially in South Africa by Sasol and Mossgas, and in Malaysia by Shell. FTS can be operated at low temperatures (LTFT) to produce a syncrude with a large fraction of heavy, waxy hydrocarbons or it can be operated at higher temperatures (HTFT) to produce a light syncrude and olefins. With HTFT the primary products can be refined to environmentally friendly gasoline and diesel, solvents, and olefins. With LTFT, the heavy hydrocarbons can be refined to speciality waxes, or if hydrocracked and/or isomerized, to produce excellent diesel, base stock for lube oils, and a naphtha that is ideal feedstock for cracking to light olefins.

FTS has been widely investigated for more than 70 years, and Fe and Co are typical catalysts. Cobalt-based catalysts are preferred because their productivity is better than Fe due to their high activity, selectivity for linear hydrocarbons, and low activity for the competing water–gas shift reaction.

The variety of composition of FT products with hundreds of individual compounds shows a remarkable degree of order with regard to class and size of the molecules. Starting from the concept of FTS as an ideal polymerization reaction, it is easily realized that the main primary products, olefins, can undergo secondary reactions and thereby modify the product distribution. This generally leads to chain length dependencies of certain olefin reaction possibilities, which are again suited to serve as a characteristic feature for the kind of olefin conversion.

Table 6.7 Typical composition of syngas

Gas	% by volume
Hydrogen	29–40
Carbon monoxide	21–32
Methane	10–15
Carbon dioxide	15–20
Ethylene	0.4–1.2
Water vapor	4–8
Nitrogen	0.6–1.2

While gasification processes vary considerably, typically gasifiers operate from 975 K and higher and from atmospheric pressure to 5 atm or higher. The process is generally optimized to produce fuel or feedstock gases. Gasification processes also produce a solid residue such as a char, ash, or slag. The product fuel gases, including hydrogen, can be used in internal and external combustion engines, fuel cells, and other prime movers for heat and mechanical or electrical power. Gasification products can be used to produce methanol, FT liquids, and other fuel liquids and chemicals. The typical composition of the syngas is given in Table 6.7.

In all types of gasification, biomass is thermochemically converted to a low or medium-energy content gas. The higher heating value of syngas produced from biomass in the gasifier is typically $10–13 \, MJ/Nm^3$. Air-blown biomass gasification results in approximately $5 \, MJ/Nm^3$ and oxygen-blown $15 \, MJ/Nm^3$ of gas and is considered a low to medium energy content gas compared to natural gas ($35 \, MJ/Nm^3$).

The process of synfuels from biomass will lower energy costs, improve the waste management, and reduce harmful emissions. This triple assault on plant operating challenges is a proprietary technology that gasifies biomass by reacting it with steam at high temperatures to form a clean burning syngas. The molecules in the biomass (primarily carbon, hydrogen and oxygen), and the molecules in the steam (hydrogen and oxygen) reorganize to form this syngas.

Reforming the light hydrocarbons and tars formed during biomass gasification also produces hydrogen. In essence, the system embodies a fast, continuous process for pyrolyzing or thermally decomposing biomass and steam reforming the resulting constituents. The entire process occurs in biomass gasifiers by reforming the light hydrocarbons (C_2H_6–C_5H_{12}). Reforming the light hydrocarbons and tars formed during biomass gasification also produces hydrogen. Steam reforming and so-called dry or CO_2 reforming occur according to the following reactions and are usually promoted by the use of catalysts:

$$C_nH_m + nH_2O \rightarrow n \, CO + (n+m/2) \, H_2 \tag{6.10}$$

$$C_nH_m + nCO_2 \rightarrow (2n) \, CO + (m/2) \, H_2 \tag{6.11}$$

The high-temperature FT technology applied by Sasol in the Synthol process in South Africa at the Secunda petrochemical site is the largest commercial scale application of FT technology. The most recent version of this technology is the Sasol Advanced Synthol (SAS) process. Although FTS was initially envisioned as a means to make transportation fuels, there has been a growing realization that the profitability of commercial operations can be improved by the production of chemicals or chemical feedstock. FTS yields a complex mixture of saturated or unsaturated hydrocarbons (C_1–C_{40+}), C_1–C_{16+} oxygenates as well as water and CO_2. Laboratory studies are often carried out in differential conditions (conversions of approx. 3%) and therefore very accurate product analysis is necessary. Hydrocarbons obtained from a SAS type reactor are: for C_5–C_{10} and C_{11}–C_{14} hydrocarbons; paraffins 13% and 15%, olefins 70% and 60%, aromatics 5% and 15%, and oxygenates 12% and 10%, respectively. From the component breakdown of the main liquid cuts it is clear that there is considerable scope for producing chemical products in addition to hydrocarbon fuels.

FTS in the supercritical phase can be developed, and its reaction behavior and mass transfer phenomenon have been analyzed by both experimental and simulation methods. The same reaction apparatus and catalysts can be used for both gas phase reaction and liquid phase FT reactions to correctly compare characteristic features of diffusion dynamics and the reaction itself, in various reaction phases. Efficient transportation of reactants and products to the inside of catalysts bed and pellet, quick heat transfer, and *in situ* product extraction from catalysts by supercritical fluid can be accomplished.

There has been an increasing interest in the effect of water on cobalt FT catalysts in recent years. Water is produced in large amounts over cobalt catalysts since one water molecule is produced for each C-atom added to a growing hydrocarbon chain and due to the low water-gas-shift activity of cobalt. The presence of water during FTS may affect the synthesis rate reversibly as reported for titania-supported catalysts, the deactivation rate as reported for alumina-supported catalysts, and water also has a significant effect on the selectivity of cobalt catalysts on different supports. The effect on the rate and the deactivation appears to depend on the catalyst system studied, while the main trends in the effect on selectivity appear to be more consistent for different supported cobalt systems. There are, however, also some differences in the selectivity effects observed. The present study deals mainly with the effect of water on the selectivity of alumina-supported cobalt catalysts, but some data on the activity change will also be reported. The results will be compared with results for other supported cobalt systems reported in the literature.

The activity and selectivity of supported Co FTS catalysts depends on both the number of Co surface atoms and on their density within support particles, as well as on transport limitations that restrict access to these sites. Catalyst preparation variables available to modify these properties include cobalt precursor type and loading level, support composition and structure, pretreatment procedures, and the presence of promoters or additives. Secondary reactions can strongly influence product selectivity. For example, the presence of acid sites can lead to the useful formation of branched paraffins directly during the FTS step. However, product water not only oxidizes Co sites making them inactive for additional turnovers, but it can inhibit secondary isomerization reactions on any acid sites intentionally placed in FTS reactors.

The iron catalysts used commercially by Sasol in the Fischer–Tropsch synthesis for the past five decades (Dry, 1981) have several advantages: (1) lower cost relative to cobalt and ruthenium catalysts, (2) high water–gas-shift activity allowing utilization of syngas feeds of relatively low hydrogen content such as those produced by gasification of coal and biomass, (3) relatively high activity for production of liquid and waxy hydrocarbons readily refined to gasoline and diesel fuels, and (4) high selectivity for olefinic C_2–C_6 hydrocarbons used as chemical feedstocks. The typical catalyst used in fixed bed reactors is an unsupported Fe/Cu/K catalyst prepared by precipitation. While having the previously-mentioned advantages, this catalyst (1) deactivates irreversibly over a period of months to a few years by sintering, oxidation, formation of inactive surface carbons, and transformation of active carbide phases to inactive carbide phases, and (2) undergoes attrition at unacceptably high rates in the otherwise highly-efficient, economical slurry bubble-column reactor. Attempts to understand and correct these problems have been impeded by the complexities of oxide and carbide phases and their transformations during pretreatment and reaction. However, it is clear from these previous efforts that the deactivation processes are an inherent consequence of unsophisticated preparation and pretreatment methods developed by trial and error without regard to a rational design of active, stable phases at the nanoscale.

It is well known that addition of alkali to iron causes an increase of both the 1-alkene selectivity and the average carbon number of produced hydrocarbons. While the promoter effects on iron has been thoroughly studied only few and at a first glance contradictive results are available for cobalt catalysts. In order to complete experimental data the carbon number distributions have been analyzed for products obtained in a fixed bed reactor under the steady state condition. Precipitated iron and cobalt catalysts with and without K_2CO_3 were used.

Activated carbon (AC) is a high surface area support with a very unique property that its textural and surface chemical properties can be changed by an easy treatment like oxidation, and these changes affect the properties of the resultant catalysts prepared with AC.

6.7.5 Supercritical Steam Gasification

Recently, the supercritical fluid treatment has been considered to be an attractive alternative in science and technology as a chemical reaction field. The molecules in the supercritical fluid have high kinetic energy like the gas and high density like the liquid. In addition, ionic product and dielectric constant of supercritical water are important parameters for chemical reaction. Therefore, the supercritical water can be realized from the ionic reaction field to the radical reaction field. For example, ionic product of the supercritical water can be increased by increasing pressure then the hydrolysis reaction field is realized. Therefore, the supercritical water is expected to be used as a solvent for converting biomass into valuable substances.

The reforming in supercritical water (SCW) offers several advantages over the conventional technologies because of the unusual properties of supercritical water. The density of supercritical water is higher than that of steam, which results in a high space–time yield. The higher thermal conductivity and specific heat of supercritical water is beneficial for carrying out the endothermic reforming reactions. In the supercritical region, the dielectric constant of water is much lower. Further, the number of hydrogen bonds is much smaller and their strength is considerably weaker. As a result, SCW behaves as an organic solvent and exhibits extraordinary solubility toward organic compounds containing large non-polar groups and most permanent gases. Another advantage of SCW reforming is that the H_2 is produced at a high pressure, which can be stored directly, thus avoiding the large energy expenditures associated with its compression. The SCWG process becomes economical as the compression work is reduced owing to the low compressibility of liquid feed when compared to that of gaseous H_2. Figure 6.13 shows the schematic set-up of the system for supercritical water liquefaction of biomass.

In supercritical water gasification, the reaction generally takes place at the temperature over 875 K and a pressure higher than the critical point of water. With temperature higher than 875 K, water becomes a strong oxidant, and oxygen in water can be transferred to the carbon atoms of the biomass. As a result of the high density, carbon is preferentially oxidized into CO_2 but also low concentrations of

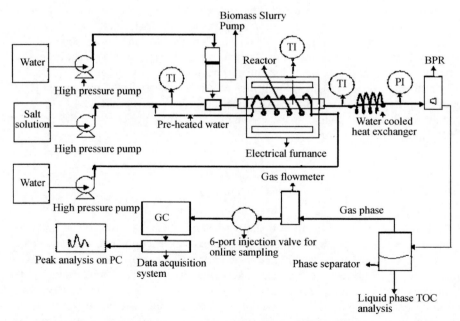

Fig. 6.13 Schematic set-up of the system for supercritical water gasification and liquefaction of biomass

CO are formed. The hydrogen atoms of water and of the biomass are set free and form H_2. The gas product consists of hydrogen, CO_2, CH_4, and CO.

A problem of general nature in SCWG is the required heat exchange between the reactor outlet and inlet streams. To achieve an acceptable thermal efficiency, it is crucial for the process that the heat of the inlet stream is utilized as far as possible to preheat the feedstock stream (mainly water) to reaction conditions. At the same time, heating of the biomass slurry in the inlet tube of a reactor is likely to cause fouling/plugging problems because the thermal decomposition (>525 K) starts already far below the desired reaction temperature (>875 K) (Gadhe and Gupta, 2007). In addition to being a high mass transfer effect, supercritical water also participates in the reforming reaction. The molecules in the supercritical fluid have high kinetic energy like the gas and high density like the liquid. Therefore, it is expected that the chemical reactivity can be high in it. In addition, the ionic product and dielectric constant of supercritical water, which are important parameters for chemical reactions, can be continuously controlled by regulating pressure and temperature. Pressure has a negligible effect on hydrogen yield above the critical pressure of water. As the temperature is increased from 600°C to 800°C the H_2 yield increases from 53% to 73% by volume, respectively. Only a small amount of hydrogen is formed at low temperatures, indicating that direct reformation reaction of ethanol as a model compound in SCW is favored at high temperatures (>700°C). With an increase in the temperature, the hydrogen and carbon dioxide yields increase, while the methane yield decreases. The water excess leads to a preference for the formation of hydrogen and carbon dioxide instead of carbon monoxide. The formed intermediate carbon monoxide reacts with water to hydrogen and carbon dioxide. The low carbon monoxide yield indicates that the water–gas-shift reaction approaches completion (Byrd et al., 2007).

Thermochemical gasification of biomass has been identified as a possible system to produce renewable hydrogen with less dependence on exhaustible fossil energy resources. Compared with other biomass thermochemical gasifications, such as air gasification or steam gasification, SCWG has high gasification efficiency at lower temperature and can deal directly with wet biomass without drying.

The capillaries (1 mm ID and 150 mm length tubular reactors) are heated rapidly (within 5 s) in a fluidized sand bed to the desired reaction temperature. Experimentation with the batch capillary method has revealed that, especially at low temperatures and high feed concentrations, char formation occurs. A fluidized bed reactor might be a good alternative to solve the problems related to this char and ash formation.

Cellulose and sawdust were gasified in supercritical water to produce hydrogen-rich gas, and Ru/C, Pd/C, CeO_2 paticles, nano-CeO_2 and nano-$(CeZr)xO_2$ were selected as catalysts. The experimental results showed that the catalytic activities were Ru/C > Pd/C > nano-$(CeZr)xO_2$ > nano-CeO_2 > CeO_2 particle in turn. The 10 wt% cellulose or sawdust with CMC can be gasified near completely with Ru/C catalyst to produce 2–4 g hydrogen yield and 11–15 g potential hydrogen yield per 100 g feedstock at the condition of 773 K, 27 MPa, 20 min residence time in supercritical water (Hao et al., 2005).

Catalysts for low-temperature gasification include combinations of stable metals, such as ruthenium or nickel bimetallics and stable supports, such as certain titania, zirconia, or carbon. Without catalyst the gasification is limited (Kruse *et al.*, 2000). Sodium carbonate is effective in increasing the gasification efficiency of cellulose (Minowa *et al.*, 1997). Likewise, homogeneous, alkali catalysts have been employed for high-temperature supercritical water gasification.

Hydrogen is a sustainable, non-polluting source of energy that can be used in mobile and stationary applications. In order to evaluate hydrogen production by SCWG of various types of biomass, extensive experimental investigations have been conducted in recent years. The supercritical water reforming of biomass materials has been found to be an effective technique for the production of hydrogen with short residence times. Hydrogen production by biomass SCW is a promising technology for utilizing high moisture content biomass.

The product gases mainly consisted of hydrogen and carbon dioxide, with a small amount of methane and carbon monoxide in SCW reforming. Hydrogen yields approaching the stoichiometric limit were obtained under catalytic supercritical water conditions. Hydrogen production is influenced by temperature, residence time, and biomass concentration. The high yield of hydrogen was obtained at high reactor temperature, low residence time, and low biomass concentration. The gas product from biomass gasification in catalytic supercritical water contains about 69% H_2 and 30% CO_2 in mole fraction. Others like CH_4 and CO exist in the gas product in less amounts.

Since a drying process is an energy-intensive operation, and the produced H_2 will be stored under high pressure, supercritical water gasification is a promising technology for gasifying biomass with high moisture content.

References

Appel, H.R., Fu, Y.C., Friedman, S., Yavorsky, P.M., Wender, I. 1971. Converting organic wastes to oil. US Burea of Mines Report of Investigation. No. 7560.

Appell, H., Fu, Y., Friedman, S., Yavorsky, P., Wender, I. 1980. Converting organic wastes to oil: A replenishable energy source. Washington, D.C. Bureau of Mines, U.S. Department of the Interior.

Babu, B.V., Chaurasia, A.S. 2003. Modeling for pyrolysis of solid particle: Kinetics and heat transfer effects. Energy Convers Mgmt 44:2251–2275.

Byrd, A.J., Pant, K.K., Gupta, R.B. 2007. Hydrogen production from glucose using $Ru/Al_2 O_3$ catalyst in supercritical water. Ind Eng Chem Res 46:3574–3579.

Balat, M. 2008. Mechanisms of thermochemical biomass conversion processes. Part 1: Reactions of pyrolysis. Energy Sources, Part A 30:620–625.

Beaumont, O. 1985. Flash pyrolysis products from beech wood. Wood Fiber Sci. 17:228–239.

Barooah, J.N., Long, V.D. 1976. Rates of thermal decomposition of some carbonaceous materials in a fluidized bed. Fuel 55:116–120.

Chen, G., Spliethoff, H., Andries, J., Glazer, M.P., Yang, L.B. 2004 Biomass gasification in acirculating fluidised bed-Part I: Preliminary xxperiments and modeling development. Energy Sources 26:485–498.

Chornet, E., Overend, R.P. 1985. Fundamentals of thermochemical biomass conversion, pp. 967–1002. Elsevier, New York.

Datta, B., McAuliffe, C. 1993. The production of fuels by cellulose liquefaction. In Proceedings of First Biomass Conference of the Americas: Energy, Environment, Agriculture, and Industry. Golden, CO: National Renewable Energy Laboratory. 931–946.

Demirbas, A. 1985. A new method on wood liquefaction. Chim. Acta Turc 13:363–368.

Demirbas, A. 1998. Kinetics for non-isothermal flash pyrolysis of hazelnut shell. Bioresource Technol 66:247–252.

Demirbas, A. 2000. Mechanisms of liquefaction and pyrolysis reactions of biomass. Energy Convers Mgmt 41:633–646.

Demirbas, A. 2007. The influence of temperature on the yields of compounds existing in bio-oils obtaining from biomass samples via pyrolysis. Fuel Proc Technol 88:591–597.

Demirbas, A. 2008. The Importance of bioethanol and biodiesel from biomass. Energy Sources Part B 3:177–185.

Desrosiers, R.E., Lin, R.J. 1984. A moving-boundary model of biomass pyrolysis. Solar Energy 33:187–196.

Dry, M E. 1981. The Fischer–Tropsch synthesis. In Anderson, J.R., Boudart, M. (eds.) Catalysis-Science and Technology, Vol. 1, p. 160. Springer, New York.

Eager, R.L., Mathews, J.F., Pepper, J.M. 1982. Liquefaction of aspen poplar wood. Canadian J Chem Eng 60:289–94.

Feng, W., van der Kooi, H.J., Arons, J.D.S. 2004. Biomass conversions in subcritical and supercritical water: driving force, phase equilibria, and thermodynamic analysis. Chem Eng Proc 43:1459–1467.

Gadhe, J.B., Gupta, R.B. 2007. Hydrogen production by methanol reforming in supercritical water: Catalysis by in-situ-generated copper nanoparticles. Int J Hydrogen Energy 2007;32:2374–2381.

Goudriaan, F., Peferoen, D. 1990. Liquid fuels from biomass via a hydrothermal process. Chem Eng Sci 45:2729–2734.

Hao, H., Guo, L., Zhang, X., Guan, Y. 2005. Hydrogen production from catalytic gasification of cellulose in supercritical water. Chem Eng J 110:57–65.

He, W., Li, G., Kong, L., Wang, H., Huang, J., Xu, J. 2008. Application of hydrothermal reaction in resource recovery of organic wastes. Res Conserv Recyc 52:691–699.

Hofmann, L., Antal, M.J, Jr. 1984. Numerical simulations of the performance of solar fired flash pyrolysis reactors. Solar Energy 33:427–440.

Inoue, S., Sawayma, S., Dote, Y., Ogi, T. 1997. Behavior of nitrogen during liquefaction of dewatered sewage sludge. Biomass Bioenergy 12:473–475.

Itoh, S., Suzuki, A., Nakamura, T., Yokoyama, S. 1994. Production of heavy oil from sewage sludge by direct thermochemical liquefaction. Proceedings of the IDA and WRPC World Conference on Desalination and Water Treatment. 98:127–133.

Jomaa, S. 2001. Combined sludge treatment and production of useful organic byproducts using hydrothermal oxidation. PhD thesis. Brisbane, Australia: Department of Civil Engineering, Queensland University of Technology.

Jomaa, S., Shanableh, A., Khalil, W., and Trebilco, B. 2003. Hydrothermal decomposition and oxidation of the organic component of municipal and industrial waste products. Adv Environ Res 7:647–53.

Koullas, D.P., Nikolaou, N., Koukkios, E.G. 1998. Modelling non-isothermal kinetics of biomass prepyrolysis at low pressure. Bioresource Technol 63:261–266.

Kranich, W.L. 1984. Conversion of sewage sludge to oil by hydroliquefaction. EPA-600/2 84-010. Report for the U.S. Environmental Protection Agency. Cincinnati, OH. EPA.

Kruse, A., Meier, D., Rimbrecht, P., Schacht, M. 2000. Gasification of pyrocatechol in supercritical water in the presence of potassium hydroxide. Ind Eng Chem Res 39:4842–2848.

Minowa, T., Ogi, T., Dote, Y., Yokoyama, S. 1994. Effect of lignin content on direct liquefaction of bark. Int Chem Eng 34:428–30.

Minowa, T., Zhen, F., Ogi, T. 1997. Cellulose decomposition in hot-compressed water with alkali or nickel catalyst. J Supercritical Fluids 13:253–259.

Mohan, D., Pittman Jr,. C.U., Steele, P.H. 2006. Pyrolysis of wood/biomass for bio-oil: A critical review. Energy Fuels 2006;20:848–889.

Molten, P.M., Demmitt, T.F., Donovan, J M., Miller, R.K. 1983. Mechanism of conversion of cellulose wastes to liquid in alkaline solution. In Klass, D.L. (ed.). Energy from biomass and wastes III. Chicago, IL: Institute of Gas Technology, p. 293.

Ogi, T., Yokoyama, S.D. 1993. Liquid fuel production from woody biomass by direct liquefaction. Sekiyu Gakkaishi 36:73–84.

Ogi, T., Yokoyama, S., Koguchi, K. 1985. Direct liquefaction of wood by alkali and alkaline earth salt in an aqueous phase. Chemical Letters 8:1199–200.

Shanableh, A., Jomaa, S. 1998. A versatile supercritical water oxidation system for hazardous Organic waste destruction. In: Fourth national hazardous and solid waste convention, CDRom record no. 1/90, Brisbane, Australia.

Stamm, A.J. 1956. Thermal degradation of wood and cellulose. Ind Engng Chem 48:413–417.

Stevens, D.J. 2001. Hot gas conditioning: Recent progress with larger-scale biomass gasification systems. National Renewable Energy Laboratory, NREL/SR-510-29952, 1617 Cole Boulevard, Golden, CO.

Suzuki, A., Yokoyama, S., Murakami, M., Ogi, T., and Koguchi, K. New treatment of sewage sludge by direct thermochemical liquefaction. Chemistry Letters CMLTAG.I 9:1425–1428.

Tester, J. W., Cline, J. A. 1999. Hydrolysis and oxidation in sub-critical and supercritical water: connecting process engineering science to molecular interactions. Corrosion 55:1088–100.

Timell, T.E. 1967. Recent progress in the chemistry of wood. Hemicelluloses. Wood Sci Technol 1:45–70.

Tran, D.Q., Charanjit, R. 1978. A kinetic model for pyrolysis of Douglas fir bark. Fuel 57:293–298.

Zhong, Z., Peters, C. J., de Swaan Arons, J. 2002. Thermodynamic modeling of biomass conversion processes. Fluid Phase Equilibria 194–197:805–815.

Warnecke, R. 2000. Gasification of biomass: comparison of fixed bed and fluidized bed gasifier. Biomass Bioenergy 18:489–497.

Chapter 7
Biofuel Economy

7.1 Introduction to Biofuel Economy

Experts suggest that current oil and gas reserves will suffice to last only a few more decades. To exceed the rising energy demand and reducing petroleum reserves, fuels such as bioethanol and biodiesel are in the forefront of the alternative technologies. Accordingly, the viable alternative for compression-ignition engines is biodiesel. The major economic factor to consider for input costs of biodiesel production is the feedstock, which is about 80% of the total operating cost. The high price of biodiesel is in large part due to the high price of the feedstock. Economic benefits of a biodiesel industry would include value added to the feedstock, an increased number of rural manufacturing jobs, an increased income taxes, and investments in plant and equipment. The production and utilization of biodiesel is facilitated firstly through the agricultural policy of subsidizing the cultivation of non-food crops. Secondly, biodiesel is exempt from oil tax.

The 20th century was the century of the petrochemical economy. However, petroleum is a finite source for fuel that is rapidly becoming scarcer and more expensive. Biorefineries will not eliminate the need for petrochemicals in the 21st century, but they will play a key role in reducing our level of dependence on imported petroleum and in making the 21st century an increasingly sustainable, domestic, and environmentally responsible biofuel economy (bioeconomy). Oil price increases have also increased the level of interest in bioenergy. Production of bioenergy, biofuel, and bioindustrial products derived from agriculture is accelerating rapidly. A strong bioeconomy may provide significant environmental benefits.

Most of the biofuel production processes developed to date are immature and have never been implemented on an industrial scale. Projections of the amount of biofuel depend on the development of the bioeconomy and society. Biofuels are generally considered as offering many priorities, including sustainability, reduction of greenhouse gas emissions, regional development, social structure and agriculture, and security of supply (Reijnders, 2006).

A. Demirbas, *Biofuels*,
© Springer 2009

There are several reasons for biofuels to be considered as relevant technologies by both developing and industrialized countries. These include energy security, environmental concerns, foreign exchange savings, and socioeconomic issues, mainly related to the rural sector. A large number of research projects in the field of thermochemical and biochemical conversion of biomass, mainly on liquefaction, pyrolysis, and gasification, have been carried out. Liquefaction is a thermochemical conversion process of biomass or other organic matters into liquid oil products in the presence of a reducing reagent, for example, carbon monoxide or hydrogen. Pyrolysis products are divided into a volatile fraction, consisting of gases, vapors, and tar components, and a carbon-rich solid residue. The gasification of biomass is a thermal treatment, which results in a high production of gaseous products and small quantities of char and ash. Bioethanol is a petrol additive/substitute. It is possible that wood, straw, and even household wastes may be economically converted to bioethanol. Bioethanol is derived from alcoholic fermentation of sucrose or simple sugars, which are produced from biomass by hydrolysis process. There has been renewed interest in the use of vegetable oils for making biodiesel due to its less polluting and renewable nature as against conventional petroleum diesel fuel. Methanol is mainly manufactured from natural gas, but biomass can also be gasified to methanol. Methanol can be produced from hydrogen-carbon oxide mixtures by means of the catalytic reaction of carbon monoxide and some carbon dioxide with hydrogen.

Biosynthesis gas (biosyngas) is a gas rich in CO and H_2 obtained by gasification of biomass. Biomass sources are preferable for biomethanol, rather than for bioethanol because bioethanol is a high-cost and low-yield product. The production of biosynfuel such as gasoline, kerosene, and diesel from biomass by the best proven techniques, involves the following complex sequence of processes: milling of the feedstock, followed by pyrolysis, followed by gasification, followed scrubbing of the biosyngas, followed by either the Mobil process, which involves an intermediate step of manufacturing methanol from biosyngas, or by the Fischer–Tropsch process where the biosyngas is reacted with a catalyst under heat and pressure. Each process requires a separate unit of equipment, and thus capital cost will be many times higher than is the case for biodiesel.

Estimation of the cost of biofuel is affected by a range of drivers that could change in direction and importance over time. These include:

1. Supply cost, market price and demand.
2. Competing, non-energy markets for biomass.
3. Preferences of farmers and woodland owners.
4. Access to market
5. Success of alternative waste recovery and recycling.

Currently, biomass-driven combined heat and power, co-firing, and combustion plants provide reliable, efficient, and clean power and heat. The production and use of biofuels are growing at a very rapid pace. Sugar cane-based ethanol is already a competitive biofuel in tropical regions. In the medium term, ethanol and high-quality synthetic fuels from woody biomass are expected to be competitive at crude oil prices above US$ 45 *per* barrel.

7.2 Biofuel Economy

The biofuel economy, and its associated biorefineries, will be shaped by many of the same forces that shaped the development of the hydrocarbon economy and its refineries over the past century. Due to its environmental merits, the share of biofuel in the automotive fuel market will grow fast in the next decade. There are several reasons for biofuels to be considered as relevant technologies by both developing and industrialized countries.

National indicative targets can only be achieved if biofuels benefit from some kind of public support system. Some support is available to encourage the supply of biofuels and their feedstocks. This includes aids for the cultivation of raw materials and for the capital cost of biofuel processing. Support systems designed to encourage demand for biofuels play a larger role. The main approaches are: (a) tax reductions/exemptions for biofuels and (b) biofuel obligations.

In the most biomass-intensive scenario, modernized biomass energy will contribute by 2050 about one half of total energy demand in developing countries (IPCC, 1997). The biomass intensive future energy supply scenario includes 385 million hectares of biomass energy plantations globally in 2050 with three quarters of this area established in developing countries (Kartha and Larson, 2000). Various scenarios have put forward on estimates of biofuel from biomass sources in the future energy system. The availability of the resources is an important factor cogenerative use of biofuel in the electricity, heat, or liquid fuel market.

Current global energy supplies are dominated by fossil fuels (388 EJ *per* year), with much smaller contributions from nuclear power (26 EJ) and hydropower (28 EJ). Biomass provides about 45 ± 10 EJ, making it by far the most important renewable energy source used. On average, in the industrialized countries biomass contributes less than 10% to the total energy supplies, but in developing countries the proportion is as high as 20–30%. In a number of countries biomass supplies 50–90% of the total energy demand. A considerable part of this biomass use is, however, non-commercial and relates to cooking and space heating, generally by the poorer part of the population. Part of this use is commercial, *i.e.*, the household fuelwood in industrialized countries and charcoal and firewood in urban and industrial areas in developing countries, but there are very limited data on the size of those markets. An estimated 9 ± 6 EJ are included in this category (WEA, 2000 and 2004).

The use of biomass for biomaterials will increase, both in well established markets (such as paper, construction) and possibly large new markets (such as biochemicals and plastics), as well as in the use of charcoal for steel making. This adds to the competition for biomass resources, in particular forest biomass, as well as land for producing woody biomass and other crops (Hoogwijk *et al.*, 2003). However, increased use of biomaterials does not prohibit the production of biofuels (and electricity and heat) *per se*. Construction wood ends up as waste wood, paper (after recycling) as waste paper, and bioplastics in municipal solid waste (Dornburg and Faaij, 2005). Such waste streams still qualify as biomass feedstock and are available often at low or even negative costs.

There are two global biomass-based liquid transportation fuels that might replace gasoline and diesel fuel. These are bioethanol and biodiesel. Transport is one of the main energy consuming sectors. It is assumed that biodiesel is used as a fossil diesel replacement and that bioethanol is used as a gasoline replacement. Biomass-based energy sources for heat, electricity, and transportation fuels are potentially carbon dioxide neutral and recycle the same carbon atoms. Due to widespread availability opportunities of biomass resources biomass-based fuel technology potentially employ more people than fossil-fuel-based technology (Kartha and Larson, 2000). The demand for energy is increasing every day due to the rapid outgrowth of population and urbanization. As the major conventional energy resources like coal, petroleum, and natural gas are at the verge of getting extinct, biomass can be considered as one of the promising environmentally friendly renewable energy options.

Agricultural energy ("green energy") production is the principal contributor in economic development of a developing country. Its economy development is based on agricultural production, and most people live in the rural areas. Implementation of integrated community development programs is therefore very necessary. It is believed that integrated community development contributes to pushing up the socio-economic development of the country.

Agriculture-(m)ethanol is at present more expensive than synthesis-ethanol from ethylene and methanol from natural gas. The simultaneous production of biomethanol (from sugar juice) in parallel to the production of bioethanol, appears economically attractive in locations where hydroelectricity is available at very low cost (~0.01 $ Kwh) and where lignocellulosic residues are available as surpluses (Grassi, 1999).

The FT synthesis-based gas to liquids (GTL) technology includes three processing steps, namely, syngas generation, syngas conversion, and hydroprocessing. In order to make the GTL technology more cost-effective, the focus must be on reducing both the capital and the operating costs of such a plant (Vosloo, 2001). For some time now the price has been up to $ 60 *per* barrel. It has been estimated that the FT process should be viable at crude oil prices of about $ 20 *per* barrel (Jager, 1990). The current commercial applications of the FT process are geared at the production of the valuable linear alpha olefins and of fuels such as LPG, gasoline, kerosene, and diesel. Since the FT process produces predominantly linear hydrocarbons, the production of high quality diesel fuel is currently of considerable interest (Dry, 2004). The most expensive section of an FT complex is the production of purified syngas, and so its composition should match the overall usage ratio of the FT reactions, which in turn depends on the product selectivity (Dry, 2002). The industrial application of the FT process started in Germany and by 1938 there were nine plants in operation having a combined capacity of about 660×10^3 t *per* year (Anderson, 1984).

7.2.1 Estimation of Biofuel Prices

The calculation of biofuel prices should be designed to maintain into the future the equilibrium between demand and supply, taking into account the costs of planned investments. They should also take into account the rest of the economy and the environment. Two very important characteristics of energy prices are equity and affordability. Biofuel prices must reflect the cost, imposed by the specific consumer category on the economy. Since energy prices based on the apparent long run marginal costs may not be sufficient to finance the development of the energy sector, and the prices should be adjusted so that the energy sector can be financed without subsidies to enhance its autonomy. In competitive markets this form of adjustment may not be possible.

Marginal biofuel is the most expensive fuel that would be used where a target is met at least cost. Fuels would be used in turn in order of price from lowest to highest. Our concern is with the most expensive fuel that would be used to meet the target, as that is the one that would be displaced by a lower cost fuel. Manufacture from wood is untested commercially at this stage, although at least one plant is expected to come into production in 2006.

Three cost factors affect the biofuel costs:

1. Operating costs
2. Distribution and blending costs
3. Capital costs.

The operating costs are defined here as all costs that are not capital or feedstock costs. The literature contains large variations in the estimation of operating costs. Where there is a range of costs reported in different sources, a midpoint has been used.

There are two components to distribution costs: The costs of delivering the feedstock to the ethanol production plant, and those of distributing the ethanol to the blending facility.

The feedstock cost consists of two parts: the conversion yield (how much ethanol is produced from 1 ton of feedstock), and the price of that feedstock. Once again, there are large variations in the literature in the estimation of conversion yield. In these cases, a midpoint has been used.

7.2.2 Biodiesel Economy

Biodiesel has become more attractive recently because of its environmental benefits. The cost of biodiesel, however, is the main obstacle to commercialization of the product. With cooking oils used as raw material, the viability of a continuous transesterification process and recovery of high quality glycerol as a biodiesel byproduct are primary options to be considered to lower the cost of biodiesel (Ma and Hanna, 1999; Zhang *et al.*, 2003). The economic performance of a biodiesel plant (*e.g.*, fixed capital cost, total manufacturing cost and the break-even price of biodiesel) can be determined once certain factors are identified, such as plant capacity, process technology, raw material cost, and chemical costs. However, the effects of these factors on the economic viability of the plant are also of concern. A sensitivity analysis involves measuring the relative magnitudes of these effects. This will also provide further information for the optimization of biodiesel production (Zhang *et al.*, 2003).

The economic advantages of biodiesel can be listed as follows: it reduces greenhouse gas emissions, it helps to reduce a country's reliance on crude oil imports and supports agriculture by providing a new labor and market opportunities for domestic crops, it enhances the lubricating property, and it is widely accepted by vehicle manufacturers (Palz *et al.*, 2002; Clarke *et al.*, 2003).

The brake power of biodiesel is nearly the same as with petrodiesel, while the specific fuel consumption is higher than that of petrodiesel. Carbon deposits inside the engine are normal, with the exception of intake valve deposits. In an earlier study, the results showed the transesterification treatment decreased the injector coking to a level significantly lower than that observed with petrodiesel (Demirbas, 2003). Although most researchers agree that vegetable oil ester fuels are suitable for use in CIE, a few contrary results have also been obtained. The results of these studies point out that most vegetable oil esters are suitable as diesel substitutes, but that more long-term studies are necessary for commercial utilization to become practical.

Biodiesel is becoming of interest to companies interested in commercial scale production, as well as the more usual home brew biodiesel user, and the user of straight vegetable oil or waste vegetable oil in diesel engines. Biodiesel is commercially available in most oilseed-producing countries. Biodiesel is a technologically feasible alternative to petrodiesel, but nowadays biodiesel cost is 1.5–3 times higher than the fossil diesel cost in developed countries. Biodiesel is more expensive than petrodiesel, though it is still commonly produced in relatively small quantities (in comparison to petroleum products and ethanol). The competitiveness of biodiesel to petrodiesel depends on the fuel taxation approaches and levels. Generally, the production costs of biodiesel remain much higher compared to the petrodiesel ones. Therefore, biodiesel is not competitive to petrodiesel under current economic conditions. The competitiveness of biodiesel relies on the prices of biomass feedstock and costs, linked to the conversion technology.

The recent increase of biodiesel potential is not only apparent in the number of plants, but also in the size of the facilities. The tremendous growth in the biodiesel industry is expected to have a significant impact on the price of biodiesel feedstocks. This growth in the biodiesel industry will increase competition. An earlier evaluation of the potential feedstocks for biodiesel by Hanna *et al.* (2005) also identified the expected price pressures on biodiesel feedstocks. Fiscal incentives for biodiesel such as reductions both in feedstock and processing costs, and tax exemptions will be the key instrument to enhance the biodiesel application as an alternative fuel for transport in the near future. Biodiesel has demonstrated a number of promising characteristics, including reduction of exhaust emissions (Dunn,

2001). The advantages offered by biodiesel have to be considered at levels beyond the agricultural, transport, and energy sectors only.

Economic benefits of a biodiesel industry would include value added to the feedstock, an increased number of rural manufacturing jobs, an increased income taxes, and investments in plant and equipment. In recent years, the importance of non-food crops increased significantly. The opportunity to grow non-food crops under the compulsory set-aside scheme is an option to increase biodiesel production. The possibility of growing non-food crops under the compulsory set-aside scheme is an opportunity for the biodiesel market, but it is not an appropriate instrument to promote non-food production.

Blends of up to 20% biodiesel mixed with petrodiesel fuels can be used in nearly all diesel equipment and are compatible with most storage and distribution equipment. Higher blends, even B100, can be used in many engines built with little or no modification. Transportation and storage, however, require special management. Material compatibility and warrantee issues have not been resolved with higher blends.

Most of the biodiesel that is currently made uses soybean oil, methanol, and an alkaline catalyst. Methanol is preferred, because it is less expensive than ethanol (Graboski and McCormick, 1998). Base catalyst is preferred in transesterification, because the reaction is quick and thorough. It also occurs at lower temperature and pressure than other processes, resulting in lower capital and operating costs for the biodiesel plant. The high value of soybean oil as a food product makes production of a cost effective fuel very challenging. However there are large amounts of low cost oils and fats, such as restaurant waste and animal fats, that could be converted to biodiesel. The problem with processing these low cost oils and fats is that they often contain large amounts of free fatty acids that cannot be converted to biodiesel using an alkaline catalyst (Canakci and Van Gerpen, 2001).

The total energy use for biodiesel production in the common method is 17.9 MJ/l biodiesel. Transesterification process alone consumes 4.3 MJ/l, while from our calculation, the supercritical methanol method requires as much as 3.3 MJ/l, or energy reduction of 1.0 MJ for each liter of biodiesel fuel. In the common catalyzed method, mixing is significant during the reaction. In our method, since the reactants are already in a single phase, mixing is not necessary. Since our process is much simpler, particularly in the purification step, which only needs a removal of unreacted methanol, it is further expected that about 20% of cost reduction can be achieved from transesterification process. Therefore, the production cost for biodiesel fuel from rapeseed oil falls to be US$ 0.59/l, compared to US$ 0.63/l for the common catalyzed method (Saka and Kusdiana, 2001).

The cost of feedstock is a major economic factor in the viability of biodiesel production. Feedstock costs typically account for 80% of the total costs of biodiesel production (Demirbas, 2006). Using an estimated process cost, exclusive of feedstock cost, of $ 0.158/l ($ 0.60/gal) for biodiesel production, and estimating a feedstock cost of $ 0.539/l ($ 2.04/gal) for refined soy oil, an overall cost of $ 0.70/l ($ 2.64/gal) for the production of soy-based biodiesel has been estimated (Haas et al., 2006). Biodiesel from animal fat is currently the cheapest option

(\$ 0.4–\$ 0.5/l), while traditional transesterification of vegetable oil is at present around \$ 0.6–\$ 0.8/l (IEA, 2007).

We believe that the ease with which a feedstock may be processed to achieve the type of fuel desired is also of great significance. Biodiesel, for example, generally utilizes the most expensive feedstocks but is produced by a relatively simple process that entails only the following processes: pressing the oil from the feedstock, the mixing the oil with methanol and an alkaline catalyst, and then the removal of the glycerol, which is the principal waste product.

A review of economic feasibility studies shows that the projected costs for biodiesel from oil seed or animal fats have a range of \$ 0.30–0.69/l, including meal and glycerin credits, and the assumption of reduced capital investment costs by having the crushing and/or esterification facility added onto an existing grain or tallow facility. Rough projections of the cost of biodiesel from vegetable oil and waste grease are, respectively, \$ 0.54–\$ 0.62/l and \$ 0.34–\$ 0.42/l. With pretax diesel priced at \$ 0.18/l in the United States and \$ 0.20–\$ 0.24/l in some European countries, biodiesel is currently not economically feasible, and more research and technological development will be needed (Bender, 1999). The cost of biodiesel production is the cause for the generally accepted view of the industry in Europe that biodiesel production is not profitable without fiscal support. Table 7.1 shows cost and return scenario for a 60,000 ton biodiesel plant (Balat, 2007; Bozbas, 2008). The average cost of biodiesel in Italy is about two to three times higher than that of diesel fuel, without considering excise tax and VAT. Since 1993, the Italian government has been promoting the use of biodiesel by abolition of the excise tax on a maximum annual amount of 125,000 tons of biodiesel. In 2001, the allowance was increased to 300,000 tons (Carraretto et al., 2004).

High petroleum price demands the study of biofuel production. Lower-cost feedstocks are needed since biodiesel from food-grade oils is not economically competitive with petroleum-based diesel fuel. Inedible plant oils have been found to be promising crude oils for the production of biodiesel.

The cost biofuel and the demand for vegetable oils can be reduced by inedible oils and used oils, instead of edible vegetable oil. In the world a large amount of inedible oil plants are available in nature.

Table 7.1 Cost and return scenario for a 60,000 tons biodiesel plant

	Million Euros
Income	
60,000 tons biodiesel (€ 617/ton)	37.03
7,500 tons 80% glycerine (€ 500/ton)	3.75
Undetermined amount of free fatty acids sold as livestock feed	40.78
Total income expenses	
60,900 tons vegetable oil (€ 520/ton)	31.67
6000 tons methanol (€ 265/ton)	1.59
Undetermined amount of NaOH and HCl included in variable costs	4.70
Variable costs equal to fixed costs	4.70
Total cost	42.66

Vegetable oil is traditionally used as a natural raw material to linoleum, paint, lacquers, cosmetics and washing powder additives. In the technical range there is a growing market in the field of lubricants, hydraulic oils and special applications. The energetic use of pure plant oil in motors is an option to replace fossil fuels. Nowadays the technique is tested and well established. Pure plant oil-fuel has the advantages of low sulfur and aromatics contents, and safer handling. With the use of cold pressed plant oil instead of fossil diesel, there is a reduction in production of the green house gas CO_2

Everybody is able to produce his own fuel. The cold-pressing process does not require complicated machinery. The characteristics of this process are low energy requirement without any use of chemical extractive agents.

The major economic factor to consider for input costs of biodiesel production is the feedstock, which is about 80% of the total operating cost. Other important costs are labor, methanol, and the catalyst, which must be added to the feedstock. In some countries, filling stations sell biodiesel more cheaply than conventional diesel.

The cost of biodiesel fuels varies depending on the base stock, geographic area, variability in crop production from season to season, the price of the crude petroleum and other factors. Biodiesel is more than double the price of petroleum diesel. The high price of biodiesel is in large part due to the high price of the feedstock. However, biodiesel can be made from other feedstocks, including beef tallow, pork lard, and yellow grease.

The cost of production of biodiesel is not necessarily the same as the market price. Market prices would be expected to equal production costs in highly competitive markets where there is competition amongst biodiesel producers. In the absence of competition, prices can rise to the costs of the marginal producer.

If biodiesel is simply competing in the market against conventional diesel then the maximum price that biodiesel can sell at is the price of the alternative diesel, assuming no quality difference. This means that if production costs were lower than for conventional diesel, the price of biodiesel would rise to the conventional diesel price. If there is a potential export market for biodiesel, this also influences price; local producers would price at least as high as the value of biodiesel in the alternative market, less transport costs.

7.2.3 Bioethanol Economy

For designing fuel bioethanol production processes, the assessment of the utilization of different feedstocks (*i.e.*, sucrose containing, starchy materials, lignocellulosic biomass) is required considering the big share of raw materials in bioethanol costs (Cardona and Sanchez, 2007). Approximately 60% of the bioethanol produced is from raw materials (Rogers *et al.*, 2005). The cost of raw material, which varies considerably between different studies (US\$ 22–US\$ 61 *per* metric ton dry matter), and the capital costs, which makes the total cost dependent on plant capacity, contribute most to the total production cost (Hahn-Hagerdal *et al.*, 2006). With these relatively high raw material costs (which includes enzyme pretreatment when starch-based crops are used), such fermentation products are currently more expensive to produce than fuels or chemicals produced from lower cost hydrocarbons (Rogers *et al.*, 2005). Pretreatment has been viewed as one of the most expensive processing steps in cellulosic biomass-to-fermentable sugar conversion with costs as high as US$ 0.3/gallon bioethanol produced (Mosier *et al.*, 2005). Enzyme pricing is assumed such that the total contribution of enzymes to production costs is about US$ 0.15/gallon of bioethanol with some variation depending upon actual bioethanol yields resulting from the particular pretreatment approach (Eggeman and Elander, 2005).

The costs of producing bioethanol were estimated for a 50 million gallons *per* year dry mill bioethanol plant using current data for corn, distillers dried grains (DDG), natural gas, enzymes, yeast and chemicals, electricity, and wage rates. A bioethanol plant of this size will produce 51.5 million gallons of denatured bioethanol annually from 18.1 million bushels of corn. In additional to bioethanol, the plant will produce 154,500 tons of DDG. The cost of producing bioethanol in a dry mill plant currently totals US$ 1.65/gallon. Corn accounts for 66% of operating costs, while energy (electricity and natural gas) to fuel boilers and dry DDG represents nearly 20% of operating costs (Urbanchuk, 2007).

Until recently, Brazil was the largest producer of bioethanol in the world. Brazil used sugarcane to produce bioethanol, which is a more efficient feedstock for bioethanol production than corn grain (Pimentel and Patzek, 2005). The costs of producing bioethanol in Brazil are the world's lowest. The production cost for bioethanol in Brazil is in the US$ 0.68 to US$ 0.95 *per* gallon range (Shapouri *et al.*, 2006). Factors contributing to Brazil's competitiveness include favorable climate conditions, low labor costs, and mature infrastructure built over at least three decades (Xavier, 2007).

Estimates show that bioethanol in the EU will become competitive when the oil price reaches US$ 70 a barrel, while in the United States it will become competitive at US$ 50–60 a barrel. For Brazil the threshold is much lower – between US$ 25 and US$ 30 a barrel. Other efficient sugar producing countries such as Pakistan, Swaziland, and Zimbabwe have production costs similar to Brazil's (Balat, 2007; Dufey, 2006).

7.2.3.1 Ethanol Supply Costs and Prices

There are a number of different production methods and feedstocks for ethanol production. It has been focused on the following feedstocks and production processes:

1. Corn/maize: starch conversion to sugars, fermentation and distillation
2. Sugar beet: fermentation and distillation
3. Wood: acid hydrolysis, fermentation and distillation.

Each of the different feedstocks and processes, and their has had their manufacturing costs should be investigated independently.

The analysis above focuses on the costs of supply. As discussed under biodiesel above, this is not necessarily the same as the market price of ethanol, which will be influenced also by world prices of ethanol.

If ethanol competes simply at market prices, its maximum price will be set by the price of the alternatives. If there is competition with other ethanol producers, this might set the price, but it can rise no higher than the price of petrol.

With biodiesel, for policy purposes, the concern is with the costs of ethanol production rather than price. The producer surplus (difference between costs and price) is not a cost to the nation, unless this surplus is expatriated profit by foreign owned firms.

7.2.4   Biorenewable Energy Costs and Biohydrogen Economy

Table 7.2 shows a cost comparison between renewable energy and other energy sources. From this it can be seen that the renewable energy sources such as biomass, geothermal, and wind are as economically usable as other common energy sources.

Table 7.3 shows a cost analysis of poultry power generation by anaerobic digestion assuming a 10 year time frame at a 10% rate of return. The primary costs included in the analysis are capital costs, operation and maintenance, and cleanout. The costs are normalized to a KWh. Transportation costs are excluded for on-site

Table 7.2   Comparison of renewable energy with other energy sources

Power Source	Production cost of 1 kWh energy (cent)	
	Minimum	Maximum
Coal	4.5	7.0
Natural gas	4.3	5.4
Geothermal	4.7	7.8
Biomass	4.2	7.9
Agricultural residues	4.5	9.8
Energy crops	10.0	20.0
Municipal solid wastes	4.2	6.3
Wind generators	4.7	7.2
Solar thermal hybrid	6.0	7.8
Solar PV	28.7	31.0
Nuclear	5.3	9.3
Large hydro	3.0	13.0
Small hydro	4.0	14.0
Hydraulic	5.2	18.9
Wave/tidal	6.7	17.2

Sources: Demirbas, 2006; Demirbas, 2007

Table 7.3 Cost analysis of poultry power generation by anaerobic digestion assuming a 10 year time frame at a 10% rate of return

Plant rating, kW	125
Probable application	on-site
Capital costs, $/kW	$ 5,612
Operation and maintenance costs, $/kW	$ 346
Capacity factor	80%
Capital costs, $/KWh	$ 0.5784
Leveling capital costs, $/KWh	$ 0.0839
Operation and maintenance costs, $/KWh	$ 0.0376
Cleanout costs, $/KWh	$ 0.0128
Cost of production, $/KWh	$ 0.1349
Turkish average electricity retailpPrice, $/KWh	$ 0.0866

Source: Demirbas, 2008

systems. Calculations show that biogas from poultry is potentially economically viable (Demirbas, 2008).

Hydrogen is a synthetic energy carrier. The synthesis of hydrogen requires energy. Production, packaging, storage, transfer, and delivery of the hydrogen gas, in essence all key components of an economy, are so energy consuming that alternatives should be considered. The production technology should be site specific and include steam reforming of methane and electrolysis in hydropower rich countries. In the long run, when hydrogen is a very common energy carrier, distribution with pipeline will probably be the preferred option. Hydrogen generated from rooftop solar electricity and stored at low pressure in stationary tanks may be a viable solution for private buildings. Hydrogen may be the only link between physical energy from renewable sources and chemical energy. It is also the ideal fuel for modern clean energy conversion devices like fuel cells or even hydrogen engines.

Hydrogen for fleet vehicles would probably dominate in the transportation sector. To produce hydrogen *via* electrolysis, and the transportation of liquefied hydrogen to rural areas with pipelines would be expensive. The cost of hydrogen distribution and refueling is very specific. However, there are some barriers to the development of hydrogen economy. They are technological, economical, supply, storage, safety, and policy barriers. Reducing these barriers is one of the driving factors in the government's involvement in hydrogen and fuel cell research and development.

References

Anderson, R.B. 1984. The Fischer–Tropsch synthesis, Academic Press, New York.
Balat, M. 2007. An overview of biofuels and policies in the European Union countries. Energy Sources, Part B 2:167–181.
Bender, M. 1999. Economic feasibility review for community-scale farmer cooperatives for biodiesel. Bioresour Technol 70:81–87.

Bozbas, K. 2008. Biodiesel as an alternative motor fuel: production and policies in the European Union. Renew Sustain Energy Rev 12:542–552.

Canakci, M., Van Gerpen, J. 2001. Biodiesel production from oils and fats with high free fatty acids. Trans ASAE 44, 1429–1436.

Cardona, C.A., Sanchez, O.J. 2007. Fuel ethanol production: Process design trends and integration opportunities. Biores Technol 98:2415–2457.

Carraretto, C., Macor, A., Mirandola, A., Stoppato, A., Tonon, S. 2004. Biodiesel as alternative fuel: Experimental analysis and energetic evaluations. Energy 29:2195–2211.

Clarke, L.J., Crawshaw, E.H., Lilley, L.C., 2003. Fatty acid methyl esters (FAMEs) as diesel blend component. In 9th Annual Fuels and Lubes Asia Conference and Exhibition, Singapore, January 21–24.

Demirbas, A., 2003. Biodiesel fuels from vegetable oils *via* catalytic and non-catalytic supercritical alcohol transesterifications and other methods: A survey. Energy Convers Mgmt 44, 2093–2109.

Demirbas, A. 2006. Biodiesel production *via* non-catalytic SCF method and biodiesel fuel characteristics. Energy Convers Mgmt 47:2271–2282.

Demirbas, A. 2008. Importance of biomass energy sources for Turkey. Energy Policy 36:834–842.

Dornburg, V., Faaij, A. 2005. Cost and CO2-emission reduction of biomass cascading: Methodological aspects and case study of SRF poplar. Climatic Change 71:373–408.

Dry, M.E. 2002. The Fischer–Tropsch process: 1950–2000. Catalysis Today 71:227–241.

Dry, M.E. 2004. Present and future applications of the Fischer–Tropsch process. Fuel Proc Technol 55:219–233.

Dufey, A. 2006. Biofuels production, trade and sustainable development: Emerging issues. Environmental Economics Programme, Sustainable Markets Discussion Paper No.2, International Institute for Environment and Development (IIED), London, September, 2006.

Dunn, R.O. 2001. Alternative jet fuels from vegetable-oils. Trans ASAE 44,1151–757.

Eggeman, T., Elander, R.T. 2005. Process and economic analysis of pretreatment technologies. Biores Technol 96:2019–2025.

Grassi, G. 1999. Modern bioenergy in the European Union. Renewable Energy 16:985–990.

Hahn-Hagerdal, B., Galbe, M., Gorwa-Grauslund, M.F., Liden, G., Zacchi, G. 2006. Bio-ethanol – The fuel of tomorrow from the residues of today. Trends Biotechnol 24:549–556.

Hanna, M.A., Isom, L., Campbell, J. 2005. Biodiesel: Current perspectives and future. J Sci Ind Res 64, 854–857.

Hoogwijk, M., Faaij, A., van den Broek, R., Berndes, G., Gielen, D., Turkenburg, W. 2003. Exploration of the ranges of the global potential of biomass for energy. Biomass Bioenergy 25:119–133.

IEA (International Energy Agency). 2007. Biodiesel statistics. IEA Energy Technology Essentials, OECD/IEA, Paris, January, 2007.

IPCC (Intergovernmental Panel on Climate Change). 1997. Greenhouse gas inventory reference manual: Revised 1996 IPCC guidelines for national greenhouse gas inventories, Report Vol. 3, p. 1.53, Intergovernmental Panel on Climate Change (IPCC), Paris, France, 1997. Available from: www.ipcc.ch/pub/guide.htm.

Jager, B. 1990. Proceedings of the 5th Natural Gas Conversion Symposium, Taormina, Italy.

Kartha, S, Larson ED. 2000. Bioenergy primer: Modernised biomass energy for sustainable Development. Technical Report UN Sales Number E.00.III.B.6, United Nations Development Programme, 1 United Nations Plaza, New York, USA.

Ma, F., Hanna, M.A. 1999. Biodiesel production: A review. Biores. Technol. 70:1–15.

Mosier, N., Wyman, C., Dale, B., Elander, R., Holtzapple, Y.Y.L.M., Ladisch, M. 2005. Features of promising technologies for pretreatment of lignocellulosic biomass. Biores Technol 96:673–86.

Palz, W., Spitzer, J., Maniatis, K., Kwant, N., Helm, P., Grassi, A., 2002. Proceeedings of 12th International European Biomass Conference; ETA-Florence, WIP-Munich: Amsterdam, The Netherlands.

Pimentel, D., Patzek, T.W. 2005. Ethanol production using corn, switchgrass, and wood; biodiesel production using soybean and sunflower. Natural Res Res 14:65–76.

Reijnders, L. 2006. Conditions for the sustainability of biomass based fuel use. Energy Policy 34:863–876.

Rogers, P.L., Jeon, Y.J., Svenson, C.J. 2005. Application of biotechnology to industrial sustainability. Proc Safety Environ Protect 83:499–503.

Saka, S., Kusdiana, D., 2001. Biodiesel fuel from rapeseed oil as prepared in supercritical methanol, Fuel 80, 225–231.

Shapouri, H., Salassi, M., Nelson, J. 2006. The economic feasibility of ethanol production from sugar in the United States, U.S. Department of Agriculture (USDA), Washington, DC, USA, July, 2006.

Urbanchuk, J.M. 2007. Economic impacts on the farm community of cooperative ownership of ethanol production. LECG, LLC, Wayne, PA, February 13, 2007. (Available from: www. verasun.com/pdf/RFA).

Vosloo, A.C. 2001. Fischer–Tropsch: A futuristic view. Fuel Process Technol 71:149–155.

WEA (World Energy Assessment). 2000. World Energy Assessment of the United Nations, UNDP, UNDESA/WEC, Published by: UNDP, New York, September.

WEA (World Energy Assessment). 2004. Overview: 2004 Update, UNDP, UNDESA and the World Energy Council.

Xavier, M.R. 2007. The Brazilian sugarcane ethanol experience. Competitive Enterprise Institute (CEI), Issue Analysis, Washington, DC, February 15, 2007. (Available from: www.cei.org/pdf/5774).

Zhang, Y., Dube, M.A., McLean, D.D., Kates, M. 2003. Biodiesel production from waste cooking oil: 2. Economic assessment and sensitivity analysis. Bioresour Technol 90:229–240.

Chapter 8
Biofuel Policy

8.1 Introduction to Biofuel Policy

Current energy policies also address environmental issues including environmentally friendly technologies to increase energy supplies and encourage cleaner, more efficient energy use, and address air pollution, greenhouse effect (mainly reducing carbon dioxide emissions), global warming, and climate change (Demirbas, 2008). In general, energy policy includes issues of energy production, distribution, and consumption. The attributes of energy policy may include international treaties, legislation on commercial energy activities (trading, transport, storage, *etc.*), incentives to investment, guidelines for energy production, conversion and use (efficiency and emission standards), taxation and other public policy techniques, energy-related research and development, energy economy, general international trade agreements and marketing, energy diversity, and risk factors of possible energy crisis. There is considerable uncertainty in the economic analysis, associated with the costs of feedstocks and oil prices, and also because of the rapid development of the technologies. Any policy to support and encourage the supply of biofuel should provide incentives for both biodiesel and ethanol supplies. Policy options include incentive payments or tax breaks. Due to rising prices for fossil fuels (especially oil, but also natural gas and to a lesser extent coal) the competitiveness of biomass use has improved considerably over time. Biomass and bioenergy are now a key option in energy policies. Security of supply, an alternative for mineral oil and reduced carbon emissions are key reasons. Targets and expectations for bioenergy in many national policies are ambitious, reaching 20–30% of total energy demand in various countries. Similarly, long-term energy scenarios also contain challenging targets.

A. Demirbas, *Biofuels*,
© Springer 2009

8.2 Biofuel Policy

Energy is an essential input driving economic development. Therefore, in developed economies energy policies constitute an important component of overall regulatory frameworks shaping the improving overall competitiveness and market integration of the private business sector. Overall competitiveness includes liberalization of the electricity and gas markets as well as separation of energy production, transportation, and distribution activities.

Biofuels are transport fuels made from organic material. The most common biofuels today are biodiesel and bioethanol. Renewable energy sources are indigenous, and can therefore contribute to reducing dependency on oil imports and increasing security of supply. The biofuel policy aims to promote the use in transport of fuels made from biomass, as well as other renewable fuels. Biofuels provide the prospect of new economic opportunities for people in rural areas in oil importer and developing countries. The central policy of biofuel concerns job creation, greater efficiency in the general business environment, and protection of the environment.

Current European Union (EU) policies on alternative motor fuels focus on the promotion of biofuels. The definition of the marginal producer depends on the policy stance on biofuels. Biofuel pricing policy should not be employed as an antiinflationary instrument. It should be applied in such a way that it does not create cross subsidies between classes of consumers. In a proposed biofuels directive the introduction of a mandatory share scheme for biofuels, including as from 2009 minimum blending shares is stated. Table 8.1 shows the shares of alternative fuels compared to the total automotive fuel consumption in the EU under the optimistic development scenario of the European Commission. The EU has set the goal of obtaining 5.75% of transportation fuel needs from biofuels by 2010 in all member states in February 2006. In the Commission's view mandating the use of biofuels will (a) improve energy supply security, (b) reduce greenhouse gas (GHG) emissions, and (c) boost rural income and employment (Jansen, 2003; Hansen *et al.*, 2005). The European Union accounted for nearly 89% of all biodiesel production worldwide in 2005. By 2010, the United States is expected to become the world's largest single biodiesel market, accounting for roughly 18% of world biodiesel consumption, followed by Germany.

Table 8.1 Shares of alternative fuels in total automotive fuel consumption in the EU under the optimistic development scenario of the European Commission

Year	Biofuel	Natural gas	Hydrogen	Total
2010	6	2	–	8
2015	7	5	2	14
2020	8	10	5	23

Source: Demirbas, 2008

The general EU policy objectives considered most relevant to the design of energy policy are (Jansen, 2003):

1. Competitiveness of the EU economy
2. Security of energy supply
3. Environmental protection.

Producing and using biofuels for transportation offers alternatives to fossil fuels that can help provide solutions to many environmental problems. Using biofuels in motor vehicles helps reduce greenhouse gases (GHG) emissions. Biodiesel and ethanol provide significant reductions in GHG emissions compared to gasoline and diesel fuel. Due to the low or zero content of pollutants such as sulfur in biofuels, the pollutant emission of biofuels is much lower than the emission of conventional fuels. Numerous low emission scenarios have demonstrated that the Kyoto Protocol cannot be achieved without establishing a large role for biofuel in the global energy economy by 2050. Low emission scenarios imply 50–70 EJ of biofuel raw material in 2050. Well-designed biofuels projects would have very significant sustainable development benefits for rural areas, including creation of rural employment, rural electricity supply, soil conservation, and environmental benefits (Demirbas, 2006).

The main biofuel opportunities where suitable land is available are in developing countries. The issue of energy security has been accorded top-most priority. Every effort needs to be made to enhance the indigenous content of energy in a time-bound and planned manner. The additional benefit of biofuel development is creation of new employment opportunities in manufacturing, construction, plant operation and servicing, and fuel supply. Rural jobs are created in fuel harvesting, transport, and maintenance of processing areas (Demirbas, 2006).

8.2.1 Biodiesel Policy

Many farmers who grow oilseeds use a biodiesel blend in tractors and equipment as a matter of policy to foster production of biodiesel and raise public awareness. It is sometimes easier to find biodiesel in rural areas than in cities. Additional factors must be taken into account, such as the fuel equivalent of the energy required for processing, the yield of fuel from raw oil, the return on cultivating food, and the relative cost of biodiesel *versus* petrodiesel. Some nations and regions that have pondered transitioning fully to biofuels have found that doing so would require immense tracts of land if traditional crops are used.

While well-intentioned, policy-makers are on the wrong track for promoting biodiesel. Biodiesel policy should reflect that theme by aggressively eliminating the government imposed barriers to its success. Biodiesel policy should be based firmly in the philosophy of freedom. Given the history of petroleum politics, it is imperative that today's policy decisions ensure a free market for biodiesel. Producers of all sizes must be free to compete in this industry, without farm subsidies, regulations, and other interventions skewing the playing field. The production and utilization of biodiesel is facilitated firstly through the agricultural policy of subsidizing the cultivation of non-food crops. Secondly, biodiesel is exempt from oil tax.

European research and testing indicate that used as a diesel fuel substitute, biodiesel can replace petroleum diesel. Biodiesel, produced mainly from rapeseed or sunflower seed, comprises 80% of Europe's total biofuel production. The European Union accounted for nearly 89% of all biodiesel production worldwide in 2005. Germany produced 1.9 billion liters, or more than half the world total. Other countries with significant biodiesel markets in 2005 included France, the United States, Italy, and Brazil. All other countries combined accounted for only 11% of world biodiesel consumption in 2005. Figure 8.1 shows the biodiesel production of the European Union (1993–2005). In Germany biodiesel is also sold at a lower price than fossil diesel fuel. Biodiesel is treated like any other vehicle fuel in the UK. In February 2006, the European Union set the goal of obtaining 5.75% of transportation fuel needs from biofuels by 2010 in all member states. Many countries have adopted various policy initiatives. Specific legislation to promote and regulate the use of biodiesel is in force in Germany, Italy, France, Austria, and Sweden. Table 8.2 shows the biodiesel production capacity of the European Union in 2003 (EBB, 2004). By 2010, the United States is expected to become the world's largest single biodiesel market, accounting for roughly 18% of world biodiesel consumption, followed by Germany. New and large single markets for biodiesel are expected to emerge in China, India, and Brazil (EBB, 2004; Pinto *et al.*, 2005; Hanna *et al.*, 2005).

Brazil and the United States have the largest programs promoting biofuels in the world. The EU is in the third rank of biofuel production world wide, behind Brazil

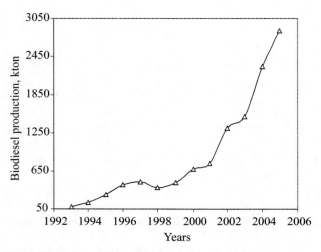

Fig. 8.1 Biodiesel production of the European Union (1993–2005)
Source: Demirbas, 2008

Table 8.2 Biodiesel production capacity of the European Union in 2003

Country	(1000 metric tons)
Germany	1025
France	500
Italy	420
Austria	50
Denmark	40
United Kingdom	5
Sweden	8
Total	2048

and the United States (Demirbas and Balat, 2006). Biofuel production in the EU is concentrated in countries hosting the major transportation fuel markets, notably Germany, France, Italy, and Spain, with substantial volumes produced also in the Czech Republic, Austria, Poland, Sweden, and Denmark (Eikeland, 2006).

Bioethanol is by far the most widely used biofuel for transportation worldwide. Global bioethanol production more than doubled between 2000 and 2005. Figure 8.2 shows the production of ethanol and biodiesel between 1980 and 2007 in the world. About 60% of global bioethanol production comes from sugarcane and 40% from other crops (Dufey, 2006).

According to the United States Department of Agriculture's 2006 figures (USDA, 2006), the United States produced 4 billion gallons (15.2 billion liters) of ethanol in 2005, up to 3.4 billion gallons in 2004. In 2005, Brazil, produced 4.2 billion gallons of ethanol, up to 4.0 billion gallons in 2004. The United States is predominantly a producer of bioethanol derived from corn, and production is con-

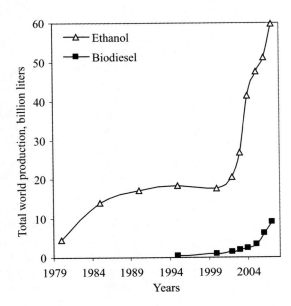

Fig. 8.2 World production of ethanol and biodiesel, 1980–2007

centrated in midwestern states with abundant corn supplies. In 2005 the United
States produced 4 billion gallons of bioethanol, which equates to about 3% of the
country's total gasoline consumption (140 billion gallons *per* year) (Asher, 2006).
Prompted by the increase in oil prices, Brazil began to produce bioethanol from
sugarcane in the 1970s and is considered the most successful example of a commercial application of biomass for energy production and use. Extensive experience in bioethanol production, suitable natural conditions for sugarcane production and low labor costs have made Brazil the most efficient bioethanol-producing
country (Dufey, 2006). Bioethanol represents approximately 1/3 of total vehicle
fuels currently used in Brazil (Eikeland, 2006).

The EU accounts for about 10% of global bioethanol production. EU bioethanol is generally produced using a combination of sugar beet and wheat (Schnepf,
2006). France is currently the front-runner in the EU's attempt to boost bioethanol
use, accounting for 2% of global production, mainly from sugar beet and wheat.
However, France is rapidly being overtaken by Germany and Spain as the EU's
largest bioethanol producers. China accounts for about 9% of global bioethanol
production, 80% of which is grain-based, mainly derived from corn, cassava, and
rice. India accounts for 4% of global bioethanol production, which is made from
sugarcane (Dufey, 2006). With all of the new government programs in America,
Asia, and Europe in place, total worldwide fuel bioethanol demand may grow to
exceed 125 billion liters by 2020 (Bohlmann, 2006).

During the 1990s the production and use of biofuels started in several European
countries and expanded significantly. Biofuels have a unique role to play in European energy policy. They are today the only direct substitute for oil in transport
that is available on a significant scale. Various policy goals, such as reducing
greenhouse gas emissions, boosting the decarbonization of transport fuels, diversifying fuel supply sources, and developing long-term replacements for fossil oil
while diversifying income and employment in rural areas, have motivated the EU
to promote the production and use of biofuels (*i.e.,* transport fuels produced from
renewable organic materials).

In the White Paper for a Community Strategy and Action Plan, entitled "Energy for the Future: Renewable Sources of Energy" (1997), the need for increasing
the share of liquid biofuels was already mentioned. It is stated that, at present,
biofuels are not competitive due to the relatively low price of crude oil. Still, it is
important to secure an increasing part of alternative fuels on the market, because
of unpredictable oil prices and long-term fossil resource depletion. The first priority that is identified in the White Paper is a decrease of the production costs of
biofuels. Other focus areas of focus are tax exemption and subsidized biomass
growing initiatives.

The general EU policy objectives considered most relevant to the design of energy policy are: (1) competitiveness of the EU economy, (2) security of energy
supply, and (3) environmental protection. All renewable energy policies should be
measured by the contributions they make to these goals. Current EU policies on
alternative motor fuels focus on the promotion of biofuels. In the Commission's
view mandating the use of biofuels will (a) improve energy supply security,

(b) reduce greenhouse gas (GHG) emissions,and (c) boost rural incomes and employment. Current regulations would preclude a notable negative impact on the rural environment (Jansen, 2003). Elements of the European biofuels policy are (EC, 2003):

- A Communication presenting the action plan for the promotion of biofuels and other alternative fuels in road transport.
- The Directive on the promotion of biofuels for transport, which requires an increasing proportion of all diesel and gasoline sold in the Member States to be biofuel.
- The biofuels taxation, which is part of the large draft directive on the taxation of energy products and electricity, proposing to allow Member States to apply differentiated tax rates in favor of biofuels.

The EU has also adopted a proposal for a directive on the promotion of the use of biofuels with measures ensuring that biofuels account for at least 2% of the market for gasoline and diesel sold as transport fuel by the end of 2005, increasing in stages to a minimum of 5.75% by the end of 2010 (Hansen *et al.*, 2005). The French Agency for Environment and Energy Management (ADEME) estimates that the 2010 objective would require industrial rapeseed plantings to increase from currently 3 million ha in the EU to 8 million ha (USDA, 2003).

In Germany, the current program of development of the biodiesel industry is not a special exemption from EU law, but rather is based on a loophole in the law. The motor fuels tax in Germany is based on mineral fuel. Since biofuel is not a mineral fuel, it can be used for motor transport without being taxed. Unlike France and Italy, where biodiesel is blended with mineral diesel, biodiesel sold in Germany is pure, or 100%, methyl ester. There is no mineral tax on biodiesel in Germany, so when diesel prices were high, and vegetable oil prices were low, biodiesel became very profitable. Additionally, there have been no restrictions on the quantity of biodiesel that can be exempted from the mineral fuel tax, so there has been a huge investment in biodiesel production capacity (USDA, 2003). In 2005 the capacity was supposed to increase to 1,600,000 tons (Brand, 2004).

8.3 Global Biofuel Projections

Projections are important tools for long-term planning and policy settings. Renewable energy sources that use indigenous resources have the potential to provide energy services with zero or almost zero emissions of both air pollutants and greenhouse gases. Renewable energy is a promising alternative solution because it is clean and environmentally safe. Currently, renewable energy sources supply 14% of the total world energy demand. Approximately half of the global energy supply will be from renewables in 2040. Photovoltaic (PV) systems and wind energy will play an important role in the energy scenarios of the future. The most

significant developments in renewable energy production are observed in photovoltaics (0.2–784 Mtoe) and wind energy (4.7–688 Mtoe) between 2001 and 2040.

Various scenarios have resulted in high estimates of biofuel in the future energy system. The availability of resources is an important factor if high shares of
biofuel penetrate the electricity, heat, or liquid fuel market. The rationale is to
facilitate the transition from the hydrocarbon economy to the carbohydrate economy by using biomass to produce bioethanol and biomethanol as replacements for
traditional oil-based fuels and feedstocks. The biofuel scenario produced equivalent rates of growth in GDP and *per capita* affluence, reduced fossil energy intensities of GDP, reduced oil imports, and gave an energy ratio. Each scenario has
advantages whether it is rates of growth in GDP, reductions in carbon dioxide
emissions, the energy ratio of the production process, the direct generation of
jobs, or the area of plantation biomass required to make the production system
feasible (Demirbas, 2006).

Renewable resources are more evenly distributed than fossil and nuclear resources, and energy flows from renewable resources are more than three orders of
magnitude higher than current global energy use. Today's energy system is unsustainable because of equity issues, as well as environmental, economic, and geopolitical concerns that have implications far into the future (UNDP, 2000).

According to International Energy Agency (IEA), scenarios developed for the
USA and the EU indicate that near-term targets of up to 6% displacement of petroleum fuels with biofuels appear feasible using conventional biofuels, given available cropland. A 5% displacement of gasoline in the EU requires about 5% of
available cropland to produce ethanol, while in the USA 8% is required. A 5%
displacement of diesel requires 13% of USA cropland, 15% in the EU. The recent
commitment by the USA government to increase bioenergy three-fold in 10 years
has added impetus to the search for viable biofuels (IEA, 2004).

The dwindling fossil fuel sources and the increasing dependency of the USA on
imported crude oil have led to a major interest in expanding the use of bioenergy.
The EU have also adopted a proposal for a directive on the promotion of the use of
biofuels with measures ensuring that biofuels account for at least 2% of the market
for gasoline and diesel sold as transport fuel by the end of 2005, increasing in
stages to a minimum of 5.75% by the end of 2010 (Hansen *et al.*, 2005).

Figure 8.3 shows the shares of alternative fuels compared to the total automotive fuel consumption in the world as a futuristic view. Hydrogen is currently
more expensive than conventional energy sources. There are different technologies presently being practiced to produce hydrogen economically from biomass.
Biohydrogen technology will play a major role in the future because it can utilize
renewable sources of energy (Nath and Das, 2003).

Biofuels are expected to reduce dependence on imported petroleum with associated political and economic vulnerability, reduce greenhouse gas emissions and
other pollutants, and revitalize the economy by increasing demand and prices for
agricultural products. Although most attention focuses on ethanol, interest in biodiesel is also increasing. Rapeseed is the primary oil used to make European biodiesel. Currently, biodiesel use is particularly strong in Germany. Biodiesel is

Fig. 8.3 Shares of alternative fuels compared to the total automotive fuel consumption in the world
Source: Demirbas, 2006

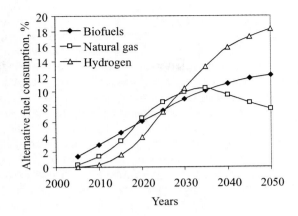

primarily produced from soybeans in the United States. The European Union has chosen biodiesel as its main renewable liquid fuel. Fuel use of ethanol in the European Union is much less important. Low European corn production and a high proportion of diesel engines compared to the United States make biodiesel a more attractive alternative in the European Union.

For fuels produced from biomass, various conversion routes are available that follow from the different types of biomass feedstocks. These routes include direct conversion processes, such as extraction of vegetable oils followed by esterification (biodiesel), fermentation of sugar-rich crops (ethanol), pyrolysis of wood (pyrolysis oil derived diesel equivalent), and hydrothermal upgrading (HTU) of wet biomass (HTU oil derived diesel equivalent). Another possibility is to produce liquid biofuels (methanol, DME, Fischer–Tropsch liquids) from synthesis gas, which results from gasification of biomass. At present, the biofuel producing countries in the European Union only have a small share in the global production of biofuels, namely a little less than 6%. Most of the global biofuel production consists of ethanol. The main ethanol producers are the USA and Brazil, whereas the share of Europe is rather small. However, Europe is the most important producer of biodiesel on the global market.

Biomass is the most used renewable energy source now and in future. The potential of sustainable large hydro is quite limited to some regions in the world. The potential for small hydropower (<10 MW) is still significant and will be more significant in future. PV systems and wind energy are technologies with annual growth rates of more than 30% during the last years that will become more significant in future. Geothermal and solar thermal sources will be more important energy sources in future. PV will then be the largest renewable electricity source with a production of 25.1% of the global power generation in 2040.

The global and regional availability has been estimated for wind energy, solar energy especially photovoltaic (PV) cells, and biomass, together with a description of factors influencing this availability. In the long term, countries with surplus biomass potential may develop into exporters of bioenergy.

Renewable energy sources or renewables contributed 2% of the world's energy consumption in 1998, including 7 exajoules from modern biomass and 2 exajoules for all other renewables (UNDP, 2000). The renewables are clean or inexhaustible and primary energy resources. Renewable technologies like water and wind power probably would not have provided the same fast increase in industrial productivity as fossil fuels did (Edinger and Kaul, 2000).

Biomass provides a number of local environmental gains. Energy forestry crops have a much greater diversity of wildlife and flora than the alternative land use, which is arable or pasture land. In industrialized countries, the main biomass processes utilized in the future are expected to be direct combustion of residues and wastes for electricity generation, bio-ethanol and biodiesel as liquid fuels, and combined heat and power production from energy crops. The future of biomass electricity generation lies in biomass integrated gasification/gas turbine technology, which offers high energy conversion efficiencies. Biomass will compete favorably with fossil mass for niches in the chemical feedstock industry. Biomass is a renewable, flexible, and adaptable resource. Crops can be grown to satisfy changing end use needs.

In the future, biomass has the potential to provide a cost-effective and sustainable supply of energy, while at the same time aiding countries in meeting their greenhouse gas reduction targets. By the year 2050, it is estimated that 90% of the world population will live in developing countries.

According to IEA, scenarios developed for the USA and the EU indicate that near-term targets of up to 6% displacement of petroleum fuels with biofuels appear feasible using conventional biofuels, given available cropland. A 5% displacement of gasoline in the EU requires about 5% of available cropland to produce ethanol, while in the USA 8% is required. A 5% displacement of diesel requires 13% of USA cropland, 15% in the EU (IEA, 2006).

References

Asher, A. 2006. Opportunities in biofuels creating competitive biofuels markets. Biofuels Australasia 2006 Conference, Sydney, Australia, November 20–22.

Bohlmann, G.M. 2006. Process economic considerations for production of ethanol from biomass feedstocks. Industrial Biotechnology 2:14–20.

Brand, R. 2004. Networks in renewable energy policies in Germany and France Berlin conference on the human dimension of global environmental change: greening of policies – Policy integration and interlinkages, Berlin, 3–4 December.

Demirbas, A. 2006. Global biofuel strategies. Energy Edu Sci Technol 17:27–63.

Demirbas, A. 2008. Biodiesel: A Realistic fuel alternative for Diesel engines. Springer, London.

Demirbas, M. F., Balat, M. 2006. Recent advances on the production and utilization trends of biofuels: A global perspective. Energy Convers Mgmt 47:2371–2381.

Dufey, A. 2006. Biofuels production, trade and sustainable development: Emerging issues. International Institute for Environment and Development (IIET), London, November, 60 p.

EBB (European Biodiesel Board). 2004. EU: Biodiesel industry expanding use of oilseeds, Brussels.

EC (European Commission). 2003. Renewable energies: an European policy. Promoting Biofuels in Europe. European Commission, Directorate – General for Energy and Transport, B-1049 Bruxelles, Belgium 2003.

Edinger, R., Kaul, S. 2000. Humankind's detour toward sustainability: Past, present, and future of renewable energies and electric power generation. Renew Sustain Energy Rev 4:295–313.

Eikeland, P.O. 2006. Biofuels – The new oil for the petroleum industry? FNI Report 15/2005, The Fridtjof Nansen Institute, Lysaker, Norway, January, 39 p.

Hanna, M.A., Isom, L., Campbell, J. 2005. Biodiesel: Current perspectives and future. J Sci Ind Res 64, 854–857.

Hansen, A.C., Zhang, Q., Lyne, P.W.L. 2005. Ethanol-diesel fuel blends – A review. Biores Technol 96:277–285.

IEA (International Energy Agency). 2006. Reference scenario projections. 75739 Paris, cedex 15, France.

Jansen, J.C. 2003. Policy support for renewable energy in the European Union. Energy Research Centre the Netherlands. Available from: www.ecn.nl/docs/library/report/2003/C03113.

Nath, K., Das, D. 2003. Hydrogen from biomass. Current Sci 85:265–271.

Pinto, A.C., Guarieiro, L.L.N., Rezende, M.J.C., Ribeiro, N.M., Torres, E.A., Lopes, W.A., Pereira, P.A.P., Andrade, J.B. 2005. Biodiesel: An overview. J Brazilian Chem Soc 16:1313–1330.

Schnepf, R. 2006. European union biofuels policy and agriculture: An overview. CRS Report for Congress, March 16.

UNDP (United Nations Development Programme). 2000. World energy assessment. Energy and the challange of sustainability.

USDA (United States Department of Agriculture). 2003. Production Estimates and Crop Assessment Division Foreign Agricultural Service. EU: Biodiesel Industry Expanding Use of Oilseeds. Available from: biodiesel. Org/resources/../reports/gen/20030920_gen330.pdf.

USDA (United States Department of Agriculture). 2006. The economic feasibility of ethanol production from sugar in the United States. Washington, DC, July. (Available from: www. usda.gov/oce/EthanolSugarFeasibilityReport3.pdf).

Index